Taxonomy, ecology and identity of conodonts

Proceedings of the Third European Conodont Symposium (ECOS III)
in Lund, 30th August to 1st September, 1982

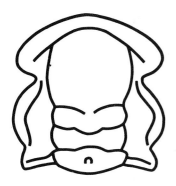

Fossils and Strata · No. 15

Universitetsforlaget · Oslo · 15th December, 1983

ECOS III

Printed by Offsetcenter AB, Uppsala
Typeset by Ord & Form AB, Uppsala
© Lethaia Foundation 1983

ISSN 0300-9491   ISBN 82-0006737-8

Contents

Foreword

The first symposium in Europe entirely devoted to conodonts was held in Marburg in 1971, the second (ECOS II) in Vienna–Prague 1980. There, we were asked to organize the third one in Lund, Sweden, in 1982. From the start these symposia have been European only in terms of location of the meetings. The topics discussed have not been exclusively European, nor has the provenance of participants been restricted. On the contrary, each of these symposia has served as an international forum, open to all. Thus, in addition to the many Europeans, ECOS III attracted scientists from as far away as Tasmania, Argentina, Japan (and five other Asian countries) as well as a large group from Canada and the USA. Altogether some 50 specialists attended.

This wide scope was evident also in the topics proposed for discussion and as the business of workshops. Anita G. Harris, Washington, organized a workshop on conodonts as palaeo-temperature indicators, Godfrey S. Nowlan, Ottawa, a workshop *cum* discussion on simple cone apparatuses, John E. Repetski, Washington, one on problems of Ordovician conodont taxonomy and biostratigraphy, Stefan Bengtson, Uppsala, a discussion of the biology and affinities of the conodont animal, and Otto H. Walliser, Göttingen, and Stig M. Bergström, Columbus, one on evolution in conodonts. A number of the talks held at the Symposium have also resulted in contributions to this volume. Other contributions are from conodontologists who had planned to, but for various reasons could not, attend the Symposium. Their articles complement the array of topics treated at the sessions in Lund [Jeppsson, L. & Löfgren, A. (eds.) 1982: Third European Conodont Symposium (ECOS III) Abstracts. *Publications from the Institutes of Mineralogy, Paleontology and Quaternary Geology University of Lund, Sweden, 238;* Jeppsson, L. 1982: Third European Conodont Symposium (ECOS III) Guide to excursion. *Publications from the Institutes of Mineralogy, Paleontology and Quaternary Geology University of Lund, Sweden, 239.*]

The ECOS III Symposium was not committed to any particular theme. The contributions to this volume are therefore a free representation of the kinds of problems that attract conodont specialists at the present time.

The main theme of the Marburg Symposium in 1971 (*Geologica et Palaeontologica,* SB 1 and Vol. 6) was transfer of taxonomy from an old mode in which taxa were based on individual elements, to one in which the basis of taxonomy is the whole apparatus. Seventeen years after the first attempts at a unified taxonomy this theme is still important; a majority of conodont taxa are still based on single elements, but usually we can employ a unified taxonomy. In the contributions to this

volume the reader can find discussions of apparatus reconstruction and/or discover different attitudes to the nomenclatorial problems that follow in the wake of taxonomic progress. The most pure contribution of this kind is that by Enrico Serpagli. Lennart Jeppsson's one-page contribution treats other aspects of that same theme.

In the Foreword to *Geologica et Palaeontologica,* SB 1, Lindström & Ziegler (1972) noted that a multielement-based taxonomy would permit us to trace lineages through time and to study their evolution. The present volume bears witness to the correctness of that prediction, since such enquiries are among the major concerns here. Many lineages can now be traced through several tens of millions of years. Contributions dealing explicitly with evolutionary problems are those by Stig M. Bergström, Thomas W. Broadhead, Thomas W. Broadhead & Ronald McComb, Pierre Bultynck, Jerzy Dzik, and Michael J. Orchard.

The third theme is as old as the study of conodonts: which group includes their closest relatives? The question has been revived following Hubert Szaniawski's demonstration of the close similarities that exist between protoconodonts and chaetognaths (Szaniawski 1982, *Journal of Paleontology 56:3*). In this volume a successor article by the same author reports on an investigation of the (micro-)structure of protoconodont elements. Thus there is at present stronger evidence of a relationship between protoconodonts and chaetognaths than there is between protoconodonts and euconodonts. Questions of conodont origin are again the principal concern in Stefan Bengtson's contribution to this volume. An event reported to the Symposium by Richard Aldridge (but published elsewhere (Briggs *et al., Lethaia 16:1*) was the discovery of a soft-bodied fossil which is the most convincing record of a conodont animal found so far.

Another classical yet still highly topical theme is presence of individual taxa in space and time and likely factors governing distribution. This is an important subject in many contributions, but it is a major one in only a few. Biostratigraphic and/or biogeographic–ecological questions are important in the contributions by Stig M. Bergström, David L. Clark & Eric Hatleberg, Jerzy Dzik, Lennart Jeppsson, Lin Bao-yu, Peep Männik, Tamara A. Moskalenko, Godfrey S. Nowlan, John E. Repetski & Thomas W. Henry, and Wang Cheng-yuan & Wang Zhi-hao.

The ECOS III Symposium could never have been carried through without much help from friends and colleagues. We wish to express sincere thanks to our co-members of the Organizing Committee, Sven Stridsberg and Eva-Marie

Widmark, as well as to Per Ahlberg, Claes Bergman, Doris Fredholm, Jan Gabrielson, Brian Holland, Louis Liljedahl, Peter Mileson, Sara Nyman, and Ewa Säll, all at the time at the Department of Historical Geology and Palaeontology in Lund, and to the Department of Geology for provision of technical help. We appreciate also the assistance of all those colleagues who acted as organizers of workshops and discussions, or as chairmen of sessions or who in any other way contributed to the Symposium.

The logotype used in this volume is a literal reconstruction of *Pygodus anserinus*, drawn by Claes Bergman for the Symposium.

The editorial work on the present volume has been greatly facilitated by the excellent services of our referees. We gratefully acknowledge help by Richard J. Aldridge, Nottingham, Christopher R. Barnes, St. John's, Stefan Bengtson, Uppsala, Thomas W. Broadhead, Knoxville, Pierre Bultynck, Brussels, David L. Clark, Madison, Maurits Lindström, Marburg, the late S. Crosbie Matthews, Uppsala, James F. Miller, Springfield, Michael A. Murphy, Riverside, Godfrey S. Nowlan, Ottawa, John E. Repetski, Washington, Carl B. Rexroad, Bloomington, Hans Peter Schönlaub, Vienna, and Walter C. Sweet, Columbus, as well as by those who prefer to remain anonymous.

The late Anders Martinsson, Uppsala, who as a friend and as Editor of *Fossils and Strata* gave us much helpful advice and unlimited support, especially during the planning of this volume, is thankfully remembered.

Special thanks are also due to Stefan Bengtson, Uppsala, who became engaged as Anders Martinsson's co-editor, but after his death alone took the responsibility for the final editing of this volume.

We would also like to thank the Swedish National Science Research Council for financial support, in the form of a grant that covered the printing of the Symposium abstracts and excursion guide and helped us pay some other costs in connection with the organizing of the Symposium. In a further notable benefaction the Council has generously met the principal part of the cost of printing this volume.

Lennart Jeppsson and Anita Löfgren

Editorial note

The Editor of *Fossils and Strata*, Anders Martinsson, died when the present volume was under preparation. The editorship then passed into my hands. At that stage the manuscripts had already been subject to thorough refereeing and revision through the efforts of Lennart Jeppsson and Anita Löfgren of the ECOS III Organizing Committee. This made the final editing of the manuscripts unusually uncomplicated. Thus the editing of the volume was the cooperative work of four persons.

The title 'Taxonomy, ecology and identity of conodonts' serves to identify the topics treated in the symposium volume, but it is not an essential part of the bibliographic identification. Thus articles in this volume may be conveniently referred to in reference lists under their serial arrangement, for example: Szaniawski, H. 1983: Structure of protoconodont elements. *Fossils and Strata 15*, 21–27.

Stefan Bengtson

The early history of the Conodonta

STEFAN BENGTSON

FOSSILS AND STRATA

ECOS III

A contribution to the Third European
Conodont Symposium, Lund, 1982

Bengtson, Stefan 1983 12 15: The early history of the Conodonta. *Fossils and Strata*, No. 15, pp. 5–19. Oslo ISSN 0300-9491. ISBN 82-0006737-8.

The slender, spine-shaped, apatitic protoconodonts appear in the fossil record near the Precambrian–Cambrian boundary and persist through the Cambrian. Recent work (Szaniawski 1982, *J. Paleont. 56*) suggests that protoconodont elements were homologous to the grasping spines of modern chaetognaths. Paraconodonts are similar to protoconodonts in their mode of growth by basal accretion. However, paraconodonts were more deeply invested in the secreting epithelium, and there are no known transitional forms between the two types. Other Cambrian conodont-like fossils have been investigated to examine the possibility that they may be homologous with paraconodonts. Some of them, such as the funnel-shaped cones of *Fomitchella*, have an internal structure which precludes homology with paraconodonts. Others, such as the cone-shaped sclerites of *Lapworthella*, might have taken on the morphological and structural characteristics of paraconodonts if they became adapted to a tooth or claw function, but there is no direct evidence to support such an interpretation. Morphological, histological and stratigraphical data indicate that euconodonts evolved from paraconodonts during the late Cambrian by acquiring a crown, a dense apatitic tissue secreted over the outer surface of the paraconodont cusp. If the paraconodont and protoconodont animals are closely related, as has been suggested earlier, the paraconodont and euconodont animals may represent a branch of the chaetognaths that had developed pharyngeal denticulation. This concept of conodont origin and early evolution fits well with the recently discovered euconodont animal in the Scottish Carboniferous (Briggs, Clarkson & Aldridge 1983, *Lethaia 16*). □*Conodonta, Chaetognatha, protoconodonts, Rhombocorniculum, Fomitchella, Lapworthella, microstructure, functional morphology, evolution, Cambrian.*

Stefan Bengtson, Department of Palaeobiology, Box 564, S-751 22 Uppsala, Sweden; 25th October, 1982 (revised 1983 06 21).

By the time of their first rise to high diversity in the Ordovician, conodonts had already attained the euconodont grade of element structure, i.e. the elements had developed a crown (term introduced by Nicoll 1977 for what was previously known as the 'conodont proper'). Most of our knowledge of conodont evolution concerns this euconodont grade of organization, which was extant from the late Cambrian (Franconian) until the final extinction of the group in the Triassic. The earlier history of the Conodonta is much less well understood, largely because Cambrian conodonts with their slow evolution and low diversity are unattractive for biostratigraphical purposes and have therefore attracted comparatively little attention, but also because the simple morphology and structure of early conodonts provide little basis for detailed phylogenetic analyses and often may even render their identification as conodonts conjectural.

The present contribution examines evidence for the early evolution of conodonts, in time from the first appearance of conodont-like fossils in the latest Precambrian to the attainment of the euconodont grade in the late Cambrian. The evidence is in some measure equivocal because of the problems of identification just mentioned. It is possible to propose evolutionary lineages between almost any two Cambrian conodont-like fossils, but very difficult to test such hypotheses. Histological characters help to channel speculation, but identifications of non-euconodonts as conodonts must nevertheless involve some degree of appeal to fairly general phylogenetic hypotheses. However, as long as the nature of the hypothesis is kept in mind, this kind of approach may be fruitful and is, indeed, often the only one available.

In this paper I shall examine the record of conodont-like fossils known from the Cambrian within the framework of a model of early conodont evolution proposed in 1976 (Bengtson 1976). The essentials of this model are: (1) conodont elements were primarily external, secreted by an epithelium, (2) an evolutionary trend during the Cambrian, represented by the sequence protoconodonts–paraconodonts–euconodonts, involved gradual retraction of the elements into pockets of this epithelium, (3) euconodonts were completely engulfed in epithelial pockets, which enabled them to grow holoperipherally and thus to assume more complex shapes that could be retained during ontogenetic growth, and (4) the secreting epithelium adhered only to the basal body of the euconodont, the crown being exposed to the aquatic medium when the apparatus was in use.

The model has a number of implications which may be tested against independent evidence: (1) The tissue of the elements was primarily derived from an ectodermal epithelium, (2) the basal body of early euconodonts is homologous with the 'cusp' of paraconodonts, (3) conodont elements functioned externally rather than as supports for soft tissue, and (4) they alternated between a retracted non-functional growth stage and a protracted functional non-growth stage.

The model was mainly based on histological structure as reflecting the mode of accretion of the elements. The present study extends the histological comparisons to conodont-like fossils additional to those that were considered in 1976. These comparisons can be used to test and develop the general model. In addition, the well-documented proposals by Jeppsson (1979) on the tooth function of euconodonts and by Szaniawski (1982; Repetski & Szaniawski 1981) on the chaetognath nature of protoconodonts may be combined with the proto–para–euconodont model to produce a more specific hypothesis of conodont evolution that sets up detailed proposals on the origin, affinities and biology of the conodont animal and on the anatomy and functional morphology of the conodont apparatus. In fact, some already published central predictions of this hypothesis have been confirmed by the recent discovery in the Scottish Carboniferous of the first convincing conodont animal with preserved soft parts (Briggs, Clarkson & Aldridge 1983), as will be discussed below.

Cambrian conodont-like fossils

The earliest known unquestionable euconodonts appear in the upper Franconian *Proconodontus tenuiserratus* Zone of western United States (Miller 1980; Miller *et al.* 1982) These forms (*P. tenuiserratus* Miller 1980 and the n. gen. n. sp. of Miller 1980:31–32) are characterized by simple-cone elements with prominent basal bodies and thin hyaline crowns. The histological evolution of crown and basal-body tissue in the late Cambrian is at present under investigation by J. F. Miller and myself and will be the subject of a later publication. Reviews of general euconodont histology have been given by Bengtson (1976), Müller (1981) and Lindström & Ziegler (1981).

Evolutionary transitions can never be *proven* to have occurred. Even if we identify fossils which are highly probably ancestors of euconodonts, there is always a remaining possibility that the ancestors of euconodonts are in fact unrepresented in the fossil record. Keeping this in mind, we may examine the record of conodont-like fossils in the Cambrian, essentially the record of phosphatic, cone-shaped fossils. The general cone shape may in each instance result from either or both of two conditions: (1) accretionary growth and (2) a function requiring a pointed shape (tooth, claw, defensive spine, etc.). Both of these conditions are common, and, consequently, so are instances of morphological convergence in simple cone-shaped fossils. In order to exclude such cases of convergence from the phylogenetic analyses it is necessary to investigate as many independent characters as possible, particularly histological. Although microstructures are not in themselves a more reliable tool for phylogenetic analysis than morphology, they may give crucial information on the origin and mode of formation of the tissue, which in turn may serve to restrict hypotheses concerning homology.

All figured specimens have been deposited at the Swedish Museum of Natural History, Stockholm (SMNH). Specimens sectioned for SEM investigation were polished and etched with 3 % HCl for 5–10 seconds before coating with gold–palladium.

Paraconodonts

The term 'paraconodont' is here used in the sense of Bengtson (1976:186) to signify conodont-like fossils with the structural

organization specified by Müller & Nogami (1971, 1972; cf. Bengtson 1976:200). The group corresponds only partly to the order Paraconodontida as defined by Miller (1981), since he also included the genera *Amphigeisina* Bengtson 1976 (only genus of the superfamily Amphigeisinacea Miller 1981), *Gapparodus* Abaimova 1978, and *Protohertzina* Missarzhevsky 1973, as well as the species '*Prooneotodus*' *tenuis* (Müller 1959) (within the superfamily Furnishinacea, family Furnishinidae Müller & Nogami 1971) which all are of the protoconodont grade of structural organization (Bengtson 1976, 1977, and herein; Szaniawski 1982), However, the terms proto-, para- and euconodont as used here are descriptive and do not necessarily imply a direct correspondence with taxonomic groups (although element histology certainly has taxonomic implications).

Morphology and histology. – The paraconodont elements are typically simple cones with more or less flaring bases. A few representatives (notably *Westergaardodina* Müller 1959) have secondary denticles at the base. Finds of bedding-plane associations and fused clusters of *Furnishina* Müller 1959 indicate that paraconodonts formed apparatuses similar to those known in simple-cone euconodonts (Szaniawski 1980a; Andres 1981). Most information about the histology of paraconodont elements derives from the work of Müller & Nogami (1971, 1972); the essential characters are: high organic content, very small apatite crystallites, thick growth lamellae, growth by basal accretion on the outside, on the basal edge, and (in some cases) on the inside of the element.

In one important respect the published accounts of paraconodont histology may have to be modified: Müller & Nogami (1971) stated that in the initial growth stages the lamellae are continuous around the tip of the element. The only specimen that has been adduced to show this is a thin section of *Problematoconites perforatus* Müller 1959 illustrated by Müller & Nogami (1971, Pl. 1:4). I have studied this section and find no evidence of apically closed lamellae – the structure that may have such an appearance in their Pl. 1:4 is formed by a dark Becke line. Other thin sections of paraconodonts in the Bonn collection (UB246, 325, 326, 328, 343–349, N922) also fail to show apically closed lamellae. The presence of apically continuous lamellae in the early growth stages of paraconodont elements has yet to be demonstrated.

Discussion. – According to the proposed evolutionary model, euconodonts were derived from paraconodonts through the addition of an outer apatitic layer (the crown) deposited from an enveloping epithelial fold. The idea was supported by the close histological similarities between paraconodont elements and the basal body of early euconodont elements. If paraconodonts did indeed have apically closed lamellae in the early growth stages, the evolutionary acquisition of a crown in euconodonts may have been due to a heterochronic process; whereas the possible absence (see above) of such lamellae in paraconodonts would postdate the appearance of the crown and make the suggested homology more exact.

Miller's (1980) unnamed new genus and species of euconodont found together with the first *Proconodontus tenuiserratus* satisfies the predictions of the hypothesized evolutionary model, for it is a form which differs visibly from associated

Fig. 1. Elements of *Protohertzina.* □A. *P. unguliformis,* Tommot, River Aldan, Yakutia, Precambrian–Cambrian boundary beds, sample 70g (V. V. Missarzhevsky). ×70. SMNH No. X2070. □B. Detail of A, base of internal cavity. ×500. □C. *P. unguliformis,* same locality as A–B, sample 70e (V. V. Missarzhevsky). ×130. SMNH No. X2081. □D. Same specimen in longitudinal section, polished and etched. Position of E indicated. ×130. □E. Detail of D. ×600. □F. *P. anabarica,* River Kotujkan, Anabar Massif, Yakutia, Nemakit-Daldyn Beds (Precambrian–Cambrian boundary beds), sample M419/12 (V. V. Missarzhevsky). Longitudinal section (perpendicular to plane of curvature), polished and etched. Position of G indicated. ×160. SMNH No. X2082. □G. Detail of F. ×1400.

paraconodonts only in the presence of a thin hyaline crown (Miller 1980:31–32). The new form may be very closely related to '*Coelocerodontus*' *rotundatus* Druce & Jones 1971 described from the Upper Cambrian of Queensland (Druce & Jones 1971); the histology of these forms and of the paraconodonts associated with them is at present under detailed investigation.

Protoconodonts

The term 'protoconodont' was introduced for slender Cambrian elements with only basal–internal growth increments (Bengtson 1976). This structure was demonstrated for Middle Cambrian *Gapparodus bisulcatus* (Müller 1959) and was later (Bengtson 1977) shown to occur also in the widespread Upper Cambrian '*Prooneotodus*' *tenuis*. The latter species is represented by numerous apparatus-like assemblages (bedding-plane associations or fused clusters) from several continents (e.g. Miller & Rushton 1973; Müller & Andres 1976; Landing 1977; Tipnis & Chatterton 1979; Abaimova 1980; Andres 1981; Szaniawski 1982). Szaniawski (1982) has shown that the patent general similarities between the '*P.*' *tenuis* apparatus and the grasping apparatus of modern chaetognaths are complemented by detailed morphological and histological similarities.

Protoconodonts are among the earliest cases of biomineralized tissues known from the fossil record. The genus *Protohertzina* is known from pre-Tommotian (i.e. uppermost Precambrian according to current stratigraphical concepts) deposits of the Siberian Platform, Kazakhstan, and Mongolia (Missarzhevsky 1973, 1982; Missarzhevsky & Mambetov 1981), and has also been reported from the base of the 'shelly' succession in China (e.g. Qian 1977; Qian, Chen & Chen 1979; Chen 1979, 1982) and the Canadian Cordillera (Conway Morris & Fritz 1980). The fine structure of *Protohertzina* is thus of particular interest and will be examined in detail here.

Morphology and histology. – Protoconodont elements are long (up to several millimetres), slender, slightly curved, commonly with one or more longitudinal keels (Figs. 1–2). The internal cavity reaches almost to the apex. Well-preserved specimens of *Gapparodus bisulcatus, Amphigeisina danica* (Poulsen 1966) and '*Prooneotodus*' *tenuis* show a three-layered wall structure: a thick, laminated middle layer bounded by thin outer and inner lamellae (Bengtson 1976, 1977; Szaniawski 1982, 1983). Organic content was originally high, and there is evidence to show that only the middle layer was to some degree mineralized with apatite (Bengtson 1976:196–197; Szaniawski 1982, 1983). This layer also demonstrates the basal–internal growth increments referred to above (not observed in *Amphigeisina*, however).

Three elements of *Protohertzina unguliformis* Missarzhevsky 1973 from the River Aldan, Yakutia, one of *P. anabarica* Missarzhevsky 1973 from the River Kotuj, Anabar Massif, and one each of *P. anabarica* and *P. cultrata* Missarzhevsky 1977 from Malyj Karatau, Kazakhstan, have been available for histological investigation. All are from Precambrian–Cambrian boundary deposits except for the *P. cultrata*, which is from the Upper Atdabanian. Two thin sections and four polished and etched sections were made. The observations indicate a consistent structure in *P. unguliformis* and *P. anabarica:* the

main layer of the wall is composed of lamellae arranged slightly obliquely to the wall in a manner indicating basal–internal accretion (Fig. 1D–G). The lamellae have a fibrous appearance, as they consist of acicular apatite crystallites with long axes parallel to the lamellae. This direction corresponds to that of the *c* axes; the structure is highly birefringent. (The fibrous structure can also be observed in an unsectioned specimen, Fig. 1B; the fibres are here probably enlarged by secondary apatite deposition.) The thin sections (one of each species) show an impersistent dark-coloured outer portion of the wall with no birefringence at all or with *c* axes perpendicular to the lamellae. This may represent recrystallized or originally less mineralized portions of the wall. There is no persistent outer layer visible; the thin crust seen in Fig. 1G is of uncertain origin.

In the specimen of *P. cultrata* the lamellae are non-fibrous (Fig. 2), as in *Gapparodus bisulcatus* and '*Prooneotodus*' *tenuis*. There is also a secondary apatitic lining of the internal cavity with surface-normal (i.e. perpendicular to the outer surface) acicular crystallites (Fig. 2C). A similar layer of surface-normal to spherulitic apatite has also been observed in the internal cavity of elements of *P. unguliformis* and *P. anabarica*.

Discussion. – Szaniawski (1982) compared the structure of grasping spines of modern *Sagitta* with that of Upper Cambrian '*Prooneotodus*' *tenuis* elements. He found a three-layered structure in the wall around the pulp cavity of *Sagitta* spines that is strikingly similar to the structure in protoconodonts. Probable growth increments, reflected on the surface as well as in the structure of the thick middle layer, are present in *Sagitta*, as in protoconodonts. The main differences are that *Sagitta* spines are unmineralized, and the middle layer is fibrous. These fibres consist of chains of highly crystalline α-chitin (Atkins, Dlugosz & Foord 1979). It is interesting to note that a fibrous structure is indeed present in the earliest protoconodonts, i.e. the *Protohertzina unguliformis* and *P. anabarica* figured here. (Although it is a protoconodont too, the younger *P. cultrata* differs considerably from the two older species both morphologically and histologically; it may in fact not be so closely related to them that it deserves placement in the same genus.) Fibres consisting of acicular apatite crystallites and chitin chains, respectively, are not likely to be strictly homologous, but the fibrous structure as such may reflect similarities in the organization of the tissues responsible for formation of the spines.

Bone, Ryan & Pulsford (in press), studying spines and teeth of *Sagitta setosa* Müller 1847, discovered complex structures (fibrils or lamellae, probably representing aggregated crystalline sheets of α-chitin) traversing the middle layer and connecting the outer and inner layers of the wall. They also found high concentrations of silicon in the tip, and of zinc in the tip and the outer and inner layers of the wall. Such features have not yet been found in protoconodont elements. However, their presence need not be expected in apatitic elements even if they are homologous to chaetognath spines, for in the latter these features probably all serve to strengthen and harden the unmineralized spines (Bone, Ryan & Pulsford, in press).

Outer and inner layers are not clearly visible on the *Protohertzina* specimens investigated here. Thin organic layers, if present, are generally not likely to be preserved, and one must be cautious when interpreting thin crusts on surfaces of fossil specimens – they may well be diagenetic, as is evident in the case of the inner apatitic crusts seen in many specimens of *Amphigeisina* (Bengtson 1976, Figs. 1A–B and 3) and *Protohertzina* (Fig. 2C herein). Such a crust is usually characterized by growth of acicular crystallites normal to the encrusted surface, with tell-tale spherulitic growth around irregularities on the surface. The fused lamellae reported as primary tissue in clusters of *'Prooneotodus' tenuis* by Tipnis & Chatterton (1979) are typical of such diagenetic structures (cf. also Repetski & Szaniawski 1981:170–171).

Although some paraconodonts (and euconodonts) have slender shapes approaching those of protoconodonts, none of the forms investigated so far could be interpreted as morphological–histological intermediates between proto- and paraconodonts (cf. Bengtson 1977:43). For this reason the evolutionary transition between these two groups is more hypothetical than the transition from para- to euconodonts.

RHOMBOCORNICULUM

Rhombocorniculum Walliser 1958 is a widely distributed fossil in the upper Lower Cambrian, first reported from Shropshire, England (Cobbold 1921; Walliser 1958). It is found as phosphatic sclerites, narrowly conical or broadly triangular, with a characteristic rhomboidal surface microsculpture (Fig. 3). The type species, *R. cancellatum* (Cobbold 1921), has been interpreted as having two kinds of elements, slender and broad, which occur in the same deposits (Walliser 1958; Landing, Nowlan & Fletcher 1980). (Mambetov 1977, however, considered the broad element to represent a separate species, *R. walliseri* Mambetov 1977.) Missarzhevsky (*in* Missarzhevsky & Mambetov 1981) has described another species, *R. insolutum* Missarzhevsky 1981, characterized by slender elements with rounded cross-section and spinose surface sculpture.

Morphology and histology. – Walliser (1958) and Landing, Nowlan & Fletcher (1980) described the morphology and histology of *R. cancellatum* in detail. The elements are up to a couple of millimetres long, usually asymmetrical, more or less twisted, with a narrow internal cavity. A slender element is figured herein (Fig. 3A) showing the characteristic surface ornament. The internal structure is coarsely fibrous (Walliser 1958, Pl. 15:6–7; Landing, Nowlan & Fletcher 1980, Figs. 3–4; Fig. 3B–C herein). The fibres are arranged obliquely to the walls, those from opposing walls converging in the apical direction, i.e. the inclination agrees with that of the fibrous lamellae in *Protohertzina unguliformis* and *P. anabarica*. The composition is apatitic (Walliser 1958), and observations under a petrographic microscope show that the fibres are composed of highly birefringent crystals with *c* axes parallel to the longitudinal direction of fibres. The presence of outer (Walliser 1958) and inner (Landing, Nowlan & Fletcher 1980) organic lamellae has been reported, but information on these structures is not sufficient to allow a more detailed determination of their nature.

Discussion. – Apart from some apparently arbitrary assignments to groups such as the blatantly polyphyletic Cambro-

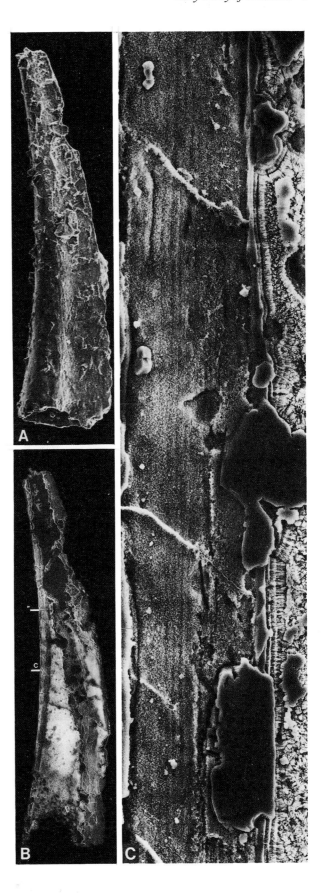

Fig. 2. Element of *Protohertzina cultrata*, Aktugaj, river Shabakty, Malyj Karatau, Shabakty Formation (upper Atdabanian), sample M52 (A. M. Mambetov). SMNH No. X2083. □A. ×65. □B. Longitudinal section, polished and etched. Position of C indicated. ×65. □C. Detail of B. ×900.

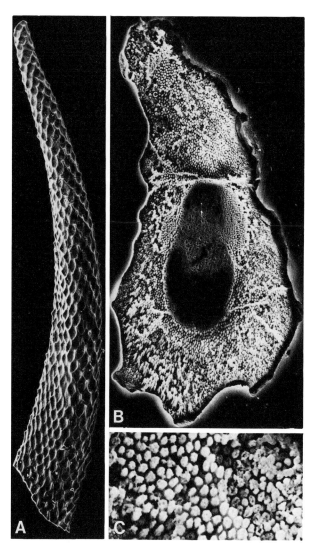

Fig. 3. Slender elements of *Rhombocorniculum cancellatum*. □A. Fortune River, Burin Peninsula, Newfoundland, Brigus Formation, ca. 10 m above base, sample Can79-30-SB. ×100. SMNH No. X2084. □B. Achchagyj-Kyyry-Taas, River Lena, Yakutia, Transitional 'Formation', 2nd Member (upper Atdabanian), sample Sib73-2-SB. Transverse section, polished and etched. Position of C indicated (upper left). ×500. SMNH No. X2085. □C. Detail of B. ×2000.

scleritida Meshkova 1974 (Meshkova 1974; Mambetov 1977; cf. Bengtson 1977:60–61) or the Halkieriidae Poulsen 1967 (Missarzhevsky 1977; Missarzhevsky & Mambetov 1981), *Rhombocorniculum* has generally been left without a suprageneric assignment. Müller (1962) originally assigned it to his order Paraconodontida together with *Problematoconites* Müller 1959 and *Pygodus* Lamont & Lindström 1957, but subsequent histological investigations led to a considerable emendation of the Paraconodontida, to include a number of Cambrian forms structurally similar to *Problematoconites*, and to exclude *Rhombocorniculum* and the euconodont *Pygodus* (Müller & Nogami 1971). Landing, Nowlan & Fletcher (1980) introduced the (seemingly expletive) term 'pseudoconodont' for *Rhombocorniculum* sclerites and discussed the general similarities with protoconodont and euconodont apparatuses, concluding that the *Rhombocorniculum* elements 'may have served a similar function and had a broadly comparable morphology in un unrelated organism'. I agree

with this conclusion, except that the fibrous structure of *Protohertzina* described herein certainly allows for the interpretation that *Rhombocorniculum* was derived from an early protoconodont. Thus a near relationship with conodonts is conceivable, but the available characters do not allow a strict assessment of this possibility. There are no known forms that show any evidence of being descendants of *Rhombocorniculum*.

FOMITCHELLA

The phosphatic cones of *Fomitchella infundibuliformis* Missarzhevsky 1969 (Fig. 4) were described by Missarzhevsky (*in* Rozanov *et al.* 1969) from the lower part of the Tommotian Stage in Siberia, i.e. basal Cambrian beds as currently understood. In 1977 Missarzhevsky described a second species, *F. acinaciformis*, from these beds, and similar forms are now known also from approximately age-equivalent beds in Malyj Karatau (Missarzhevsky & Mambetov 1981), eastern Massachusetts (Landing & Brett 1982) and southeastern Newfoundland (Bengtson & Fletcher 1983). [Other reports of *Fomitchella* appear questionable on the evidence of published illustrations, viz. the *Fomitchella* sp. of He (*in* Yin *et al.* 1980, Pl. 19:7), the *F. rugosa* and *F.* cf. *infundibuliformis* of Jiang (*in* Luo *et al.* 1982, Pl. 17:14–15 and Pl. 17:16, respectively), and the *F. yankonensis* of Yuan & Zhang (1983, Pl. 1:6); also the *Paraformichella* [sic!] *orientalis* of Qian & Zhang (1983, Pl. 2:15–17).] *Fomitchella infundibuliformis* is very conodont-like and has even been suggested as the stock from which all later conodonts, including paraconodonts, arose (Dzik 1976, Fig. 1).

Morphology and histology. – Specimens of *F. infundibuliformis* have also been illustrated by Missarzhevsky (*in* Rozanov *et al.* 1969, Pl. 6:12, 15, 16), Meshkova (1969, Pl. 56:1–5, as '*Oneotodus?* sp.'; *in* Repina *et al.* 1974, Pl. 17:5) and Matthews & Missarzhevsky (1975, Pl. 3:8). They are hollow cones, usually slightly flattened laterally, with widely flaring bases and narrow, drawn-out tips. The internal cavity is large and the walls consequently very thin (down to 5 μm) but the cavity does not extend into the narrow tip. Outer and inner surfaces are practically smooth; a faint radial ornament is sometimes visible on the outside.

In longitudinal section (Fig. 4B–D) the cones are seen to consist of very fine (0.5–2 μm) lamellae parallel to the outer surface of the cones. These lamellae wedge out towards the inner side, so that the innermost lamellae are restricted to the apical part. The lamellae appear finely granular under the SEM, and the substance shows very weak birefringence in polarized light, suggesting that individual crystallites are very small with a *c* axis orientation which is only to a degree preferential (tangentially in the cross-sectional plane). There may be fine (about 1 μm in diameter) tubules piercing the lamellae (Fig. 4D, top), but these may have been produced *post mortem* by boring organisms.

Discussion. – If the lamellae are taken to represent successive growth increments, their arrangement shows that secretion took place on the *outside* of the cone, i.e. that the outer side was covered with secretory tissue. This is similar to the condition in the euconodont crown (cf. Fig. 5), but fundamentally different

Fig. 4. Fomitchella infundibuliformis. River Fomich, Anabar Massif, Zone of *Aldanocyathus sunnaginicus – Tiksitheca licis*; sample M314/4 (V. V. Missarzhevsky). □A. ×100. SMNH No. X2086. □B. Longitudinal thin section in transmitted, plane-polarized light. ×600. SMNH No. X2087. □C. Longitudinal section, polished and etched. Position of D indicated. ×200. SMNH No. X2088. □D. Detail of C. ×1 500.

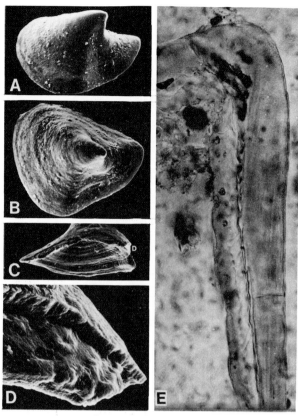

Fig. 5. Elements of *Pseudooneotodus* cf. *mitratus* (Moskalenko 1973), 'elf caps' of Winder (1976). Colbourne Quarry, Ontario, upper Trenton Group (Caradoc), coll. C. G. Winder. □A–B. ×100. SMNH No. X2089. □C. View from below showing basal body and thin crown. Position of D indicated. ×100. SMNH No. X2090. □D. Detail of specimen in C, showing partly exfoliated crown on basal body. ×800. □E. Thin section in transmitted, plane-polarized light, showing crown (top and right) and basal body. ×400. SMNH No. X2091.

from that in the Cambrian paraconodonts. In fact, there seems to be no possibility that the elements of, e.g., *Furnishina furnishi* Müller 1959 and *Prooneotodus gallatini* (Müller 1959) could be derived from the morphologically somewhat similar, but histologically different, cones of *Fomitchella infundibuliformis*. There is a radical difference in mode of secretion which hardly permits the structures to be interpreted as homologous.

This similarity between *Fomitchella* and euconodonts presents a challenge to the hypothesis that euconodonts evolved from paraconodonts in the late Cambrian. There are four possible interpretations:

(1) Late Cambrian euconodonts did not evolve from paraconodonts but from *Fomitchella infundibuliformis* or a close ancestor in common with *Fomitchella*.

(2) Euconodonts did in fact evolve from paraconodonts, but considerably earlier than supposed, in the late Precambrian rather than the late Cambrian. *Fomitchella* is an early euconodont. The stratigraphic order of appearance of the other forms misrepresents their evolutionary order of appearance.

(3) The euconodont grade of organization was attained more than once and at different times, *Fomitchella* representing an early such example. The euconodont animals are polyphyletic and the crown not truly homologous between different euconodonts.

(4) *Fomitchella* is not a member of the conodont stock, and the similarity to euconodonts is convergent or fortuitous.

The first interpretation would reduce the similarities beween paraconodont elements and the basal bodies of euconodonts documented by Müller & Nogami (1971, 1972; cf. Bengtson 1976) to convergence or chance, and it would negate the hypothesis that euconodonts evolved from paraconodonts by the acquisition of a crown. Against this interpretation there stands the orderly stratigraphic succession in the Upper Cambrian of paraconodonts, euconodonts with thin crown and prominent basal body (e.g. *Proconodontus* Miller 1969) and euconodonts with thicker crown and smaller basal body (e.g. *Teridontus* Miller 1980, *Eoconodontus* Miller 1980, *Cambroistodus* Miller 1980) as documented by, e.g., Miller (1969, 1980; Miller *et al.* 1982). The complete absence of a basal body in the well-preserved Tommotian *Fomitchella* contrasts with the prominent basal bodies present in the earliest appearing late Cambrian euconodonts (*Proconodontus* and the n. gen. et sp. of Miller 1980:31–32).

The second interpretation is also at variance with the observed stratigraphical succession. Whereas such stratigraphical misrepresentations are theoretically possible, they are not very likely in the present case of commonly abundant microfossils whose distribution has been investigated in detail in key sections in North America (e.g. J.F. Miller 1969, 1978, 1980; Derby, Lane & Norford 1972; R. H. Miller & Paden 1976; R. H. Miller *et al.* 1981; Fåhraeus & Nowlan 1978; Landing, Ludvigsen & von Bitter 1980; J. F. Miller *et al.* 1982), northern Europe (Müller 1959; Szaniawski 1971, 1980b; Bednarczyk 1979), Kazakhstan (Abaimova & Ergaliev 1975; Abaimova *et al.* 1978; Abaimova 1978; Dubinina 1982), Iran (Müller 1973), China (Xiang *et al.* 1981), and Australia (Druce & Jones 1971; Druce, Shergold & Radke 1982). The lack of a trace of a basal body in *Fomitchella* is also inconsistent with this interpretation.

The third interpretation is somewhat strained, again because of the lack of a basal body in *Fomitchella*.

The fourth interpretation fits the available data, but at the price of an *ad hoc* assumption that *Fomitchella* has nothing to do with conodonts. This assumption can be partly defended by the fact that *Fomitchella*, apart from the anomalous stratigraphic position, differs in some respect from early euconodonts. First, in spite of the good preservation (no sign of etching, recrystallization, mechanical abrasion or heating) and the deep internal cavity there is no trace of a basal body. Second, the lamellae consist of finely granular apatite without preferred crystallographic orientation, whereas euconodont crowns typically show a strong preferred orientation of crystallographic *c* axes, usually in connection with acicular crystallites. None of these differences can be taken to prove that *Fomitchella* is not a euconodont, but it must be remembered that the remaining similarities to the euconodont crown are of a simple and general nature, making convergence or fortuitous resemblance a not unlikely possibility.

In summary, *Fomitchella* is not a paracondont or a paraconodont ancestor, but its possible euconodont nature cannot be ruled out by its characters alone. In view of the total stratigraphical and histological evidence on the origin of euconodonts, however, acceptance of *Fomitchella* as a conodont strains credulity much more than the alternative assumption that it is no close relative of conodonts.

If not a conodont element, the phosphatic cone of *Fomitchella* may be interpreted as a supporting cup for an organism having secreting soft tissue extending over the outside of the mineralized structure. This organism may have been a colonial animal, the cones secreted by a coenenchyme, but there is no direct comparison with the superficially similar laminated phosphatic linings of some bryozoan zooecia (Martinsson 1965) because the lamination indicates that these were secreted by the zooids themselves (Martinsson 1965; Eisenack 1964).

LAPWORTHELLA

The phosphatic cones of *Lapworthella* Cobbold 1921 are not strictly conodont-like, but a case could be made for a *Lapworthella*-like animal as an ancestor of paraconodonts; this possibility and its implications will be dealt with here.

Species of *Lapworthella* are known from Lower Cambrian (and some possibly Middle Cambrian) deposits in northwestern Europe, eastern North America, the Siberian Platform, Mongolia, South China, and South Australia (species list in Bengtson 1980; new occurrences reported by Landing, Nowlan & Fletcher 1980; Yin *et al.* 1980; Landing & Brett 1982; Voronin *et al.* 1982; Luo *et al.* 1982; Bengtson & Fletcher 1983; Qian & Zhang 1983). Sclerites typically show very large morphological variation within samples (this has led to some taxonomic oversplitting). Several lines of evidence suggest that the sclerites were parts of a composite exoskeleton similar to that inferred for mitrosagophorans (Bengtson 1977:58–60); *Lapworthella* is generally considered to be closely related to the mitrosagophorans and some other Cambrian animals with composite phosphatic exoskeletons, such as *Dailyatia* Bischoff 1976 and the Kelanellidae Missarzhevsky & Grigor'eva 1981. The choice of *Lapworthella* for the present comparison with paraconodonts is mostly for the sake of illustrating a point; its sclerites are most similar to paraconodonts, but because detailed morphology may not be relevant at this level of comparison, any of these other forms could have been chosen.

Morphology and histology. – Lapworthella sclerites (Fig. 6A–B; see also Matthews 1973 for examples of morphological variability) range from broadly pyramidal to narrowly conical. In most species there are very pronounced annulations as well as finer growth lines present (Fig. 6C). A section through a specimen of *L. dentata* Missarzhevsky 1969 (Fig. 6D–E) shows that the fine growth lines correspond to a fine lamination, confirming that growth took place by basal–internal accretion. Under a petrographic microscope, the walls in this species are seen to be made up of highly birefringent apatite with *c* axes aligned parallel to the growth axis of the sclerite. The etching pattern (Fig. 6E) suggests very small (in the size order of tenths of micrometres) isodiametric crystallites; there is no sign of acicular structures.

Discussion. – In general terms, *Lapworthella* sclerites resemble paraconodont elements in gross morphology, composition and inferred external mode of formation. The main differences are the pronounced surface sculpture and lack of basal growth

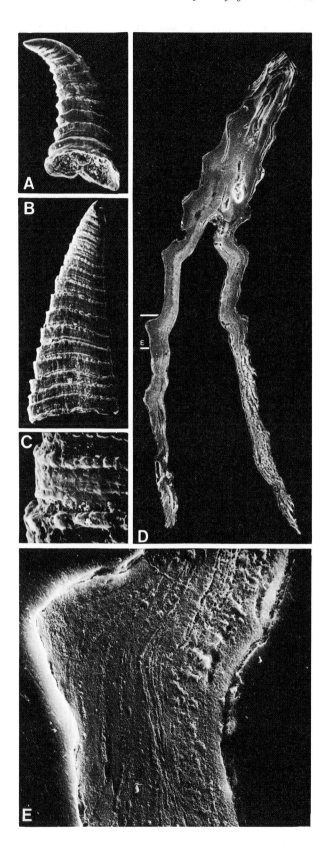

Fig. 6. Sclerites of *Lapworthella dentata* Missarzhevsky, Achchagyj-Kyyry-Taas, River Lena, Yakutia, Transitional 'Formation', 3rd Member (upper Atdabanian or lower Lenian). □A. Sample M49/106 (V. V. Missarzhevsky). ×60. SMNH No. X2092. □B. Same sample as A. ×30. SMNH No. X2093. □C. Detail of B. ×200. □D. Sample Sib73-1-SB. Longitudinal section, polished and etched. Position of E indicated. ×200. SMNH No. X2094. □E. Detail of D. ×1 500.

increments deposited on the outside. None of the similarities are necessarily indicative of close relationship (they are common features of external sclerites), and the differences indicate a significant difference in function at least: *Lapworthella* sclerites were probably passive protective devices (the sculpture would have interfered with an active function as teeth or graspers) whereas paraconodont elements, judging both from their morphology and their probable homology with euconodont elements, most likely were mouth parts.

Nevertheless, at first an evolutionary transition between the two does not seem entirely unlikely – it could be regarded as a process analogous to the odontode–tooth transition in vertebrates (cf. Ørvig 1977). In the absence of plausible transitional forms leading to paraconodonts (contrast the state of the evidence bearing on the paraconodont–euconodont transition) both the hypothesis of a protoconodont ancestry and that of a *Lapworthella* ancestry (as well as other alternative hypotheses) of paraconodonts must be evaluated along less direct lines of evidence. Such indirect evidence, however, seems to favour protoconodont rather than *Lapworthella* ancestry:

(1) Although in terms of shift of growth loci the adjustment to the paraconodont mode is about the same from both protoconodonts and *Lapworthella*, the derivation of a paraconodont apparatus from a protoconodont apparatus would be considerably simpler in terms of arrangement and probable function than its derivation from the sclerite armour of *Lapworthella*.

(2) Various lines of evidence suggest that the euconodont animal was an active nektic or planktic predator (e.g. Seddon & Sweet 1971; Jeppsson 1979; Briggs, Clarkson & Aldridge 1983). This is much more in keeping with a derivation from the chaetognath-like protoconodonts (Szaniawski 1982) than from *Lapworthella*, which with its heavy mail of sclerites was probably a sluggish or sessile benthic organism.

These considerations would not contradict a more distant relationship between *Lapworthella* with relatives on one side and the conodonts on the other, for example the sharing of a close common ancestor. However, at the level of comparison the similarities only amount to such general characters as the presence of external phosphatic sclerites, and at present there seems to be no practical possibility of testing any such hypothesis.

Other conodont-like fossils

The remaining Cambrian conodont-like fossils to be discussed here are too poorly known to play a role in the evaluation of the proto–para–euconodont evolutionary model. They indicate areas where more investigation is needed and also serve as a reminder that more fossils with some bearing on the conodont problem may yet turn up in the Cambrian.

Odontogriphus omalus Conway Morris 1976 is represented by a single specimen with preserved soft-tissue remains in the Middle Cambrian Burgess Shale. The specimen has been described in great detail by Conway Morris (1976) who interpreted it as a paraconodont animal in keeping with the hypothesis advanced by Lindström (1973, 1974), that conodont elements were tentacle supports in a lophophorate-like animal. The presumed conodont elements in the specimen are, however, too poorly preserved to be identified as

paraconodonts or even conodont-like structures (composition and histology are unknown, and the structures may owe their apparent shape largely to their position between the lobate structures interpreted by Conway Morris as remains of tentacles). The evidence for tentacles is also equivocal. Until we can with some confidence refute or confirm the presence of conodont elements, *Odontogriphus* is of very dubious relevance to questions of the history and nature of the Conodonta. (See also discussions on the interpretation of *Odontogriphus* by Landing 1977:1082–1083; Conway Morris 1980; and Jeppsson 1980.)

Mongolodus rostriformis Missarzhevsky 1977 was described from the base of the Lenian Stage of Western Mongolia (Missarzhevsky 1977). It is represented by claw-shaped, curved, laterally compressed sclerites with a base strongly expanded in the plane of curvature. The internal cavity is large and the walls thin. There is a suggestion of 'cone-in-cone' lamellation indicating growth by basal–internal accretion, but the histology is not known in detail. *Mongolodus* is possibly related to the protoconodonts, but the available evidence does not permit a firm evaluation of this possibility.

Yunnanodus doleres Wang & Jiang 1980 was described (*in* Jiang 1980) from phosphorite beds in the upper part of the Meishucunian Stage in eastern Yunnan. The sclerites of this species have a straight 'cusp' attached to a basal plate set with irregularly arranged smaller denticles. The original composition and histology are not known. Jiang (1980) suggested conodont affinity. Although *Yunnanodus* is vaguely conodont-like, it does not show any distinctive conodont features, and certainly none that suggest any affinity with known Cambrian conodonts. Some attention should be given to the possibility that it is related instead to the Zhijinitidae Qian 1978 (=Cambroclavitidae Mambetov 1979), known from the upper Atdabanian Stage of Kazakhstan (Mambetov & Repina 1979; Missarzhevsky & Mambetov 1981) and the upper Meischucunian Stage of China (Qian 1978; Qian, Chen & Chen 1979; Luo *et al.* 1982). The sclerites of this enigmatic group are composed of a straight or curved 'cusp' attached to a distinct basal plate. Finds of articulated sclerites show that they united to form a complex palisade-like aggregate (Mambetov & Repina 1979, Pl. 14:6, 8, 9) with no resemblance to a conodont apparatus.

In a recent publication, Chen (1982) described some new presumed conodonts from the Meishucunian of Emei-shan, Szechuan. Of these, *Emeidus primitivus* Chen 1982 appears morphologically similar to *Protohertzina*, whereas the two serrate species of *Paracanthodus* Chen 1982 are more difficult to place in a high-level taxonomic group on the basis of the published illustrations. The fragment of a curved rod with three processes illustrated as 'Conodont Form A gen. et sp. indet.' appears to have little claim to recognition as a conodont.

A reassessment of the evolutionary model

The search among Cambrian conodont-like fossils has not yielded a more likely ancestor for the euconodonts than a paraconodont, or a more likely ancestor for the paraconodonts than a protoconodont. *Rhombocorniculum* may be envisaged as derived from a protoconodont, but not itself as a paraconodont ancestor. *Fomitchella* cannot be envisaged as a protoconodont or a paraconodont ancestor, and hardly even as an ancestor of the simplest euconodonts (*Proconodontus* and Miller's new genus) – and if it were an ancestor of other conodonts this would falsify not only the evolutionary model under assessment, but also the identification of almost all Cambrian conodonts, including the earliest euconodonts. *Lapworthella* (or another animal with external phosphatic protective sclerites) can be envisaged as a paraconodont ancestor, but much less easily so than any protoconodont.

Thus the proto–para–euconodont model has passed the test in good condition: any alternative interpretation of the phylogenetic relationships of Cambrian conodont-like fossils appears less probable than the one shown in Fig. 7. The remaining possibilities that either euconodonts or paraconodonts (or both) evolved directly from other, perhaps unknown, forms, are more difficult to assess. That euconodonts should not be derived from paraconodonts seems very unlikely, but the possibility will be further appraised in the current work on the histology of the apparent transitional forms. Paraconodont derivation from something other than a protoconodont is more of a possibility. At present it can only be assessed indirectly, through the way in which the implications of the protoconodont model are compatible with new evidence. these implications can be made considerably more specific and testable by reference to the proposed close affinity between Cambrian protoconodonts and modern chaetognaths (Szaniawski 1982).

It should be noted that a simple list of character similarities between two groups does not suffice for systematic purposes. Comparisons with other groups, both those closely related (to exclude similarities due to symplesiomorphy) and those more distantly related (to rule out similarities due to convergence) are required. In the case of grasping spines it has not yet been shown whether structures similar to those found in chaetognaths and protoconodonts may also occur by convergence in cuticular spines or teeth of other groups (e.g. priapulids, acanthocephalans, gymnosomatous pteropods). However, the available evidence certainly supports Szaniawski's proposal, and the fibrous structure of *Protohertzina* reported herein serves to strengthen the case.

Chaetognath grasping spines sit laterally on the head of the animal, well outside the mouth. The action of the grasping apparatus is well integrated with the action of the whole head in feeding (Kuhl 1932), and it is obvious that the morphologically more complex conodonts could not have functioned in an identical way. Nevertheless, the evolutionary model under consideration postulates that no radical change in function is necessary to explain the burgeoning of conodont element morphology in the early Ordovician (Bengtson 1976:202). Jeppsson (1979) has pointed to a number of analogies between euconodont elements and teeth (*sensu lato*) in various groups of animals (and included reference to structures that had previously been quoted as arguments *against* a tooth function of conodont elements), which suggest similarities of function. (See Conway Morris 1980; Jeppsson 1980; Bengtson 1980, 1983; and Briggs, Clarkson & Aldridge 1983 for a recent discussion of the tooth model *versus* the tentacle-support model of conodont function.)

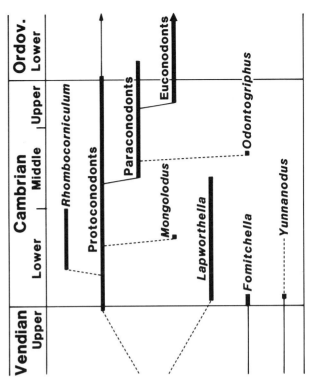

Fig. 7. Proposed phylogenetic relationships of Cambrian conodont-like fossils discussed in the text. Proto-, para- and euconodonts informally grouped as taxa. Bars show known stratigraphic ranges of 'taxa'. Unbroken lines indicate proposed evolutionary lineages; broken lines indicate possible but insufficiently corroborated lineages. *Fomitchella* and *Yunnanodus* are thus shown as independently derived from Vendian ancestors, whereas the possibility of a near common ancestry for *Lapworthella* (and other related forms) and protoconodonts is indicated by the divergence of broken lines.

Three central predictions of the hypotheses in question appear to have been verified by the recent discovery in the Scottish Carboniferous of a conodont animal with preserved soft parts (Briggs, Clarkson & Aldridge 1983):

(1) Contrary to customary reconstructions (e.g. Rhodes & Austin 1981), an apparatus with ramiform and pectiniform elements should have the ramiform, not the pectiniform, elements anteriormost (Jeppsson 1971:101, 120, Fig. 4).

(2) The conodont apparatus, when not in function, should be stored in a retracted resting position (Bengtson 1976:203; Jeppsson 1979:167–168).

(3) The euconodont animal should show chaetognath affinities (Repetski & Szaniawski 1981).

The first two predictions have been confirmed by the Scottish animal (Bengtson 1983). The suggestion of verification of the third one can be challenged, in that the chaetognath-like characters seen in the Scottish animal are not exclusively indicative of chaetognath affinity, and there is one feature – the oblique (possibly V-shaped) regularly repeated structures in the posterior part of the trunk – that cannot easily be matched with any known chaetognath character. In a chordate model they could represent myotomes (Briggs, Clarkson & Aldridge 1983:10); in a chaetognath model they could possibly be cuticular bandings or gonad structures, but both of these interpretations offer problems. Further finds are not likely to solve this problem directly (although gonad structures may be expected to show varying development in different individuals depending on the stage of maturity), but might, one would hope, provide an answer to the important question of the orientation of the animal. If the Scottish animal has a dorso-ventrally flattened chaetognath-type tail, further finds should show the symmetry of the (lateral) fins; if it has a laterally flattened chordate-type tail, the asymmetry of the (sagittal) fins is likely to show up as a consistent pattern. It should be kept in mind, however, that the conodont animals may yet show significant deviations from the body plan of their closest living relative; thus fin morphology may not be conclusive evidence against either chordate or chaetognath affinity.

The question of conodont affinity is thus unresolved even by the find in the Carboniferous of Scotland, but there are now sufficient grounds for developing the proto–para–euconodont model around the possibility of chaetognath affinity. The model can then be reformulated as follows:

(1) Predatory chaetognaths first appeared no later than at the Vendian–Cambrian transition. They were equipped with a grasping apparatus of essentially modern aspect. Individual spines, the protoconodont elements, were at least in some cases mineralized with calcium phosphate.

(2) No later than early middle Cambrian a branch of the chaetognath stock evolved, characterized by a robust denticulation around the mouth or in the pharynx. The individual denticles, the paraconodont elements, were derived either from the lateral spines or from a weaker denticulation around the mouth of the ancestor (cf. the teeth of modern chaetognaths). The denticles were deeply invested in the epithelium of the mouth–pharynx, and grew during the course of the animal's ontogeny by secretion of mineralized tissue from this epithelium. They were partly or wholly enveloped in epithelial folds when the animal was not feeding and the pharynx was in a constricted resting position.

(3) In the middle late Cambrian there appeared forms in which the secretionary activity of the pharyngeal epithelium in the folds around the denticles created favourable conditions for deposition of a thin apatite crust on the free surface of the denticle during periods when the pharynx was not expanded for feeding. Denticles so invested were the first euconodont elements.

(4) Soon after the first appearance of this new tissue – the crown – it began to be utilized for more complex constructional needs. Initially it formed cutting edges (keels), then it took over more and more of the construction of the cusp. This involved increasing biological control of the formation of the tissue, probably through an increasingly structured organic matrix. As the plastic potential of this new mode of growth began to be realized (Bengtson 1976:202), the euconodont elements grew more complicated, and the enfolding of the pharynx into the rest position became increasingly complex. This change was accompanied by a diversification of feeding habits and life modes, but the denticles, the euconodont elements, continued to function as teeth (*sensu lato*; Jeppsson 1979).

In regard to the systematic position of the Conodonta, the model outlined allies them closely to the Phylum Chaetognatha, but as formulated here it does not touch on the interesting possibility of a close relationship between chaetognaths and chordates. If one accepts that taxonomic groups may be paraphyletic (i.e. if one is not an ardent cladist) it is possible to argue for a separate paraconodont–euconodont Phylum Conodonta, derived from the protoconodont Phylum Chaetognatha. But with regard to the total phenotype, the presence or absence of pharyngeal denticulation is only a detail, and if further finds of well-preserved conodont animals were to confirm close adherence to a chaetognath body plan, the Conodonta would be better placed as a subphylum or class within the Phylum Chaetognatha.

This model is sufficiently specific to be both informative and testable. Unfortunately for the latter aspect, however, it suggests that there may be an evolutionary discontinuity between the protoconodont and paraconodont elements, in that the latter may not have evolved from the former, but from histologically similar but morphologically different structures nearer to the mouth. Thus there may be no prospect of finding an evolutionary series between these two types of elements (and there is no evidence that any species was provided with both protoconodont grasping spines and paraconodont denticles). Nevertheless, in other respects the model allows clear inferences on the early evolution of conodont elements and their histogenesis. It precludes the possibility that the primary mineralized tissue of conodont elements was formed by cells homologous to vertebrate osteogenic cells; any mesodermally derived mineralized tissue would have had to be secondarily added to the primarily ectodermal elements. (The recent suggestion by Barskov, Moskalenko & Starostina 1982 that vertebrate bony tissue is present in Ordovician conodonts would need to be strengthened – with regard to the conodont nature of their specimens of *Coleodus* as well as to the morphology and identity of the alleged cavities after osteocytes, osteoblasts and osteoclasts – before it would offer any threat to the idea that conodont elements were derived from an ectodermal epithelium.) It further suggests that element morphology in euconodonts may be analyzed specifically in terms of (1) function in the manipulation of prey, (2) storage within a collapsed pharynx, and (3) growth within epithelial folds.

Questions concerning the origin and early evolution of conodonts can thus be seen to be in continuous association with the wide spectrum of problems of euconodont biology. Many answers to the current problems of the early history of conodonts, including the further testing of the evolutionary model presented here, are likely to come from improved insights into the nature of post-Cambrian conodonts.

Acknowledgements. – My work has been financed through grants from the Swedish Natural Science Research Council. I thank Vladimir V. Missarzhevsky (Moscow), Amanbek M. Mambetov (Frunze) and Gordon C. Winder (London, Ontario) for providing material indispensable for this investigation. Quentin Bone (Plymouth), Derek E. G. Briggs (London) and Hubert Szaniawski (Warsaw) kindly gave me access to unpublished manuscripts. Of the many colleagues who have helped me with information and critique I would especially like to mention Simon Conway Morris (Milton Keynes), Lennart Jeppsson (Lund) and Klaus J. Müller (Bonn). Simon Conway Morris, Anders Martinsson (Uppsala), S. Crosbie Matthews (Uppsala) and James F. Miller (Springfield, Missouri) read the manuscript and provided many helpful suggestions; it is with grief that I record that Anders Martinsson and Crosbie Matthews both died shortly before this volume went to the press.

References

Abaimova, G. P. (Абаимова, Г. П.) 1978: Позднекембрийские конодонты центрального Казахстана. [Late Cambrian conodonts from central Kazakhstan.] *Палеонтологический журнал 1978:4*, 77–87.

Abaimova, G. P. (Абаимова, Г. П.) 1980: Аппараты кембрийских конодонтов из Казахстана. [Apparatuses of Cambrian conodonts from Kazakhstan.] *Палеонтологический журнал 1980:2*, 143–146.

Abaimova, G. P. & Ergaliev, G. Kh. (Абаимова, Г. П. & Ергалиев, Г. Х.) 1975: О находке конодонтов в среднем и верхнем кембрии Малого Каратау. [Finds of conodonts in the Middle and Upper Cambrian of Malyj Karatau.] *Труды Института Геологии и Геофизики СО АН СССР 333*, 390–394.

Abaimova, G. P., Ergaliev, G. Kh., Koneva, S. P. & Bajtorina, G. B. (Абаимова, Г. П., Ергалиев, Г. Х., Конева, С. П. & Байторина, Г. Б.) 1978: Конодонты и другие группы среднего–верхнево кембрия Малого Каратау. [Conodonts and other faunal groups from the Middle–Upper Cambrian of Malyj Karatau.] *Известия АН СССР, Серия Геологическая 1978:6*, 128–131.

Andres, D. 1981: Beziehungen zwischen kambrischen Conodonten und Euconodonten (Vorläufige Mitteilung). *Berliner Geowissenschaftliche Abhandlungen A 32*, 19–31.

Atkins, E. D. T., Dlugosz, J. & Foord, S. 1979: Electron diffraction and electron microscopy of crystalline α-chitin from the grasping spines of the marine worm *Sagitta*. *International Journal of Biological Macromolecules 1*, 29–32.

Barskov, I. S., Moskalenko, T. A. & Starostina, L. P. (Барсков, И. С., Москаленко, Т. А. & Старостина, Л. П.) 1982: Новые доказательства принадлежности конодонтофорид к позвоночным. [New evidence for the vertebrate affinity of the conodontophorids.] *Палеонтологический журнал 1982:1*, 80–86.

Bednarczyk, W. 1979: Upper Cambrian to Lower Ordovician conodonts of Łeba elevation, NW Poland, and their stratigraphic significance. *Acta Geologica Polonica 29:4*, 409–442.

Bengtson, S. 1976: The structure of some Middle Cambrian conodonts, and the early evolution of conodont structure and function. *Lethaia 9:2*, 185–206.

Bengtson, S. 1977: Aspects of problematic fossils in the early Palaeozoic. *Acta Universitatis Upsaliensis. Abstracts of Uppsala Dissertations from the Faculty of Science 415.* 71 pp.

Bengtson, S. 1980: Conodonts: the need for a functional model. *Lethaia 13:4*, 320.

Bengtson, S. 1983: A functional model for the conodont apparatus. *Lethaia 16:1*, 38.

Bengtson, S. & Fletcher, T. P. 1983: The oldest sequence of skeletal fossils in the Lower Cambrian of southeastern Newfoundland. *Canadian Journal of Earth Sciences 20:4*, 525–536.

Bone, Q., Ryan, K. P. & Pulsford, A. (in press): The structure and composition of the teeth and grasping spines of chaetognaths. *Journal of the Marine Biological Association of the United Kingdom.*

Briggs, D. E. G., Clarkson, E. N. K. & Aldridge, R. J. 1983: The conodont animal. *Lethaia 16:1*, 1–14.

Chen Meng-e 1979: [Some skeletal fossils from the phosphatic sequence, early Lower Cambrian, south China.] *Scientia Geologica Sinica 1979:2*, 187–189. (In Chinese, with an English summary.)

Chen Meng-e 1982: [The new knowledge of the fossil assemblages from Maidiping section, Emei County, Sichuan with reference to the Sinian–Cambrian boundary.] *Scientia Geologica Sinica 182:3*, 253–262. (In Chinese, with an English summary.)

Cobbold, E. S. 1921: The Cambrian horizons of Comley (Shropshire) and their Brachiopoda, Pteropoda, Gasteropoda, etc. *Quarterly Journal of the Geological Society of London 76:4*, 325–386.

Conway Morris, S. 1976: A new Cambrian lophophorate from the Burgess Shale of British Columbia. *Palaeontology 19:2*, 199–222.

Conway Morris, S. 1980: Conodont function: fallacies of the tooth model. *Lethaia 13:1*, 107–108.

Conway Morris, S. & Fritz, W. H. 1980: Shelly microfossils near the Precambrian–Cambrian boundary, Mackenzie Mountains, northwestern Canada. *Nature 286*, 381–384.

Derby, J. R., Lane, H. R. & Norford, B. S. 1972: Uppermost Cambrian – basal Ordovician faunal succession in Alberta and correlation with similar sequences in the western United States. *24th International Geological Congress, Section 7*, 503–512. Montreal.

Druce, E. C. & Jones, P. J. 1971: Cambro-Ordovician conodonts from the Burke River structural belt, Queensland. *Bureau of Mineral Resources, Geology and Geophysics, Bulletin 110.* 158 pp.

Druce, E. C., Shergold, J. H. & Radke, R. M. 1982: A reassessment of the Cambrian–Ordovician boundary section at Black Mountain, western Queensland, Australia. *In* Bassett, M. G. & Dean, W. T. (eds.): The Cambrian–Ordovician boundary: sections, fossil distributions, and correlations, 193–209. *National Museum of Wales, Geological Series 3.*

Dubinina, S. V. (Дубинина, С. В.) 1982: Конодонтовые ассоциации пограничных отложений кембрия и ордовика Малого Каратау (южный Казахстан). [Conodont associations from Cambrian–Ordovician boundary beds of Malyj Karatau (southern Kazakhstan).] *Известия АН СССР, Серия Геологическая 1982:4*, 47–54.

Dzik, J. 1976: Remarks on the evolution of Ordovician conodonts. *Acta Palaeontologica Polonica 21:4*, 395–455.

Eisenack, A. 1964: Mikrofossilien aus dem Silur Gotlands. Phosphatische Reste. *Paläontologische Zeitschrift 38:3–4*, 170–179.

Fåhræus, L. & Nowlan, G. S. 1978: Franconian (Late Cambrian) to early Champlanian (Middle Ordovician) conodonts from the Cow Head Group, western Newfoundland. *Journal of Paleontology 52:2*, 444–471.

Jeppsson, L. 1971: Element arrangement in conodont apparatuses of *Hindeodella* type and in similar forms. *Lethaia 4:1*, 101–123.

Jeppsson, L. 1979: Conodont element function. *Lethaia 12:2*, 153–171.

Jeppsson, L. 1980: Function of the conodont elements. *Lethaia 13:3*, 228.

Jiang Zhi-wen 1980: [The Meishucun Stage and fauna of the Jinning County, Yunnan.] *Bulletin of the Chinese Academy of Geological Sciences, Series 1, 2:1*, 75–92. (In Chinese, with an English summary.)

Kuhl, W. 1932: Untersuchungen über die Bewegungsphysiologie der Fangorgane am Kopf der Chätognathen. *Zeitschrift für Morphologie und Ökologie der Tiere 24:3–4*, 526–575.

Landing, E. 1977: 'Prooneotodus' tenuis (Müller, 1959) apparatuses from the Taconic allochthon, eastern New York: construction, taphonomy and the protoconodont 'supertooth' model. *Journal of Paleontology 51:6*, 1072–1084.

Landing, E. & Brett, C. E. 1982: Lower Cambrian of eastern Massachusetts: microfaunal sequence and the oldest known borings. *Geological Society of America, Abstracts with Programs 14:1–2*, 33.

Landing, E., Ludvigsen, R. & von Bitter, P. H. 1980: Upper Cambrian to Lower Ordovician conodont biostratigraphy and biofacies, Rabbitkettle Formation, District of Mackenzie. *Royal Ontario Museum, Life Sciences Contributions 126.* 42 pp.

Landing, E., Nowlan, G. S. & Fletcher, T. P. 1980: A microfauna associated with Early Cambrian trilobites of the *Callavia* Zone, northern Antigonish Highlands, Nova Scotia. *Canadian Journal of Earth Sciences 17:3*, 400–418.

Lindström, M. 1973: On the affinities of conodonts. *In* Rhodes, F. H. T. (ed.): Conodont paleozoology. *Geological Society of America, Special Paper 141*, 85–102.

Lindström, M. 1974: The conodont apparatus as a food-gathering mechanism. *Palaeontology 17:4*, 729–744.

Lindström, M. & Ziegler, W. 1981: Surface micro-ornamentation and observations on internal composition. *In* Robison, R. A. (ed.): Treatise on Invertebrate Paleontology, W, Miscellanea, Supplement 2, Conodonta, W41–W52. Geological Society of America, Boulder, Colorado, and University of Kansas, Lawrence, Kansas.

Luo Hui-lin et al. 1982: [The Sinian–Cambrian Boundary in Eastern Yunnan, China.] 265 pp. Yunnan Institute of Geological Sciences, Kunming. (In Chinese, with an English summary.)

Martinsson, A. 1965: Phosphatic linings in bryozoan zooecia. *Geologiska Föreningens i Stockholm Förhandlingar 86:4*, 404–408.

Mambetov, A. M. (Мамбетов, А. М.) 1977: К ревизии рода *Helenia*. [To the revision of the genus *Helenia*.] *Палеонтологический журнал 1977:1*, 96–102.

Mambetov, A. M. & Repina. L. N. (Мамбетов, А. М. & Репина, Л. Н.) 1979: Нижний кембрий Таласского Ала-Тоо и его корреляция с разрезами Малого Каратау и Сибирской платформы. [The Lower Cambrian of Talasskij Ala-Too and its correlation with sections of Malyj Karatau and the Siberian Platform.] *In* Журавлева, И. Т. & Мешкова, Н. П. (eds.): Биостратигпафия и палеонтология нижнего кембрия Сибипи, 98–138. *Труды Института Геологии и Геофизика СО АН СССР 406.*

Matthews, S. C. 1973: Lapworthellids from the Lower Cambrian *Strenuella* limestone at Comley, Shropshire, *Palaeontology 16:1,* 139–148.

Matthews, S. C. & Missarzhevsky, V. V. 1975: Small shelly fossils of late Precambrian and early Cambrian age: a review of recent work. *Journal of the Geological Society 131,* 289–304.

Meshkova, N. P. (Мешкова, Н. П.) 1969: К вопросу о палеонтологической характеристике нижнекембрийских отложений Сибирской платформы. [To the question of the palaeontological characteristics of the Lower Cambrian sediments of the Siberian Platform.] *In* Журавлева, И. Т. (ed.): *Биостратиграфия и палеонтология нижнего кембрия Сибири и Дальнего Востока,* 158–174. Наука, Москва.

Meshkova, N. P. (Мешкова, Н. П.) 1974: Cambroscleritida incertae sedis – новый отряд кембрийских ископаемых. [Cambroscleritida incertae sedis – a new order of Cambrian fossils.] *In* Журавлева, И. Т. & Розанов, А. Ю. (eds.): *Биостратиграфия и палеонтология нижнего кембрия Европы и Северной Азии,* 190–193. Наука, Москва.

Miller, J. F. 1969: Conodont fauna from the Notch Peak Limestone (Cambro-Ordovician), House Range, Utah. *Journal of Paleontology 43:2,* 413–439.

Miller, J. F. 1978: Upper Cambrian and lowest Ordovician conodont faunas of the House Range, Utah. *In* Miller, J. F. (ed.): Upper Cambrian to Middle Ordovician conodont faunas of western Utah, 1–33. *Southwest Missouri State University Geoscience Series 5.*

Miller, J. F. 1980: Taxonomic revisions of some Upper Cambrian and Lower Ordovician conodonts with comments on their evolution. *The University of Kansas Paleontological Contributions 99.* 39 pp.

Miller, J. F. 1981: Order Paraconodontida. *In* Robison, R. A. (ed.): *Treatise on Invertebrate Paleontology, W, Miscellanea, Supplement 2, Conodonta,* W111–W115. Geological Society of America, Boulder, Colorado, and University of Kansas, Lawrence, Kansas.

Miller, J. F. & Rushton, A. W. A. 1973: Natural conodont assemblages from the Upper Cambrian of Warwickshire. *Geological Society of America, Abstracts with Programs 5,* 338–339.

Miller, J. F., Taylor, M. E., Stitt, J. H., Ethington, R. L., Hintze, L. F. & Taylor, J. F. 1982: Potential Cambrian–Ordovician boundary stratotype sections in the western United States. *In* Bassett M. G. & Dean, W. T. (eds.): The Cambrian–Ordovician boundary: sections, fossil distributions, and correlations, 155–180. *National Museum of Wales, Geological Series 3.*

Miller, R. H. & Paden, E. A. 1976: Upper Cambrian stratigraphy and conodonts from eastern California. *Journal of Paleontology 50:4,* 590–597.

Miller, R. H., Sundberg, F. A., Harma, R. H. & Wright, J. 1981: Late Cambrian stratigraphy and conodonts of southern Nevada. *Alcheringa 5:3,* 183–196.

Missarzhevsky, V. V. (Миссаржевский, В. В.) 1973: Конодонтообразные организмы из пограничных слоев кембрия и докембрия Сибирской платформы и Казахстана. [Conodont-shaped organisms from the Precambrian and Cambrian boundary strata of the Siberian Platform and Kazakhstan.] *In* Журавлева, И. Т. (ed.): *Проблемы палеонтологии и Биостратиграфии нижнего кембрия Сибири и Дальнего востока,* 53–57. Наука, Новосибирск.

Missarzhevsky, V. V. (Миссаржевский, В. В.) 1977: Конодонты(?) и фосфатные проблематики кембрия Монголии и Сибири. [Conodonts(?) and phosphatic problematica from the Cambrian of Mongolia and Siberia.] *In* Татаринов, Л. П. *et al.* (eds.): *Беспозвоночные палеозоя Монголии,* 10–19. Наука, Москва.

Missarzhevsky, V. V. (Миссаржевский, В. В.) 1982: Расчленение и корреляция пограничных толщ докембрия и кембрия по некоторым древнейшим группам скелетных организмов. [Subdivision and correlation of the Precambrian–Cambrian boundary beds using some groups of the oldest skeletal organisms.] *Бюллетень Московского Общества Испытателей Природы, Отдел Геологический, 57:5,* 52–67.

Missarzhevsky, V. V. & Mambetov, A. M. (Миссаржевский, В. В. & Мамбетов, А. М.) 1981: Стратиграфия и фауна пограничних слоев кембрия и докембрия Малого Каратау. [Stratigraphy and fauna of the Precambrian–Cambrian boundary beds of Malyj Karatau.] *Труды Геологического Института АН СССР 326.* 90 pp.

Müller, K. 1959: Kambrische Conodonten. *Zeitschrift der Deutschen Geologischen Gesellschaft 111:2,* 434–485.

Müller, K. J. 1962: Supplement to systematics of conodonts. *In* Moore, R. C. (ed.): *Treatise on Invertebrate Paleontology, W, Miscellanea,* W246–W249. Geological Society of America, Boulder, Colorado, and University of Kansas Press, Lawrence, Kansas.

Müller, K. J. 1973: Late Cambrian and Early Ordovician conodonts from northern Iran. *Geological Survey of Iran, Report 30.* 77 pp.

Müller, K. J. 1981: Micromorphology of elements. Internal structure. *In* Robison, R. A. (ed.): *Treatise on Invertebrate Paleontology, W, Miscellanea, Supplement 2, Conodonta,* W20–W41. Geological Society of America, Boulder, Colorado, and University of Kansas, Lawrence, Kansas.

Müller, K. J. & Andres, D. 1976: Eine Conodontengruppe von *Prooneotodus tenuis* (Müller, 1959) in natürlichem Zusammenhang aus dem Oberen Kambrium von Schweden. *Paläontologische Zeitschrift 50:3–4,* 193–200.

Müller, K. J. & Nogami, Y. 1971: Über den Feinbau der Conodonten. *Memoirs of the Faculty of Science, Kyoto University, Series of Geology and Mineralogy 38:1.* 87 pp.

Müller, K. J. & Nogami, Y. 1972: Growth and function of conodonts. *24th International Geological Congress, Section 7,* 20–27. Montreal.

Nicoll, R. S. 1977: Conodont apparatuses in an Upper Devonian palaeoniscid fish from the Canning Basin, Western Australia. *BMR Journal of Australian Geology and Geophysics 2,* 217–228.

Ørvig, T. 1977: A survey of odontodes ('dermal teeth') from developmental, structural, functional, and phyletic points of view. *In* Mahala Andrews, S., Miles, R. S. & Walker, A. D. (eds.): *Problems in Vertebrate Evolution,* 53–75. *Linnean Society Symposium Series 4.*

Qian Yi 1977: [Hyolitha and some problematica from the Lower Cambrian Meishucun Stage in central and SW China.] *Acta Palaeontologica Sinica 16:2,* 255–278. (In Chinese, with an English abstract.)

Qian Yi 1978: [The early Cambrian hyolithids in central and southwest China and their stratigraphical significance.] *Memoirs of Nanjing Institute of Geology and Palaeontology, Academia Sinica, 1978:11,* 1–38.

Qian Yi, Chen Meng-e & Chen Yi-yuan 1979: [Hyolithids and other small shelly fossils from the Lower Cambrian Huangshandong Formation in the eastern part of the Yangtze Gorge.] *Acta Palaeontologica Sinica 18:3,* 207–230. (In Chinese, with an English summary.)

Qian Yi & Zhang Shi-ben 1983: [Small shelly fossils from the Xihaoping Member of the Tongying Formation in Fangxian County of Hubei Province and their stratigraphical significance.] *Acta Palaeontologica Sinica 22:1,* 82-94. (In Chinese, with an English summary.)

Repetski, J. E. & Szaniawski, H. 1981: Paleobiologic interpretation of Cambrian and earliest Ordovician conodont natural assemblages. *In* Taylor, M. E. (ed.): Short Papers for the Second International Symposium on the Cambrian System 1981. *U. S. Geological Survey Open-File Report 81-743,* 169–172.

Repina, L. N. *et al.* (Репина, Л. Н. *et al.*) 1974: *Биостратиграфия и фауна нижнего кембрия Хараулаха (хр. Туора-Сис). [Biostratigraphy and fauna of the Lower Cambrian of Kharaulakh (Tuora-Sis Range).]* 300 pp. Наука, Москва.

Rhodes, F. H. T. & Austin, R. L. 1981: Natural assemblages of elements: interpretation and taxonomy. *In* Robison, R. A. (ed):

Treatise on Invertebrate Paleontology, W, Miscellanea, Supplement 2, Conodonta, W68–W78. Geological Society of America, Boulder, Colorado, and University of Kansas, Lawrence, Kansas.

Rozanov, A. Yu. *et al.* (Розанов, А. Ю. *et al.*) 1969: Томмотский ярус и проблема нижней границы кембрия. [The Tommotian Stage and the problem of the lower boundary of the Cambrian.] *Труды Геологического Института АН СССП 206.* 380 pp. (Translated into English by Amerind Publishing Co. in 1981.)

Seddon, G. & Sweet, W. C. 1971: An ecologic model for conodonts. *Journal of Paleontology 45:5,* 869–880.

Szaniawski, H. 1971: New species of Upper Cambrian conodonts from Poland. *Acta Palaeontologica Polonica 16:4,* 401–413.

Szaniawski, H. 1980a: Fused clusters of paraconodonts. *In* Schönlaub, H. P. (ed.): Guidebook, abstracts, Second European Conodont Symposium (ECOS II), 211. *Abhandlungen der Geologischen Bundesanstalt 35.*

Szaniawski, H. 1980b: Conodonts from the Tremadocian Chalcedony Beds, Holy Cross Mountains (Poland). *Acta Palaeontologica Polonica 25:1,* 101–121.

Szaniawski, H. 1982: Chaetognath grasping spines recognized among Cambrian protoconodonts. *Journal of Paleontology 56:3,* 806–810.

Szaniawski, H. 1983: Structure of protoconodonts. *Fossils and Strata 15,* 21–27.

Tipnis, R. S. & Chatterton, B. D. E. 1979: An occurrence of the apparatus of 'Prooneotodus' (Conodontophorida) from the Road River Formation, Northwest Territories. *Current Research B, Geological Survey of Canada, Paper 79-1B,* 259–262.

Voronin, Yu. I. *et al.* (Воронин, Ю. И. *et al.*) 1982: Граница докембрия и кембрия в геосинклинальных областях (опорный разрез Саланы-Гол, МНР). [The Precambrian–Cambrian boundary in the geosynclinal areas (the reference section of Salany-Gol, MPR).] *Совместная советско-монгольская палеонтологическая экспедиция, Труды 18.* 150 pp.

Walliser, O. H. 1958: *Rhombocorniculum comleyense* n. gen., n. sp. (Incertae sedis, Unterkambrium, Shropshire). *Paläontologische Zeitschrift 32:3–4,* 176–180.

Winder, C. G. 1976: Enigmatic objects in North American Ordovician carbonates. *In* Bassett, M. G. (ed.): *The Ordovician System: Proceedings of a Palaeontological Association Symposium, Birmingham, September 1974,* 645–657. University of Wales Press and National Museum of Wales, Cardiff.

Xiang Li-wen *et al.* 1981: [The Cambrian System of China. *Stratigraphy of China 4.*] 210 pp. Geological Publishing House, Beijing. (In Chinese.)

Yin Ji-cheng, Ding Lian-fang, He Ting-gui, Li Shi-lin & Shen Li-juan 1980: [*The Palaeontology and Sedimentary Environment of the Sinian System in Emei–Ganluo Area, Sichuan.*] 231 pp. People's Republic of China. (In Chinese, with an English summary.)

Yuan Ke-xing & Zhang Sen-gui 1983: [Discovery of the *Tommotia* fauna in SW China.] *Acta Palaeontologica Sinica 22:1,* 31–41. (In Chinese, with an English Summary.)

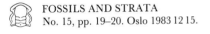

FOSSILS AND STRATA
No. 15, pp. 19–20. Oslo 1983 12 15.

A contribution to the Third European Conodont Symposium, Lund, 1982   ECOS III

Patterns and processes in conodont evolution: a prospectus

THOMAS W. BROADHEAD

Since their discovery, conodonts have become highly regarded as useful biostratigraphic tools, and correlation may always be the principal *raison d'être* of conodont work. More recently, multielement taxonomic concepts and examinations of paleoecologic controls seem to have lent truly organismal status to conodonts, now climaxed by the discovery of the new conodont animal (Briggs, Clarkson & Aldridge 1983). Examinations of conodont evolution have been largely restricted to the determination of phylogenetic relationships, usually as a codominant or subordinate theme to biostratigraphy (e.g. Sweet & Bergström 1981; Bergström 1982; Fåhraeus 1982).

Abundant stratigraphic and geographic distributional data are available for conodonts in many parts of the geologic column. These data lend themselves to a closer examination of conodont evolution, namely recognition of patterns and processes of morphologic change.

Such recognition, however, is heavily dependent on several kinds of data, including (1) precise stratigraphic limits of range, (2) ontogenetic development and (3) range and kind of intraspecific variation.

Stratigraphic data must be precise to document the extent of local teilzones and show physical stratigraphic relations between supposed ancestor and descendant species. The answerable questions become: Is there stratigraphic overlap? At all occurrences? The answers suggest patterns of lineage branching, if present, and possible dispersal gradients of newly developed morphotypes (see, e.g., Fåhraeus 1982). Single collections 'characteristic' of entire formations or large intervals ('bag biostratigraphy' of W. C. Sweet, 1982, personal communication) are grossly inadequate.

Detailed knowledge of ontogenetic development of discrete conodont elements is vital to the recognition, not only of ancestor–

descendant relationships, but also of the timing of natural selection favoring morphologic change. Heterochrony (see Gould 1977 for an historical review and examples) is the temporal displacement (in terms of ontogeny) of morphologic features recognized at other stages of development in ancestors or descendants. Recapitulation commonly produces an increase in morphologic complexity by selection for features developing in later ontogenetic stages (e.g., serration–denticulation in *Histiodella*, McHargue 1982; the fourth row of nodes in *Pygodus anserinus*). A high degree of resemblance may mark the two taxa related in this manner.

Larger-scale morphologic 'shifts' more likely may be due to paedomorphosis, heterochrony that tends to preserve juvenile morphologic features throughout ontogeny. The derivation of *Icriodus* from *Pedavis* (Broadhead & McComb 1983) presents a striking example of paedomorphosis, also possibly shown in the origin of *Rhachistognathus minutus* from *R. muricatus* in the Carboniferous (Lane & Baesemann 1982). Another possible paedomorphic development, but one which evolved several times in only distantly related lineages, is the development of widely spaced discrete denticles of ramiform elements from ancestors with more closely spaced or even confluent denticles. Because selection can become vigorous beginning at early ontogenetic stages to select against adult elaborative features, successful paedomorphs may develop within only one or a few generations, presenting a virtual instantaneous 'jump' in morphology definitive of a generic or higher-level taxon.

Even without an abundance of ontogenetic data, we can still propose phylogenetic relationships between species, genera and so on, but to be confident we must ask the answerable: When does morphologic change originate during ontogeny?

Much has been said about the manner in which species originate and by which evolutionary change proceeds. Such topics can be approached from studies of conodonts, but not without a firm knowledge of the kinds of intraspecific variation (subspecies as utilized by most biostratigraphers mostly fall outside biologic definitions of the term). Morphologic variation may be considered to be either discrete or continuous, both of which lend themselves to measurement and thus statistical manipulation.

Morphologic characteristics that exhibit continuous intraspecific variation will tend to be the only features observed to evolve phyletically. Even in extremely closely stratigraphically spaced samples, statistical means may show a continuous directional change. Continuously varying features include amount of lateral expansion and depths of basal cavities, lengths of processes, lateral and longitudinal dimensions of all elements (e.g., platform length in *Gondolella mombergensis*, Dzik & Trammer 1980; height/length ratio in *Histiodella* elements, McHargue 1982) and heights of cusps and denticles.

Although features that exhibit continuous variation may evolve continuously (phyletically), they may also evolve in a punctuational mode (*sensu* Eldredge & Gould 1972). Determination of evolutionary mode for aspects of a character's continuous variation demands extreme precision and close spacing in stratigraphic sampling in addition to thorough documentation of geographic distribution.

The discrete or discontinuous variation exhibited by conodont elements is best expressed in terms of presence and absence of features. Denticle number might at first appear to be part of the range of continuous variation because variation within a sample may show a complete and continuous range of integral values for denticle number (e.g. in *Gondolella mombergensis*, Dzik & Trammer 1980). The discontinuous nature of this variation, however, commonly involves the presence or absence of one or more denticles but may exhibit continuous change where changes in denticle number are associated

with increasing/decreasing height at added/removed denticle positions. Additionally, other features that may more commonly vary continuously (e.g. process length) may also be observed to vary discontinuously.

Evolution of discretely varying features will manifest itself as step-like morphologic shifts of various magnitudes that are of different significance to the taxonomist. Although these changes are essentially punctuated, those that represent a change in the mean of widely varying populations may exhibit a shift through a large number of successive generations that appears to be essentially phyletic. In contrast, larger-scale change that occurs in one or a small number of generations has a larger apparent punctuated component. Successful modifications (e.g. *Icriodus woschmidti*) may then have achieved virtually instantaneous dispersal enhancing even their biostratigraphic utility.

Conodonts have the potential to provide examples of evolutionary patterns and processes that may surpass all other fossil metazoans. Benefits accruing from careful examination of evolutionary patterns and processes extend not only to understanding why so many species are useful in biostratigraphy but also to interpretations of the effects of recognized environmental changes on morphology.

References

Bergström, S. M. 1982: Evolutionary relationships and biostratigraphic significance of Ordovician platform conodonts. *In* Jeppsson, L. & Löfgren, A. (eds.): Third European Conodont Symposium (ECOS III) Abstracts. *Publications from the Institutes of Mineralogy, Paleontology and Quaternary Geology, University of Lund, Sweden No. 238*, 7–8.

Briggs, D. E. G., Clarkson, E. N. K. & Aldridge, R. J. 1983: The conodont animal. *Lethaia 16*, 1–14.

Broadhead, T. W. & McComb, R. 1983: Paedomorphosis in the Icriodontidae (Conodonta) and the evolution of *Icriodus*. *Fossils and Strata 15*, 149–154.

Dzik, J. & Trammer, J. 1980: Gradual evolution of conodontophorids in the Polish Triassic. *Acta Palaeontologica Polonica 25*, 55–89.

Eldredge, N. & Gould, S. J. 1972: Punctuated equilibria: an alternative to phyletic gradualism. *In* Schopf, T. J. M. (ed.): *Models in Paleobiology*, 82–115. Freeman, Cooper & Co., San Francisco, California.

Fåhraeus, L. A. 1982. Allopatric speciation and lineage zonation exemplified by the *Pygodus serrus – P. anserinus* transition (Conodontophorida, Ordovician). *Newsletters in Stratigraphy 11*, 1–7.

Gould, S. J. 1977: *Ontogeny and Phylogeny*. 501 pp. Belknap, Cambridge, Massachusetts.

Lane, H. R. & Baesemann, J. F. 1982: A mid-Carboniferous boundary based on conodonts and revised intercontinental correlations. *In* Ramsbottom, W. H. C., Saunders, W. B. & Owens, B. (eds.): *Biostratigraphic data for a mid-Carboniferous boundary*, 6–11. Subcommission on Carboniferous Stratigraphy, Leeds.

McHargue, T. R. 1982: Ontogeny, phylogeny, and apparatus reconstruction of the conodont genus *Histiodella*, Joins Fm., Arbuckle Mountains, Oklahoma. *Journal of Paleontology 56*, 1410–1433.

Sweet, W. C. & Bergström, S. M. 1981: Biostratigraphy and evolution. *In* Robison, R. A. (ed.): *Treatise on Invertebrate Paleontology W, Miscellanea, Supplement 2 Conodonta*, W92–W101. Geological Society of America and University of Kansas, Lawrence, Kansas.

Thomas W. Broadhead, Department of Geological Sciences, University of Tennessee, Knoxville, Tennessee 37996-1410, U.S.A.; 23rd November, 1982.

Structure of protoconodont elements

HUBERT SZANIAWSKI

FOSSILS AND STRATA

ECOS III

A contribution to the Third European
Conodont Symposium, Lund, 1982

Szaniawski, Hubert 1983 12 15: Structure of protoconodont elements. *Fossils and Strata*, No. 15, pp. 21–27. Oslo. ISSN 0300–9491. ISBN 82-0006737-8.

The internal structure of the elements of *'Prooneotodus' tenuis* (Müller 1959) and of some other protoconodont elements has been investigated under the optical microscope, scanning electron microscope and transmission electron microscope. The elements are constructed of three layers: thin outer organic cover, thick laminated organo-phosphatic middle layer and thin, faintly laminated organic inner layer. The original structure of the organic matrix is not well preserved. The outer layer is interpreted as a cuticle and the inner as a secreting layer of epidermis. Comparison of structural details of protoconodont elements with grasping spines of Recent Chaetognatha shows their great similarity although there is no definite evidence of their identity. It is suggested that it is not the whole element of protoconodonts but only their inner or/and middle layer that should be homologized with the basal plate of euconodonts. □*Conodonta, protoconodonts, Chaetognatha, structure, Upper Cambrian.*

Hubert Szaniawski, Department of Paleobiology, Polish Academy of Sciences, Al. Zwirki i Wigury 93, 02-089 Warszawa, Poland; 10th November, 1982.

Protoconodonts are an informal group of marine animals of which only the organo-phosphatic, spinose elements are known as microfossils, commonly occurring from the Upper Precambrian to the Lower Ordovician. Protoconodonts are usually assigned to Paraconodontida Müller but their elements differ from typical elements of paraconodonts in the mode of growth (Bengtson 1976). Because the group is distinguished on the base of structural features, only those species can be regarded as protoconodonts in which such features have been checked by means of special structural investigations. Therefore at present only *Amphigeisina danica* (Poulsen 1966), *Gapparodus bisulcatus* (Müller 1959), *'Prooneotodus' tenuis* (Müller 1959), *Gapparodus heckeri* (Abaimova 1978), *'Prooneotodus' savitzkyi* (Abaimova 1978), *Protohertzina anabarica* Missarzhevsky 1973, *P. unguliformis* Missarzhevsky 1973, and *P. cultrata* Missarzhevsky 1977 can be referred to the group. However, it is most probable that many more species belong to it.

The probable relationship of protoconodonts to Recent Chaetognatha (Szaniawski 1980a, 1982; Repetski & Szaniawski 1981; Bengtson 1983) and also their possible affinity with paraconodonts and euconodonts (Bengtson 1976, 1977) makes this group of microfossils very interesting. A better understanding of its systematic position may possibly be reached by structural and comparative-anatomical studies.

Published information on the structure of protoconodont elements is based on optical microscope and SEM investigations (Müller 1959; Müller & Nogami 1971, 1972; Bengtson 1976, 1977, 1983; Landing 1977; Repetski 1980; Szaniawski 1980a, 1982; Andres 1981). In the present study TEM has also been applied in order to produce more information on the organic matrix. TEM investigations of protoconodont elements are technically very difficult, but nevertheless easier than in the case of conodont elements because they contain much more organic material. The results of the new investigations have helped to provide for more detailed comparison of protoconodont elements and the grasping spines of modern chaetognaths.

Material and methods

The studies were made mostly on *'Prooneotodus' tenuis* (Müller) material obtained from the Upper Cambrian in borehole cores from northern Poland. Some specimens of *Gapparodus* sp. from the same locality and specimens of *Gapparodus heckeri* Abaimova and *'Prooneotodus' savitzkyi* (Abaimova) from the Upper Cambrian of Central Kazakhstan (material kindly made available for study by Galina P. Abaimova, Novosibirsk) were also examined.

Most of the specimens from Poland are black in colour. They were processed from black or dark bituminous limestones formed in an anoxic environment. The specimens possess well-preserved organic material and retain their shape even after complete demineralization in mineral acids. In lighter-coloured rocks one usually finds lighter-coloured elements containing much less organic matter. Such elements are much more easily destroyed in acid. The investigated specimens are comparatively well preserved, although their original structure is somewhat changed by recrystallization and secondary mineralization.

The following preparation procedures have been applied:

For optical microscope observations: a – thin sections of specimens embedded in original limestone matrix, b – thin sections of orientated specimens embedded in an epoxy resin.

For SEM studies: a – fractured untreated specimens, b – fractured and etched specimens, c – etched sections of specimens embedded in epoxy resin, d – etched sections made

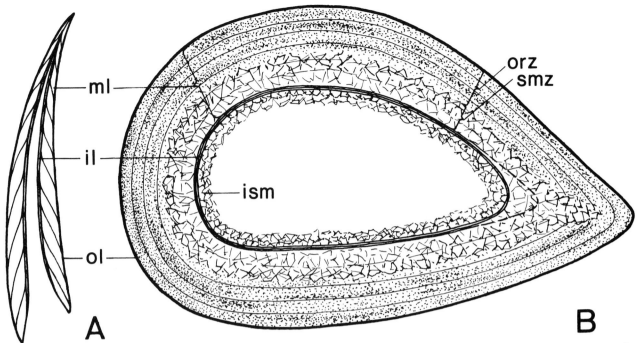

Fig. 1. Schematic longitudinal (A) and cross (B) sections of an element of *'Prooneotodus' tenuis* (Müller) with secondary mineralization as usually preserved in the Upper Cambrian limestones of the Baltic region; il – inner layer, ism – internal secondary mineralization, ml – middle layer, ol – outer layer, orz – organic-rich zone of the middle layer, smz – secondarily mineralized zone of the middle layer.

with an ultramicrotome. For etching, different methods were used depending on the state of preservation.

For TEM studies: oriented ultramicrotome sections of demineralized specimens, prepared with a diamond knife and not stained. Different methods of staining were tried but none gave good results. Demineralization was done with strong HNO_3 and HF.

The collection investigated is housed in the Institute of Paleobiology of the Polish Academy of Sciences (abbreviated as ZPAL).

Description of the preserved structure

Earlier investigations have shown that protoconodont elements were made of phosphate and organic material, possess a three-layered structure, grew by basal–internal accretion and have their main, middle layer laminated (Bengtson 1976, 1977; Szaniawski 1982). New studies make it possible to present more detailed structural descriptions, especially of the relationship between the organic and mineral components. However, the original structure of the organic matrix is still not well known because of diagenetic changes. The generalized structure of *'Prooneotodus' tenuis* elements as preserved in the Upper Cambrian limestone of the Baltic region is illustrated in Fig. 1.

The outer layer is composed of compact structureless organic matter. It is comparatively thin and its thickness seems to be approximately constant along the entire length of the element. The outer organic layer is rarely well preserved because it is less resistant than the rest of the specimen and thus detaches from the underlying layer easily (Figs. 3A, 5). In some specimens the outer layer is secondarily mineralized

with phosphate or pyrite. Often it is covered or replaced by a secondary phosphatic overgrowth (Fig. 2B).

The middle layer is constructed of phosphatic crystallites and organic matrix (Figs. 5, 6, 7). The thickness of the layer rapidly decreases in the basal part of the specimens. The layer is faintly laminated. This lamination is clearly visible in thin sections studied under the optical microscope and in etched sections observed with SEM (Fig. 4B) but it can hardly be noticed in ultra-thin sections studied in TEM. This is because the interlamellar spaces known in the elements of euconodonts do not exist here. Also the lamination is not caused by alternation of pure organic and phosphatic laminae but rather only by laminae more and less rich in organic matter (Fig. 6B). The new laminae are added at the basal–internal surface (Bengtson 1976, 1977). Phosphatic crystallites are very fine, usually irregular in shape, and irregularly distributed in the organic matrix (Figs. 5, 7). However, in some specimens the crystallites are more regularly elongated (Fig. 4A) and are arranged sub-parallel to the layer surface (Fig. 6C). The organic matrix too is more regular in some specimens and forms thin sheets and enveloping individual crystallites. In most of the studied elements from the Baltic region the middle layer can be divided into two zones (Figs. 2C, 5A, 6A). The outer zone is usually the better preserved, contains much more organic material, and is more clearly laminated. The inner zone is strongly secondarily mineralized with calcite, phosphate and pyrite. In some specimens grains of amorphous matter also occur there. Crystallites developed during the mineralization are comparatively coarse (Fig. 6A). The transition from the outer to the inner zone is usually gradual. Development of the inner zone depends very much on the preservation of the particular specimen involved.

The inner layer is very thin but nevertheless thicker than the

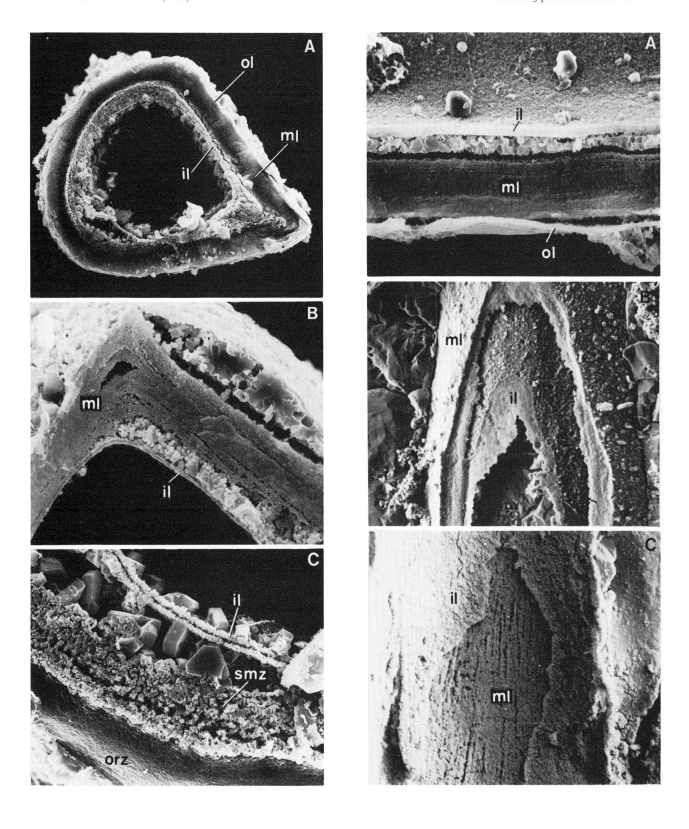

Fig. 2. Broken and etched elements of *'Prooneotodus' tenuis;* explanations as for Fig. 1; SEM. □A. Cross section of the element; ZPAL C. IV/153; ×480. □B. Fragment of a cross section of an element with secondary phosphatic overgrowth; ZPAL C. IV/1593; ×1000. □C. Fragment of a cross section showing lamination of the inner layer and secondary mineralization on both sides; ZPAL C. I/1562; ×2000.

Fig. 3. Etched sections of *'Prooneotodus' tenuis* elements; explanations as for Fig. 1; SEM. □A. Fragment of a longitudinal section of a broken and etched element; ZPAL C.IV/1597. ×900. □B. Fragment of an oblique section of a polished and etched element in its original limestone matrix; ZPAL C.IV/1558; ×450. □C. Fragment of an element embedded in epoxy resin, longitudinally sectioned and etched; part of the inner layer removed to show structure of the middle layer; ZPAL C.IV/1615; ×1000.

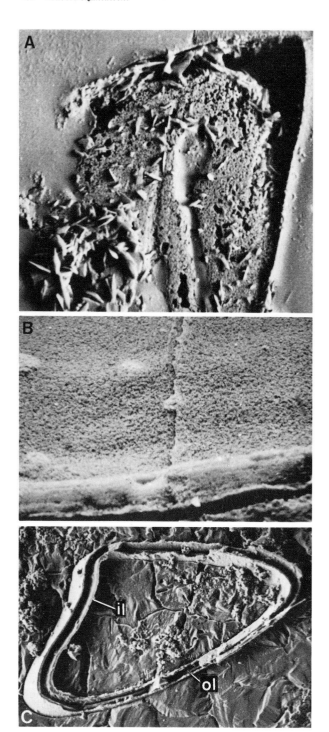

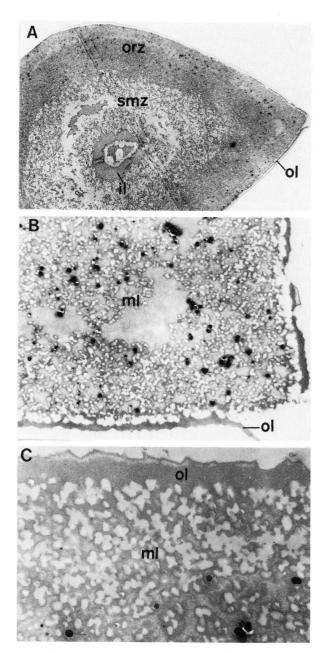

Fig. 4. Different preparations of *'Prooneotodus' tenuis* elements; explanations as for Fig. 1; SEM. □A. Incomplete cross section through distal part of an element embedded in epoxy resin. The element was sectioned with an ultramicrotome and etched. Note elongate apatite crystallites pulled out of the organic matrix; ZPAL C.IV/1607; ×1500. □B. Fragment of an etched internal surface of the middle layer showing its lamination and porosity. Inner layer removed; ZPAL C. IV/1616; ×1000. □C. Cross section of an element polished and etched in its original limestone matrix. Middle layer dissolved; ZPAL C.IV/ 1560; ×300.

Fig. 5. Ultramicrotome cross sections through distal part of a demineralized element of *'Prooneotodus' tenuis;* white – empty space, dark – undissolved amorphous material (mainly organic), black spots – undissolved mineral crystals (pyrite?); explanations as for Fig. 1; TEM. □A. Almost complete cross section of the element. Internal cavity partly filled with amorphous material; ×1200. □B. Detail of A, ×7000. □C. Fragment of another cross section of the same element, ×18,000.

outer one (Figs. 2, 3A, 5). Its thickness decreases slightly toward the base. This layer also is composed of organic matter and phosphate crystallites, but organic matter predominates (Figs. 5A, 6A) and most probably the layer was originally exclusively organic. The layer is laminated but has only a few

laminae which are so faint that usually they cannot be recognized under the optical microscope. They are best seen in etched sections observed under SEM (Figs. 2A, C, 3B). On both the inner and the outer side the inner layer is usually covered with secondary mineralization, which is often pyritic (Fig. 2C). It is probably because of this mineralization that the inner layer is the most resistant part of the whole element. It is often the only remnant of a specimen.

The internal cavity extends almost all the way to the tip of an element. Usually it is filled with sediment and often with pyrite and secondary phosphate.

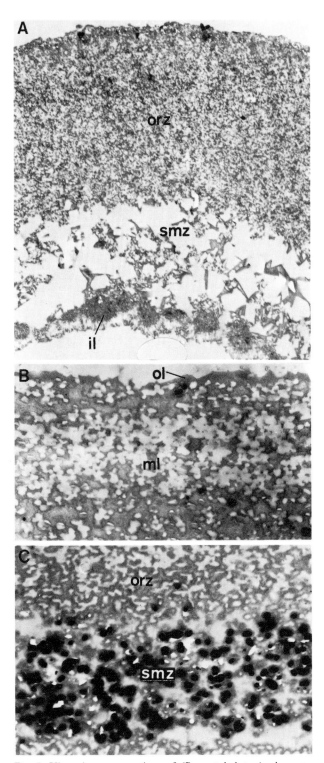

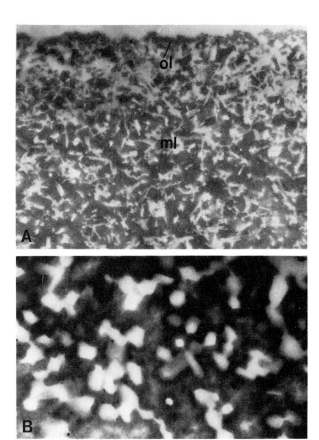

Fig. 7. Ultramicrotome sections of *'Prooneotodus' tenuis* elements; explanations as for Fig. 1; TEM. □A. Fragment of a longitudinal section showing irregular shape and arrangement of apatite crystallites in the middle layer. Outer layer not completely preserved; ×13,500. □B. Fragment of the middle layer in cross section showing distribution of apatite crystallites and their outline in section; ×45,000.

Fig. 6. Ultramicrotome sections of *'Prooneotodus' tenuis* elements; explanations as for Fig. 1; TEM. □A. Fragment of a cross section. Outer layer not preserved. Middle and inner layers secondarily mineralized; ×4000. □B. Fragment of a longitudinal section showing lamination of the middle layer. Outer layer not completely preserved; ×13,500. □C. Fragment of a longitudinal section showing elongation of apatite crystallites in an organic-rich zone of the middle layer and undissolved mineral crystals (pyrite?) in the secondarily mineralized zone of the middle layer; ×14,000.

The described state of preservation of the structure of *'P.' tenuis* elements refers to specimens preserved having black colour. Lighter-coloured specimens contain much less organic matter. After brief etching of such specimens in 2–3 % HCl

their middle layer usually dissolves (Fig. 4C). At many localities where *'P.' tenuis* elements occur, only the secondarily phosphatized inner and outer layers, separated by a gap filled with sediment, are preserved. Such a mode of preservation is seen in the clusters described by Tipnis & Chatterton (1979).

Comparisons and interpretations

It is now possible to make a somewhat more detailed comparison of protoconodont elements with on the one hand, grasping spines of Recent Chaetognatha and, on the other, elements of para- and euconodonts.

It has previously been thought that the main difference between protoconodont elements and chaetognath spines is in their chemical compositions, because the modern spines are chitinous and the fossil forms are phosphatic. Although Clark & Miller (1969) established by means of electron-probe analyses that elements of *Gapparodus* contain an admixture of organic material, and Bengtson (1976) suggested that the admixture in primitive protoconodont elements is high, it has not been demonstrated until now that the organic matter was originally the main constituent of protoconodont elements. However, the original relationship of organic material to phosphate is still not well known.

Chaetognath spines consist of three layers: a thin cuticular

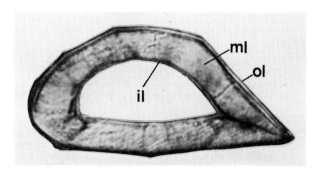

Fig. 8. Microtome cross section of a grasping spine of the Recent chaetognath *Sagitta maxima* Conant; explanations as for Fig. 1; light microscope photograph (the same specimen was illustrated in Szaniawski 1982:809); ZPAL C.IV/1700; ×1000.

outer layer, a thick, fibrous, laminated middle layer and an again thin epidermal layer (Fig. 8). The chemical compositions of the layers are most probably distinct because the outer cuticle is comparatively easily damaged with KOH solution whereas the fibrous layer is not. The cuticular layer, as far as it is known, is structureless, as is the outer layer of protoconodont elements. Homology of the layers seems very probable. The middle layer of chaetognath grasping spines has fibrous structure but the fibrils are arranged in fine lamellae. Definite fibrils have not been found in 'P.' *tenuis* elements, although some preparations suggest their possible presence (Fig. 3C). According to Bengtson (1983) the elements of early protoconodonts, *Protohertzina unguliformis* and *P. anabarica*, have a fibrous structure, but the fibrils consist of acicular apatite crystallites. It is possible that the middle layer of 'P.' *tenuis* elements originally possessed fibrous structure which has been damaged during the diagenesis. The secondarily mineralized zone of the middle layer probably developed as a result of contraction of the organic inner layer and filling of the empty space with mineral solutions which later impregnated also the inner part of the middle layer.

The inner layer of chaetognath grasping spines is formed of a stratified secreting layer of epidermis. This layer is probably homologous with the inner layer of protoconodont elements.

The pyritic mineralization of protoconodont elements and especially of their inner layer is understandable when one takes into account the fact that the pyrite usually forms as organic material decays in anoxic conditions.

A possible structural difference between chaetognath spines and protoconodont elements lies in the fact that the middle layer in chaetognath spines is interrupted along their edge (Fig. 8), whereas in cross sections of the majority of the 'P.' *tenuis* elements investigated the middle layer appears as a complete ring. However, some sections of 'P.' *tenuis* elements suggest that their middle layer originally had interruptions of the same kind as seen in modern forms (Fig. 2A). Some early protoconodonts, e.g. *Amphigeisina danica*, have all three layers interrupted in the basal part of the element (Bengtson 1976).

Structural investigations completed at present do not provide definite proof of a protoconodont–chaetognath relationship, however, they do supply some arguments supporting this hypothesis and, what is also important, they have produced nothing that would contradict such a hypothesis.

A thorough comparison of protoconodont elements with the elements of para- and euconodonts was made by Bengtson (1976). Here only some questions of possible homologies will be discussed. All euconodont elements are composed of two parts, the element proper and a basal plate. Both parts grew by outer accretion of continuously developed laminae (Müller & Nogami 1971). The basal plate, which contains much more organic matter, usually has been considered as a probable homologue of the complete protoconodont and paraconodont elements (Bengtson 1976; Szaniawski 1980b:116; Andres 1981; Müller 1981). Now, knowing that the protoconodont elements are constructed of three different layers, it seems doubtful that the entire elements can be strictly homologous to the one-layered basal plate of the euconodont elements. If any homology between proto- para- and euconodont element exists, it seems more probable that it is between euconodont basal plates, paraconodont elements without their outer organic cover, and the inner or/and middle layer of protoconodont elements.

Acklnowledgements. – I would like to thank Galina P. Abaimova, Novosibirsk, for providing some specimens for study, Stefan Bengtson, Uppsala, Anita Löfgren and Lennart Jeppsson, Lund, for valuable comments to the manuscript, the late S. Crosbie Matthews, Uppsala, and John Repetski, Washington, for linguistic corrections.

References

Abaimova, G. P. (Абаимова, Г. П) 1978: Позднекембрийские конодонты центрального Казахстана. [Late Cambrian conodonts from central Kazakhstan.] *Палеонтологический журнал 1978:4*, 77–87.

Andres, D. 1981: Beziehungen zwischen Kambrischen Conodonten und Euconodonten (Vorläufige Mitteilung). *Berliner geowiss. Abh., A, 32*, 19–31.

Bengtson, S. 1976: The structure of some Middle Cambrian conodonts, and the early evolution of conodont structure and function. *Lethaia 7*, 185–206.

Bengtson, S. 1977: Aspects of problematic fossils in the early Palaeozoic. *Acta Univ. Ups. Abstr. Uppsala Diss. Fac. Sci. 415*, 71 pp.

Bengtson, S. 1983: The early history of the Conodonta. *Fossils and Strata 15*, 5–19.

Clark, D. L. & Miller, J.F. 1969: Early evolution of conodonts. *Geol. Soc. Am. Bull. 80:1*, 125–134.

Landing, E. 1977: '*Prooneotodus*' *tenuis* (Müller, 1959) apparatuses from Taconic allochton, eastern new York: construction, taphonomy and the protoconodont 'supertooth' model. *J. Paleontol. 51:6*, 1072–1084.

Missarzhevsky, V. V. (Миссаржевский, В. В.) 1973: Конодонтообразные организмы из пограничных слоев кембрия и докембрия Сибирской платформы и Казахстана. [Conodont-shaped organisms from the Precambrian and Cambrian boundary strata of the Siberian Platform and Kazakhstan.] *In* Журавлева, И. Т. (ed.): Проблемы палеонтологии и биостратиграфии нижнего кембрия Сибири и Дальнего Востока, 53–57. Наука, Новосибирск.

Missarzhevsky, V. V. (Миссаржевский, В. В.) 1977: Конодонты (?) и фосфатные проблематики кембрия Монголии и Сибири. [Conodonts (?) and phosphatic problematica from the Cambrian of Mongolia and Siberia.] *In* Татаринов, Л. П. *et al.* (eds.): Безпозвоночные палеозоя Монголии, 10–19. Наука, Москва.

Müller, K. J. 1959: Kambrische Conodonten. *Z. deutsch. geol. Ges. 111*, 434–485.

Müller, K. J. 1981: Internal structure. *In* Robison, R.A. (ed.): *Treatise on Invertebrate Paleontology, W. Miscellanea: Supplement 2, Conodonta*, W20–W44. Geol. Soc. Am. and University of Kansas Press.

Müller, K. J. & Nogami, Y. 1971: Über den Feinbau der Conodonten. *Mem. Fac. Sci., Kyoto Univ., Ser. Geol. Miner., 38:1.* 87 pp.

Müller, K. J. & Nogami, Y. 1972: Growth and function of conodonts. *24th Int. Geol. Congr., Sect. 7,* 20–27. Montreal.

Poulsen, V. 1966: Early Cambrian distacodontid conodonts from Bornholm. *Biol. Medd. Dan. Vidensk. Selsk. 23:15.* 10 pp.

Repetski, J. E. 1980: Early Ordovician conodont clusters from the western United States. *In* Schönlaub, H. P. (ed.): *Second European Conodont Symposium, Guidebook, Abstracts. Abh. Geol. B. – A. (Austria) 35,* 207–209.

Repetski, J. E. & Szaniawski, H. 1981: Paleobiologic interpretation of Cambrian and earliest Ordovician conodont natural assemblages. *In* Taylor, M. E. (ed.): *Short Papers for the Second International Symposium on the Cambrian System 1981. U.S. Geol. Surv. Open-File Report 81-743,* 169–172.

Szaniawski, H. 1980a: Fused clusters of paraconodonts. *In* Schönlaub, H. P. (ed.): *Second European Conodont Symposium, Guidebook, Abstracts, Abh. Geol. B. – A. (Austria) 35,* 211–213.

Szaniawski, H. 1980b: Conodonts from the Tremadocian chalcedony beds, Holy Cross Mountains (Poland). *Acta Paleont. Pol. 25,* 101–121.

Szaniawski, H. 1982: Chaetognath grasping spines recognized among Cambrian protoconodonts. *J. Paleontol. 56,* 806–810.

Tipnis, R. S. & Chatterton, B. D. E. 1979: An occurrence of the apparatus of '*Prooneotodus*' (Conodontophorida) from the Road River Formation, Northwest Territories. *Curr. Res. (B), Geol. Surv. Canada, Pap. 79–1B,* 259–262.

Review of conodont biostratigraphy in China

WANG CHENG-YUAN AND WANG ZHI-HAO

FOSSILS AND STRATA

ECOS III

A contribution to the Third European
Conodont Symposium, Lund, 1982

Wang Cheng-yuan & Wang Zhi-hao 1983 12 15: Review of conodont biostratigraphy in China.
Fossils and Strata, No. 15, pp. 29–33. Oslo. ISSN 0300-9491. ISBN 82-0006737-8.

A preliminary conodont sequence from Cambrian to Triassic has been recognized in China in
recent years, totalling 104 conodont Zones or Assemblage Zones (Cambrian–9, Ordovician–22,
Silurian–9, Devonian–26, Carboniferous–11, Permian–8, Triassic–19). Primitive Lower and
Middle Cambrian conodonts have been found. *Monocostodus sevierensis* or *Cordylodus intermedius* may
be selected as a marker for the beginning of the Ordovician. Two endemic conodont Zones in the
Lower Silurian have been established. Devonian conodont Zones are exactly the same as those in
Europe. *Polygnathus costatus partitus* and *Palmatolepis disparilis* have been proposed as defining the
beginning of the Middle and Upper Devonian, respectively. There are still many gaps in our
knowledge of the Carboniferous conodont sequence. The Changhsing Limestone is considered to
be the highest horizon of the Permian. *Anchignathodus parvus* may be a marker for the beginning of
the Triassic. □*Conodonta, biostratigraphy, Cambrian, Ordovician, Silurian, Devonian, Carboniferous,
China.*

*Wang Cheng-yuan & Wang Zhi-hao, Nanjing Institute of Geology and Palaeontology, Academia Sinica,
Nanjing, China; 11th April 1982 (revised 1983 03 01).*

The serious study of conodonts in China has begun only in the
last decade. However, the geological significance of conodonts
has been fully realized by geologists and palaeontologists
throughout China, who pay more attention to these microfossils than ever before. At present, there are more than a
hundred conodont workers in China, most of whom work on
geological surveying teams and in oil exploration, using
conodonts for dating strata. Some encouraging achievements,
especially in conodont biostratigraphy, have been made, and a
preliminary conodont sequence of the Cambrian to Triassic in
China has been established. The recognition of the sequence
lays a foundation for further research. At present, a comprehensive study of conodonts, including conodont classification, palaeoecology, palaeobiogeography and the application
of the CAI (Conodont Color Alteration Index) etc., is being
carried out. Based on the data available, this paper outlines
the conodont sequence of the Cambrian–Triassic in China
(including 104 conodont zones). The present authors summarized the conodont sequence in China and published it in
Chinese (1981a). Some new information has been added in
this paper as some achievements in conodont study in China
have been made in the past two years.

Cambrian

The first report on Cambrian conodonts in China was that of
Nogami (1966, 1967), who described the conodonts recovered
from the Kushan Formation (lower Upper Cambrian) in
Shandong and the Fengshan Formation in Liaoning. Recently
our knowledge of Cambrian conodonts in China has been
rapidly advanced. In the Lower Cambrian some protoconodonts, such as *Protohertzina anabarica* and *P. robustus*, have been
found in the Meishucun Formation in Central and Southwest
China by Qian (1977) and Chen (1982). Middle and Upper

Cambrian conodont zones have been established by An
(1982), Yao (1982) and Wang Z.-h. (1983, MS):

Upper Cambrian
  Fengshan Formation
    9 *Cordylodus proavus* Zone
    8 *Proconodontus* Zone
      (c) *Cambrooistodus minutus* Subzone
      (b) *Eoconodontus notchpeakensis* Subzone
      (a) *Proconodontus muelleri* Subzone
    7 *Proconodontus muelleri* Zone
  Changshan Formation
    6 *Westergaardodina* aff. *fossa – Prooneotodus rotundatus* Assemblage
      Zone
    5 *Muellerodus? erectus* Zone
  Gushan Formation
    4 *Westergaardodina matsushitai* Zone
    3 *Westergaardodina orygma* Zone
Middle Cambrian
  Zhangxia Formation
    2 *Shandongodus priscus* Zone
    1 *Laiwugnathus laiwuensis* Zone

The problem concerning the Cambrian–Ordovician boundary remains a dispute in China. There are two alternative
suggestions that the Cambrian–Ordovician boundary should
be drawn either at the first appearence of *Monocostodus
sevierensis* (Miller) or the first appearance of *Cordylodus intermedius* Furnish. These questions have been discussed by, inter
alia, An & Yang (1980), An (1981), and Wang Z.-h. & Luo
(1982).

Ordovician

A review of the Ordovician System of China was published by
Lai *et al.* (1979). Ordovician conodonts are extremely abundant and diversified both in North and South China. In South
China the following conodont zones have been found by An *et
al.* (1981), An & Ding (1982) and Ni (1981):

Upper Ordovician
  Mufeng Formation (no conodont zone erected)
  Linxiang Formation (no conodont zone erected)
  Baota Formation
    22 *Protopanderodus inflectus* Zone
    21 *Icriodella baotaensis* Zone
    20 *Hamarodus europaeus* Zone
Middle Ordovician
  Miaopo Formation
    19 *Amorphognathus* cf. *tvaerensis* Zone
    18 *Pygodus anserinus* Zone
    17 *Pygodus serra* Zone
  Guniutan Formation
    16 *Eoplacognathus reclinatus* Zone
    15 *Eoplacognathus foliaceus* Zone
    14 *Eoplacognathus suecicus* Zone
    13 *Eoplacognathus variabilis* Zone
    12 *Eoplacognathus pseudoplanus* Zone
  Lower part of Guniutan Formation and Upper part of Dawan Formation
    11 *Amorphognathus antivariabilis* Zone
Lower Ordovician
  Dawan Formation
    10 *Baltoniodus* aff. *navis* Zone
    9 *Paroistodus originalis* Zone
    8 *Periodon flabellum – Oistodus* aff. *cornutiformis* Assemblage Zone
    7 *Oepikodus evae* Zone
  Hunghuayuan Formation
    6 *Serratognathus diversus* Zone
  Fenxiang Formation
    5 *Drepanodus deltifer* Zone
  Nanjinguan Formation
    4 *Scolopodus paucicostatus* Zone
    3 *Scolopodus quadraplicatus* Zone
    2 *Acanthodus costatus* Zone
  Upper part of Sanyoudong Formation and lower part of Nanjinguan Formation
    1 *Monocostodus sevierensis* Zone

According to An (1983), Zhou Z.–y. *et al.* (1983) and Wang Z.-h. (1983, MS), the Ordovician conodont zones in North China can be recognized as follows:

Upper Ordovician
  Taoqupo Formation
    17 *Yaoxianognathus yaoxianensis* Zone
  Yaoxian Formation
    16 *Tashmanognathus shichuanheensis* Zone
    15 *Tashmanognathus borealis – T. gracilis* Assemblage Zone
Middle Ordovician
  Fenfeng Formation
    14 *Tashmanognathus sishuiensis – Microcoelodus* Assemblage Zone
  Majiagou Formation
    13 *Aurilobodus serratus* Zone
    12 *Plectodina onychodonta* Zone
    11 *Eoplacognathus suecicus – Acontiodus? linxiensis* Assemblage Zone
    10 unnamed conodont zone
Lower Ordovician
  Beianzhuang Formation
    9 *Tangshanodus tangshanensis* Zone
    8 *Aurilobodus leptosomatus – Loxodus discectus* Assemblage Zone
  Liangjiashan Formation
    7 *Paraserratognathus paltodiformis* Zone
    6 *Serratognathus extensus* Zone
    5 *Serratognathus bilobatus* Zone
    4 *Scalpellodus tersus* Zone
  Yehli Formation
    3 *Chosonodina herfurthi – Scolopodus quadraplicatus* Assemblage Zone
    2 *Cordylodus angulatus – C. rotundatus* Assemblage Zone
    1 *Cordylodus intermedius* Zone

Silurian

A review of the Silurian System of China was published by Lin (1979). Zhou X.-y., Zhai & Xian (1981) have published an article on Lower Silurian conodonts from Guizhou Province. They have established three conodont assemblage zones and two interval zones. Some Middle to Upper Silurian conodonts were recovered from Gansu, Yunnan and Tibet by Wang C.-y. (1980, 1982) and Wang C. -y. & Ziegler (1983a). The Silurian conodont zonation in China can be summarized as follows:

Upper Silurian
    9 *Ozarkodina remscheidensis eosteinhornensis* Zone
    8 *Ozarkodina crispa* Zone
Middle Silurian
    7 *Kockelella variabilis* Zone
    6 *Kockelella ranuliformis* Zone
    5 *Pterosphathodus amorphognathoides* Zone (=Interval Zone B of Zhou, Zhai & Xian 1981)
Lower Silurian
    4 *Pterosphathodus celloni* Zone
    3 *'Spathognathodus' parahassi – 'S' quizhouensis* Assemblage Zone
    2 Interval Zone A
    1 *'Spathognathodus' obesus* Zone

Devonian

The Devonian conodont sequence of China has been investigated by Wang C.-y. & Wang Z.-h. (1978a, b, 1981a), Bai *et al.* (1980) and Wang. C.-y. & Ziegler (1983b). All conodont zones, except the *Icriodus woschmidti* Zone which was found in Sichuan and Yunnan (Qin & Gan 1976; Wang C.-y. 1981a, b.; Tan, Dong & Qin 1982), were recognized in Guangxi. Many problems concerning the subdivision and correlation of different facies of Devonian strata in South China have been solved by means of conodonts (Wang C.-y. *et al.* 1979). Recently, in the geosynclinal region of North China, Devonian conodonts have been found (Wang C.-y. & Ziegler 1981), providing reliable evidence for correlation of Devonian strata in North and South China. The Devonian conodont zonation in China can be summarized in the following table (Wang C.-y. & Ziegler 1983b: Wang C.-y. 1983, MS). It is exactly the same as in Europe and in North America. Because of the absence of upper Gedinnian to lower Siegenian marine deposits in Guangxi, conodonts from this interval have not been found. *Polygnathus costatus partitus* and *Palmatolepis disparilis* have been proposed as index fossils for the Lower – Middle and Middle – Upper Devonian boundaries, respectively (Wang C.-y. & Ziegler 1983b). A review of the Devonian System of China was given by Hou (1979).

Upper Devonian
  Famennian
    26 *Protognathodus* Fauna
    25 *Bispathodus costatus* Zone
    24 *Polygnathus styriacus* Zone
    23 *Scaphignathus velifer* Zone
    22 *Palmatolepis marginifera* Zone
    21 *Palmatolepis rhomboidea* Zone
    20 *Palmatolepis crepida* Zone
  Frasnian
    19 *Palmatolepis triangularis* Zone
    18 *Palmatolepis gigas* Zone
    17 *Ancyrognathus triangularis* Zone
    16 *Polygnathus asymmetricus* Zone
    15 *Palmatolepis disparilis* Zone

Middle Devonian
  Givetian
    14 *Schmidtognathus hermanni – P. cristatus* Zone
    13 *Polygnathus varcus* Zone
    12b *Polygnathus xylus ensensis* Zone (upper part)
  Eifelian
    12a *Polygnathus xylus ensensis* Zone (lower part)
    11 *Tortodus kockelianus kockelianus* Zone
    10 *Tortodus kockelianus australis* Zone
    9 *Polygnathus costatus costatus* Zone
    8 *Polygnathus costatus partitus* Zone
Lower Devonian
  Emsian
    7 *Polygnathus costatus patulus* Zone
    6 *Polygnathus serotinus* Zone
    5 *Polygnathus inversus* Zone
    5 *Polygnathus inversus* Zone
    4 *Polygnathus perbonus* Zone
    3 *Polygnathus dehiscens* Zone
  Siegenian
    2 *Eognathodus sulcatus* Zone
  Gedinnian
    1 *Icriodus woschmidti woschmidti* Zone

Carboniferous

The first report on Carboniferous conodonts in China was written by Wang C.-y. (1974). Afterwards, Wang C.-y. & Wang Z.-h. (1978b), Wang C.-y. (1979), Zhao Z.-x. (1979, MS), Zhao S.-y. (1981), Wang Z.-h. & Li (1982, MS), Wang Z.-h. & Wang C.-y. (1983), Wang Z.-h. & Zhang (1983, MS). Wang C.-y. & Ziegler (1983a) and Yang (1983, MS) have done research on Carboniferous conodonts from Guizhou, Shanxi, Henan, Guangxi and Tibet. On the basis of somewhat limited data, the following conodont Zones or Assemblage Zones can be recognized:

Upper Carboniferous
  11 *Streptognathodus elongatus* Zone
  10 *Streptognathodus oppletus – S. elegantulus* Assemblage Zone
  9 *Streptognathodus parvus – S. suberectus* Assemblage Zone
  8 *Idiognathodus delicatus – Neognathodus bassleri* Assemblage Zone
  7 *Idiognathoides corrugatus – Polygnathodella? ouachitensis* Assemblage Zone
  6 *Idiognathoides noduliferus* Zone
Lower Carboniferous
  5 *Gnathodus bilineatus – G. nodosus* Assemblage Zone
  4 *Gnathodus cuneiformis – Gnathodus typicus* Assemblage Zone
  3 *Polygnathus bischoffi* Zone
  2 *Siphonodella duplicata* Zone
  1 *Siphonodella sulcata* Zone

Permian

Ching (1960) made the first important report on Lower Permian conodonts in China. In 1974, Wang C.-y described and figured some Permian conodonts from the Chihsia and Maokou Formations of Sichuan. Wang Z.-h. (1978) published an article on Permian conodonts from the Maokou and Wuchiaping Formations of Shaanxi Province, providing the first report on *Sweetognathus hanzhongensis*, *Neogondolella liangshanensis*, *N. aserrata* and other important Permian conodont species. Later, Wang C.-y. & Wang Z.-h. (1981a, b) published on the conodonts from the Changhsing Formation of Zhejiang Province; the Changhsing Limestone is considered to be the highest horizon of the Permian. As a result of this work, a

preliminary Permian conodont succession has been established, though it is rather incomplete (Wang C. -y. & Wang Z.-h. 1981c).

Upper Permian
  Changhsing Stage
    8 *Neogondolella deflecta – N. subcarinata changxingensis* Assemblage Zone
    7 *Neogondolella subcarinata subcarinata – N. s. elongata* Assemblage Zone
  Wuchiaping Stage
    6 *Neogondolella orientalis* Zone
    5 *Neogondolella liangshanensis – N. bitteri* Assemblage Zone
Lower Permian
  Maokou Stage
    4 *Sweetognathus hanzhongensis – N. aserrata* Assemblage Zone
  Chihsia stage
    3 *Neogondolella idahoensis – N. serrata* Assemblage Zone
    2 *Neostreptognathodus pequopensis – N. sulcaplicatus* Assemblage Zone
    1 *Neogondolella bisselli – Sweetognathus whitei* Assemblage Zone

Triassic

The first report on Triassic conodonts in China was made by Wang C.-y. & Wang Z.-h. (1976). Later, Wang Z.-h. (1978, 1980, 1982), Wang Z.-h. & Dai (1981), Wang Z.-h. & Cao (1981) published articles on the Triassic conodonts from Sichuan, Hubei, Shaanxi, and Guizhou Provinces. Recently, some Upper Triassic conodonts have been found in West Yunnan by Wang Z.-h. & Dong (1983, MS). On the basis of present information, the Triassic conodont sequence in China can be summarized briefly as follows:

Upper Triassic
  Rhaetian
    No conodonts have been found
  Norian
    19 *Epigondolella postera* Zone
    18 *Epigondolella abneptis* Zone
    17 *Epigondolella* cf. *primitia* Zone
  Carnian
    16 *Neogondolella polygnathiformis* Zone
Middle Triassic
  Ladinian
    15 *Neogondolella excelsa* Fauna
  Anisian
    14 *Neogondolella constricta* Zone
    13 *Neospathodus germanicus – N. kockeli* Assemblage Zone
    12 *Neogondolella regale* Zone
Lower Triassic
  Olenikian
    11 *Neospathodus timorensis* Zone
    10 *Pachycladina – Neospathodus homeri* Assemblage Zone
    9 *Neogondolella jubata* Zone
    8 *Neospathodus? collinsoni* Zone
    7 *Platyvillosus costatus* Zone
    6 *Neospathodus waageni* Zone
    5 *Neospathodus pakistanensis* Zone
  Indusian
    4 *Neospathodus cristagalli* Zone
    3 *Neospathodus dieneri* Zone
    2 *Isarcicella isarcica* Zone
    1 *Anchignathodus parvus* Zone

Anchignathodus parvus has been found at the base of the Feihsienkuan Formation of the Lower Triassic in Sichuan and Yuannan Provinces. It may be a good marker for defining the Permian–Triassic boundary.

References

An T.-x. 1981: Recent progress in Cambrian and Ordovician conodont biostratigraphy. *Geol. Soc. Am. Spec. Paper* 187, 209–226.

An T.-x. 1982: Study on the Cambrian conodonts from North and Northeast China. *Sci. Rep., Inst. Geosci., Univ. Tsukuba, Sec. B, 3*, 113–159.

An. T.-x. 1983: *The conodont sequence of the Cambrian and Ordovician in China*, 1–2. Dept. Geology, Beijing University. (In Chinese.)

An T.-x. & Ding L.-s. 1982: Preliminary studies and correlations on Ordovician conodonts from the Ningzhen Mountains, Nanjing, China. *Acta Petrolei Sinica 1982:4*, 1–12. (In Chinese with English abstract.)

An T.-x., Du G.-q., Gao Q.-q., Chen Q.-b. & Li W.-t. 1981: Conodont biostratigraphy of the Ordovician System of Yichang, China. *Selected Papers 1st Convention Micropaleont. Soc. China*, 105–113. (In Chinese.)

An T.-x. & Yang C.-s. 1980: Cambro-Ordovician conodonts in North China, with special reference to the boundary between the Cambrian and Ordovician Systems. In *Scientific Papers on Geology for International Exchange Prepared for the 26th International Geological Congress, Part 4, Stratigraphy and Paleontology*, 7–14. Beijing, Publishing House of Geology. (In Chinese with English abstract.)

Bai S.-l., Jin S.-y., Ning Z.-s., He J.-h. & Han Y.-j. 1980: Devonian conodonts and tentaculitids of Kwangsi, their zonation and correlation. *Acta Scientiarum Naturalium Universitatis Pekinensis 1*, 99–117. (In Chinese with English abstract.)

Chen M.-e 1982: The new knowledge of the fossil assemblages from Maidiping section, Emei County, Sichuan with reference to the Sinian–Cambrian boundary. *Scientia Geologica Sinica 3*, 253–262. (In Chinese with English abstract.)

Ching Y.-k. 1960: Conodonts from the Kufeng Suite (Formation) of Lungtan, Nanjing. *Acta Palaeont. Sinica 8:3*, 230–248. (In Chinese and English.)

Hou H.-f., Wang S.-t., Gao L.-d., Xian S.-y., Bai S.-l., Cao X.-d, P'an, K., Liao W.-h. et al. 1979: The Devonian system of China. In Chinese Academy of Geological Sciences (ed.): *Stratigraphy of China. Papers submitted to the Second All-China Stratigraphic Congress, 1979*, 19–22. Beijing.

Lai C.-g., Wang X.-f., Fu K., An T.-x., Yi Y.-e., Qiu H.-r., Yang J.-z., Zhang W.-t., Liu D.-y. & Chen T.-e. 1979: The Ordovician system of China. In Chinese Academy of Geological Sciences (ed.): *Stratigraphy of China. Papers submitted to the Second All-China Stratigraphic Congress, 1979*, 10–13. Beijing.

Lin B.-y. 1979: The Silurian System of China. *Acta Geologica Sinica. 53:3*. 190–200. (In Chinese with English abstract.)

Ni S.-z. 1981: Discussion on some problems of Ordovician stratigraphy by means of conodonts in eastern part of Yangtze Gorge Region. *Selected Papers, 1st Convention Micropaleont. Soc. China*, 127–134. (In Chinese.)

Nogami, Y. 1966: Kambrische Conodonten von China, Teil 1, Conodonten aus den oberkambrischen Kushan-Schichten. *Kyoto Univ. Coll. Sci., Mem., Ser. B, 32:4*, 351–367.

Nogami, Y. 1967: Kambrische Conodonten von China. Teil 2, Conodonten aus dem hoch oberkambrischen Yencho-Schichten. *Kyoto Univ., Coll. Sci., Mem. Ser. B. 33:4*, 211–219.

Qian Y. 1977: Hyolitha and some problematica from the Lower Cambrian Meishucun stage in Central and S.W. China. *Acta Palaeont. Sinica 16:2*, 255–278. (In Chinese with English abstract.)

Qin F. & Gan Y.-y. 1976: The Palaeozoic stratigraphy of Western Qin Ling Range. *Acta Geologica 1976:1*, 74–97. (In Chinese with English abstract.)

Tan X.-c., Dong Z.-z & Qin D.-h. 1982: Lower Devonian of the Baoshan area, Western Yunnan and the boundary between Silurian and Devonian systems. *Journal of Stratigraphy 6:3*, 199–208. (In Chinese.)

Wang C.-y. 1974: Carboniferous conodonts from Guangxi and Permian conodonts from Sichuan. In: *Handbook of the Stratigraphy and Palaeontology in S.W. China*, 383–384. (In Chinese.)

Wang C.-y. 1979: A progress report of Carboniferous conodonts in China. *Paper for the 9th International Congress of Carboniferous Stratigraphy and Geology*.

Wang C.-y. 1980: Upper Silurian conodonts from Qujing District,

Yunnan. *Acta Palaeont. Sinica 19:5*, 369–378. (In Chinese with English abstract.)

Wang C.-y. 1981a: Conodonts from the Devonian Ertang Formation in Central Guangxi. *Acta Palaeont. Sinica 20:5*, 400–405. (In Chinese with English abstract.)

Wang C.-y. 1981b: Lower Devonian conodonts from the Xiaputonggou Formation at Zoige, N.W. Sichuan. *Bull. Xian Inst. Geol. M., R., Chinese Acad. Geol. Sci. 3*, 76–84. (In chinese with English abstract.)

Wang C.-y. 1982: Upper Silurian and Lower Devonian conodonts from Lijiang of Yunnan. *Acta Palaeont. Sinica 21:4*, 436–448. (In Chinese with English abstract.)

[Wang C.-y. 1983: Devonian conodonts of Guangxi (MS).]

Wang C.-y., Ruan Y.-p., Mu D.-c., Wang Z.-h., Rong J.-y., Yin B.-a., Kuang G.-d. & Su Y.-b. 1979: Subdivision and correlation of the Lower and Middle Devonian series in different facies of Guangxi. *Acta Stratigraphica Sinica 3:4*, 305–311. (In Chinese.)

Wang C.-y & Wang Z.-h. 1976: Triassic conodonts from the Mount Jolmo Lungma Region. *A Report of Scientific Expedition in the Mount Jolmo Lungma Region (1966–1968) (Palaeontology), Fase II*, 387–419. (In Chinese.)

Wang C.-y. & Wang Z.-h. 1978a: Early and Middle Devonian conodonts of Kwangsi and Yunnan. In *Symposium on the Devonian System of south China (1974)*, 334–345. (In Chinese.)

Wang C.-y. Wang Z.-h. 1978b: Upper Devonian and Lower Carboniferous conodonts from Southern Guizhou. *Memoirs of Nanjing Institute of Geology and Palaeontology, Academia Sinica 11*, 51–96. (In Chinese with English abstract.)

Wang C.-y. & Wang Z.-h. 1981a: Conodont sequence in China (Cambrian–Triassic). *12th Anm. Conf. Palaeont. Soc. China. Selected Papers*, 105–115. (In Chinese.)

Wang C.-y. & Wang Z.-h. 1981b: Permian conodonts from the Longtan Formation and Changhsing Formation of Changxing, Zhejiang and their stratigraphical and palaeoecological significance. *Selected Papers 1st Convention Micropaleont. Soc. China (1979)*, 114–120.(In Chinese.)

Wang C.-y. & Wang Z.-h. 1981c: Permian conodont biostratigraphy of China. *Geol. Soc. am. Spec. Paper 187*, 227–236.

Wang C.-y. & Ziegler, W. 1981: Middle Devonian conodonts from Xiguitu Qi, Inner Mongolia Autonomous Region, China. *Senckenbergiana lethaea 62*, 125–139.

Wang C.-y. & Ziegler, W. 1983a: Conodonts from Tibet. *N. Jahrb. Geol. Paläontol, Monatsh. 1983:2*, 69–79.

Wang C.-y. & Ziegler, W. 1983b: Devonian conodont biostratigraphy of Guangxi and its correlation with Europe. *Geologica et Palaeontologica 17* (in press).

Wang Z.-h. 1978: Permian–Lower Triassic conodonts of the Liangshan area, Southern Shaanxi. *Acta Palaeont. Sinica 17:2*, 213–232. (In Chinese with English abstract.)

Wang Z.-h. 1980: Outline of Triassic conodonts in China. *Rev. Ital. Paleont. 85:3–4*, 1221–1226.

Wang, Z.-h. 1982: Discovery of Early Triassic *Neospathodus timorensis* Fauna in Ziyun of Guizhou. *Acta Palaeont. Sinica 21:5*, 584–587. (In Chinese with English abstract.)

[Wang Z.-h. 1983: Cambro-Ordovician conodonts from North and Northeast China with additional remarks on Cambro-Ordovician boundary (MS).]

Wang Z.-h. & Cao Y.-y. 1981: Early Triassic conodonts from Lichuan, Western Hubei. *Acta Palaeont. Sinica 20:4*, 363–375. (In Chinese with English abstract.)

Wang Z.-h. & Dai J.-y 1981: Triassic conodonts from the Jiangyou–Beichuan area. Sichuan Province. *Acta Palaeont. Sinica 20:2*, 138–150. (In Chinese with English abstract.)

[Wang Z.-h. & Dong Z.-z. 1983: Upper Triassic conodonts from West Yunnan (MS).]

[Wang Z.-h. & Li R.-l. 1982: Upper Carboniferous conodonts of the Taiyuan Formation in Taiyuan, Shanxi (MS).]

Wang Z.-h. & Luo K.-q. 1982: Cambro-Ordovician conodonts from the marginal areas of Ordos Platform, China.

Wang Z.-h. & Wang C.-y 1983: The Carboniferous conodonts from the Jingyuan formation of Gansu province. Acta Palaeont. Sinica 22:4, 437–446. (In Chinese with English abstract.)

[Wang Z.-h. & Zhang W.-s. 1983: Discovery of conodont fauna at the upper part of the Taiyuan Formation, Henan (MS).]

Xian S.-y., Wang S.-d., Zhou, X.-y., Xiong J.-f. & Zhou T.-r. 1980: *Nandan Typical Stratigraphy and Paleontology of Devonian in South China*, 1–159. Publishing House of Guizhou. (In Chinese.)

[Yang J.-z. 1983: Tentative discussion on the Mid-Carboniferous boundary of China (MS).]

Yao L.-q 1982: Upper Cambrian conodont fauna and stratigraphy of Fengshan formation in Northern Anhui. *Journal of Tonji University 1982:3*, 24–42. (In Chinese with English abstract.)

Zhao S.-y. 1981: Late Carboniferous conodonts from the Qinshui Basin in Shanxi Province. *Bull. Taijin Inst. Geol. Min. Res. 4*, 97–108. (In Chinese with English abstract.)

[Zhao Z.-x. 1979: Carboniferous and Permian conodonts from the North slope of the Kunlun Mountain, Xingjiang (MS).]

Zhou X.-y., Zhai Z.-q. & Xian S.-y. 1981: On the Silurian conodont biostratigraphy, new genera and species in Guizhou Province. *Oil & Gas Geology 2:2*, 123–140. (In Chinese with English abstract.)

[Zhou Z.-y., Chen J.-y., Lin Y.-k., Wang Z.-h., Xu J.-t. & Zhang J.-l. 1983: New observation on the Ordovician strata of the Tangshan area, Hebei (MS).]

Biogeography, evolutionary relationships, and biostratigraphic significance of Ordovician platform conodonts

STIG M. BERGSTRÖM

FOSSILS AND STRATA

Bergström, Stig M. 1983 12 15: Biogeography, evolutionary relationships, and biostratigraphic significance of Ordovician platform conodonts. *Fossils and Strata*, No. 15, pp. 35–58. Oslo. ISSN 0300-9491. ISBN 82-0006737-8.

Fewer than 15 of the nearly 100 currently recognized Ordovician conodont genera have platform elements in the apparatus, but these are of special interest because they represent the oldest such types known and they include some types of platform developments not known in younger strata. Eight of these genera have their origin and main occurrence in the North Atlantic Province, three in the Midcontinent Province, and three are present in both provinces. The data at hand suggest that especially during Middle Ordovician time, some North Atlantic Province platform taxa invaded parts of the Midcontinent Province, but there is much less evidence of such migrations in the opposite direction. Five evolutionary lineages are recognized and discussed (those of *Amorphognathus*, *Cahabagnathus* nom. nov., *Eoplacognathus*, *Icriodella*, and *Pygodus*) and other genera (*Complexodus*, *Nericodus*, *Polonodus*, *Prattognathus* n. gen., *Rhodesognathus*, *Sagittodontina*, *Scyphiodus*, and *Serratognathus*) are dealt with in less detail. Species of *Amorphognathus*, *Cahabagnathus*, *Eoplacognathus*, and *Pygodus* are of major biostratigraphic significance but most other taxa are either too long-ranging or too restricted in their distribution to be useful biostratigraphically. Two new generic designations (*Cahabagnathus*, *Prattognathus*) and three new species (*Cahabagnathus chazyensis*, *C. carnesi*, *Pygodus anitae*) are proposed. □*Conodonta, platform conodonts, biostratigraphy, biogeography, evolution, Ordovician.*

ECOS III

A contribution to the Third European Conodont Symposium, Lund, 1982

Stig M. Bergström, Department of Geology & Mineralogy, The Ohio State University, 125 S. Oval Mall, Columbus, Ohio 43210, USA; 20th October, 1982.

A decade ago, at the 1971 *Marburg Symposium on Conodont Taxonomy,* I reviewed some new data bearing on the morphology and phylogenetic relationships of Middle and Upper Ordovician platform conodonts (Bergström 1971c). For various reasons, especially the need to get additional supportive materials, most of that information has remained unpublished, and on the basis of additional collections and other data now at hand, some of my 1971 interpretations have been modified. The purpose of the present study is to summarize data that bear on the biogeography, evolution, mutual relationships, and biostratigraphic significance of Ordovician platform conodonts. It may be particularly appropriate to attempt such a summary at this time because the recently published revised conodont volume of the *Treatise on Invertebrate Paleontology* (Clark *et al.* 1981) does not address most of the matters to be discussed below.

Different authors have given a somewhat different scope to the admittedly somewhat vague term 'platform conodont'. 'Platform' was recently (Sweet 1981) defined as a 'laterally produced shelflike structure flanking a process . . .', and in the present contribution, I use the term 'platform conodont' in a rather broad sense for conodonts having platformlike elements; thus I include both *Nericodus* Lindström 1955 (which has no distinct processes) and *Serratognathus* Lee 1970 (which is based on alate ramiform elements with platformlike lateral processes) among the platform conodonts. On the other hand, I do not regard *Prioniodus* (*Baltoniodus*) Lindström 1971 as a

platform conodont genus although it includes some species (e.g. *P.* (*B.*) *gerdae* Bergström 1971) in which one element has a platformlike posterior process. All in all, I here recognize 12 previously named genera of Ordovician platform conodonts, namely *Amorphognathus* Branson & Mehl 1933, *Complexodus* Dzik 1976, *Eoplacognathus* Hamar 1966, *Icriodella* Rhodes 1953, *Nericodus* Lindström 1955, *Polonodus* Dzik 1976, *Polyplacognathus* Stauffer 1935, *Pygodus* Lamont & Lindström 1957, *Rhodesognathus* Bergström & Sweet 1966, *Sagittodontina* Knüpfer 1967, *Scyphiodus* Stauffer 1935, and *Serratognathus* Lee 1970. In addition, I am aware of the existence of several other types of platform conodonts that appear distinctive enough to merit generic recognition, and two of these, *Cahabagnathus* nom. nov. and *Prattognathus* n. gen., are named in this contribution.

One problem confronting everyone who is carrying out studies related to the vertical distribution of Ordovician fossils on a world-wide scale is the current uncertainty of the precise chronostratigraphic relations between units in the successions on different continents. Although considerable progress has been made in recent years in establishing long-distance correlations of Ordovician rocks, alternate interpretations are still being proposed, especially in the case of the lower half of the system, and it may take years before conclusive evidence becomes available to settle some of these problems. In the present contribution, I have followed the graptolite–conodont biostratigraphy set forth in a recent paper (Bergström 1983), and I use the terms Lower, Middle, and Upper Ordovician in the sense of the Baltoscandic Oelandian, Viruan, and Harjuan Series. For convenience, the British series designations have been employed; the base of the Middle Ordovician is taken to be in the middle Llanvirnian, and that of the Upper Ordovician in the upper Caradocian.

Scope of study

Extensive studies during the last three decades have made Ordovician conodonts reasonably well known in North America and northwestern Europe, but Asian, Australian, and South American forms have received far less study. However, the data at hand do not suggest that there are numerous undescribed Ordovician platform conodonts. The interpretations and descriptions presented below are based largely on specimens in my own collections, which have been assembled since 1958 from a considerable number of localities in Europe and North America, and a few in North Africa and The People's Republic of China. I have also had full access to large collections of Ordovician conodonts from North America kept at the Department of Geology and Mineralogy at the Ohio State University, and I have examined most type collections containing Ordovician platform conodonts in Europe and North America. Unfortunately, I have not had the opportunity to study specimens of a few Siberian platform conodonts dealt with by Moskalenko (1970, 1977) as well as those in other collections in the USSR, and the currently available descriptions and illustrations are insufficient for a proper evaluation of some of these forms, most of which appear to be unrelated to the taxa dealt with herein. Although the oldest platform conodonts known are from strata as old as the Tremadocian (Lindström 1955, 1964), this type of conodont became common and diversified only in the Llanvirnian and younger parts of the Ordovician. Because of this, and the fact that the available collections of Tremadocian and Arenigian platform conodonts are small and not very informative, my discussion will center on post-Arenigian forms.

Provincial differentiation

The vast majority of the currently known Ordovician platform conodonts were originally described from Europe, particularly from the Baltoscandic–Polish region, and this area appears to have a greater diversity of such conodonts than any other region with Ordovician deposits. In terms of the two commonly recognized Ordovician conodont faunal provinces (Sweet & Bergström 1974), North American Midcontinent Province and the North Atlantic Province, the Ordovician platform conodont genera are strikingly unequally distributed. That is, *Cahabagnathus*, *Complexodus*, *Eoplacognathus*, *Nericodus*, *Polonodus*, *Prattognathus*, *Pygodus*, and *Sagittodontina* are virtually restricted to the North Atlantic Province, whereas only *Scyphiodus*, *Serratognathus*, and *Polyplacognathus* are Midcontinent Province genera. *Amorphognathus*, *Icriodella*, and *Rhodesognathus* are known from numerous localities in both provinces, and are almost cosmopolitan in their distribution; however, *Amorphognathus* and *Icriodella* are most diversified,

and have their evolutionary origin, in the North Atlantic Province and may be considered more typical of the latter province than of the Midcontinent Province. Thus, there is no question that at the generic level, the North Atlantic Province has the greatest diversity of platform conodonts. That this diversity trend is even more evident at the species level is illustrated by the fact that more than 30 platform conodont species are known from the North Atlantic Province and fewer than five are restricted to the other province. The reasons for this difference are obscure and may be very complex. The Midcontinent Province appears to have included primarily, but not exclusively, shallow-water environments in the tropical zone whereas the North Atlantic Province seems to have occupied mainly temperate and arctic latitudes where its conodont biotas became preserved in rocks representing a very wide range of depth environments. Interestingly, the greatest diversity of platform conodonts in the latter province was in off-shore subtidal environments, in many cases on continental-shelf margins, and these environments are also characterized by a high diversity of platform conodonts in some other geologic periods, for instance, the Devonian. Quite clearly, the data at hand support the idea that the Ordovician platform-condont distribution pattern is a conspicuous exception to the well-known rule that tropical faunas generally are characterized by a considerably higher species diversity than the temperate–arctic ones.

Migration patterns

A discussion of the distribution of Ordovician platform conodonts in time and space should clearly address the question of distribution centers and patterns of migration at both the species and genus levels. Regrettably, the current lack of a reliable base map of the Ordovician regional paleogeography prevents proper illustration of these matters but pertinent data are discussed below.

As noted below, the stratigraphically oldest known platform conodont, *Ambalodus* n. sp. of Lindström (1955) from the upper Tremadocian of Sweden, is similar to an element of *Eoplacognathus* and it may well be interpreted as an ancestor of the latter genus. Be this as it may, the data at hand suggest that the center of diversification of *Eoplacognathus* was in the Baltoscandic area, where at least seven species appeared during Llanvirnian through Llandeilian time, and representatives of these are regularly, and in many cases quite commonly, represented in most lower Middle Ordovician samples. The only known other area with an early (Arenigian–Llanvirnian) diversification of *Eoplacognathus* comparable to that of the Baltoscandic region is Hubei Province in The People's Republic of China (Sheng 1980; An 1981). Unfortunately, the Chinese taxa are not yet known in enough detail to permit firm conclusions regarding their precise relations to the Baltoscandic forms. Clearly, however, they are very similar to and occur with many other conodont species characteristic of Baltoscandic faunas. Outside these areas, occurrences of representatives of *Eoplacognathus* are far more scattered but some species exhibit a very wide, in some cases almost cosmopolitan, distribution. For instance, *E. suecicus* is known from Nevada (Harris *et al.* 1979) and forms similar to both *E.*

foliaceus and *E. reclinatus* have been found, however in most cases only rarely, in Tennessee (Bergström 1971a; Bergström & Carnes 1976), central Nevada, eastern New York and Oklahoma (Harris *et al.* 1979), and Alabama (Bergström 1971a; listed as *E. foliaceus*). Similar forms are also known from south-central China (Sheng 1980; An 1981; Ni 1981) and the Canning Basin of Western Australia (McTavish & Legg 1976). In contrast, representatives of *E. robustus*, although found locally in Tennessee (Bergström 1973) and possibly in northeastern Newfoundland (Bergström *et al.* 1974; Fåhraeus & Hunter 1981), are very rare outside the Baltoscandic–Polish region. A far wider distribution is exhibited by *E. lindstroemi*, and especially by *E. elongatus*; in addition to its many occurrences in Baltoscandia, the latter species is known from Wales (Bergström 1981a), many localities in the Appalachians (Bergström 1971a; Bergström & Carnes 1976; Nowlan 1981), Oklahoma, Arkansas (Repetski & Ethington 1977), central Nevada (Harris *et al.* 1979), Poland (Dzik 1978), and western USSR (Drygant 1974). This species may possibly be the most widespread of all those currently referred to *Eoplacognathus* suggesting a maximum dispersal of the genus just before its extinction.

Although the ancestor of *Cahabagnathus* has not been identified with certainty, it appears that the genus had a center of evolution in southeastern and southern USA, where it is common in the lower Middle Ordovician and all known forms are present. Representatives of the stratigraphically next oldest species, *C. friendsvillensis*, occur also outside this area in North America, for instance in New York and Quebec (Roscoe 1973), Northwest Territories in western Canada (Tipnis *et al.* 1978), and central Nevada (Harris *et al.* 1979). As far as I know, this form is still unknown in Europe but a closely similar species was recently reported from The People's Republic of China (An 1981). The geographically most widespread species of *Cahabagnathus* is likely to be *C. sweeti*, which is known not only from localities in much of the Appalachians but also from Texas (Bergström 1978), Arkansas (Repetski & Ethington 1977), southwestern Scotland (Bergström 1971a), and, from rare specimens in a very limited interval, in Norway and Sweden (Bergström 1971a) as well as in Volyn in western USSR (Drygant 1974), and Siberia (Moskalenko 1977). This distribution pattern clearly shows a considerable expansion of the range of the genus in Llandeilian time. The stratigraphically youngest species of the *Cahabagnathus* lineage (*C. carnesi*) has not as yet been found outside Tennessee, and the genus might have had a very restricted geographic range before becoming extinct.

The presence in the Baltoscandic region of not only its apparent ancestors but also primitive species of *Pygodus* may be taken as an indication that this area was a center of diversification of that important genus. The virtually cosmopolitan distribution of *P. serra* and *P. anserinus*, especially in geosynclinal regions in North America (Sweet & Bergström 1974), Scotland (Bergström 1971a), Wales (Bergström 1981a), the Urals (Nasedkina & Pushkov 1979), The People's Republic of China (Sheng 1980; An 1981; Ni 1981), New Zealand (Simes 1980), and Australia (Nicoll 1980) shows power of dispersal over huge regions. In view of this distribution, and the common occurrence of forms of *Pygodus* also in black shale facies, the suggestion recently put forward by

Fåhraeus & Hunter (1981) that these conodonts were benthic, seems rather improbable. Interestingly, it appears that the stratigraphically youngest species of the *Pygodus* lineage had a wider geographic distribution than any other species of the genus.

Another area of origin is apparently represented by *Icriodella*, the oldest representatives of which are known from the British Isles and Brittany in strata of latest Llandeilian to earliest Caradocian age (Lindström *et al.* 1974). This genus is not known from the Baltoscandic region in strata older than the late Caradocian, and the genus is only rarely represented in the Ashgillian there. In North America, the stratigraphically oldest species is *I. superba*, which appears in Rocklandian strata (Schopf 1966; Roscoe 1973), that is, in beds coeval with a portion of the lower Caradocian. This species migrated into the North American Midcontinent, where it is quite widespread, during middle to late Caradocian time and it persisted there to the late Ashgillian (Richmondian). I am aware of only one reported occurrence of Ordovician representatives of *Icriodella* outside Europe and North America, namely in The People's Republic of China (Sheng 1980), but it would not be surprising if occurrences were also found in other parts of Asia and in Australia.

From available evidence (see pp. 38–40), it appears that the evolutionary origin of *Amorphognathus* was in the Baltoscandic area, where there are some Arenigian–Llanvirnian species not known with certainty elsewhere. However, An (1981) has recently reported primitive forms from the People's Republic of China that are associated with conodonts of Baltoscandic type. Unfortunately, those platform conodonts are still not known in enough detail to permit close comparison with Baltoscandic forms. The same applies to some forms from the Urals described by Nasedkina & Pushkov (1979). The Llandeilian species *A. inaequalis* is currently best known from Wales (Rhodes 1953; Bergström 1964) and Brittany (Lindström *et al.* 1974) but has also been recorded from Poland (Dzik 1976). It may also be present in Baltoscandia but most middle Middle Ordovician specimens there are clearly *A. tvaerensis*. The Siberian form referred to as *A.* cf. *inaequalis* by Moskalenko (1977, Pl. 8:14, 15) is clearly not conspecific, and may not even be congeneric, with Rhodes' species. Unlike *A. inaequalis*, representatives of *A. tvaerensis* are widely distributed also in North America where the species has been recorded in the Appalachians (Bergström 1971a; Bergström & Carnes 1976; Nowlan 1981), Oklahoma, central Nevada (Harris *et al.* 1979), New York State and the Midcontinent (Sweet & Bergström 1971). The stratigraphically younger species *A. superbus* and *A. ordovicicus* have an even wider distribution in North America and Europe, and the latter species may be regarded as essentially cosmopolitan. From distribution data now available, it seems that *Amorphognathus* gradually expanded its geographic range during late Middle and Late Ordovician time.

Most of the other Ordovician platform genera have a very limited geographic range and exhibit no very clear migration patterns. Yet, representatives of *Complexodus* are known from Wales (Bergström 1981a), the Baltoscandic–Polish–Volyn region (Dzik 1976), and The People's Republic of China (An 1981) but all these occurrences are in a narrow stratigraphic interval in the *Pygodus anserinus* to lower *Amorphognathus tvaerensis* Zone. Representatives of *Rhodesognathus* first appear in the lower Caradocian of the British Isles (Bergström 1971a) and in equivalent strata in New York State and Vermont (Schopf 1966; Roscoe 1973), and the genus expanded its range into much of the North American Midcontinent and the Baltoscandic region (Bergström, J. *et al.* 1968) during middle to late Caradocian time. At least in North America, it exhibits a distribution pattern similar to that of *Icriodella superba*. Also, representatives of *Rhodesognathus* almost invariably occur with specimens of *Amorphognathus*, which is far more widespread than *Rhodesognathus* but also exhibits a striking progressive evolution in several features that has no counterpart in the morphologically stable and conservative representatives of *Rhodesognathus*.

Summarizing the above observations, it appears that *Amorphognathus*, *Eoplacognathus*, and *Pygodus* had their evolutionary origin in Baltoscandia in Early and early Middle Ordovician time, and subsequently spread from that region into particularly the British Isles and marginal areas of North America. Whereas the last two genera became extinct in the early Caradocian, *Amorphognathus* later expanded its range and in the Late Ordovician became essentially cosmopolitan. The distribution of these three genera in time and space also suggests close faunal interchange between Baltoscandia and The People's Republic of China and, possibly, New Zealand and the Canning Basin of Western Australia. *Icriodella* and *Rhodesognathus* may have originated in the British Isles – Brittany region, and representatives of these genera subsequently migrated into large areas of the North American plate. These genera are comparatively poorly represented in Baltoscandia where most occurrences are in a limited interval in the middle to upper Caradocian. Interestingly, many of these occurrences are associated with the invasion of North American megafossils discussed by Jaanusson & Bergström (1980:92–93), and it seems appropriate to regard both *Icriodella* and *Rhodesognathus* as exotic elements in the Baltic conodont faunas. *Cahabagnathus* seems to have had a center of evolution in eastern North America and it never became common or widespread outside the North American plate although rare specimens of one species (*C. sweeti*) have been found in Scotland, Baltoscandia, Volyn in western USSR, and Siberia. Again, it is of interest to note that in Baltoscandia this species occurs in an interval that also has representatives of the unusual but widespread genus *Complexodus* and the Midcontinent Province genus *Phragmodus* (Bergström 1971a:106), which suggests a brief period of long-distance faunal exchange. Another, and markedly stronger contemporaneous expression of this may be the 'flood' of North Atlantic Province elements, including *Pygodus* and *Prioniodus* (*Baltoniodus*), which in early Middle Ordovician time invaded the continental-plate margins in several parts of the world, especially around North America.

Evolutionary lineages

Based on their vertical occurrence and evolutionary relationships, one can distinguish three groups of Ordovician platform-conodont genera. One includes *Nericodus*, *Prattognathus*, and *Serratognathus*, each of which appears suddenly in the succession without any obvious ancestral forms. Another, which comprises *Complexodus*, *Icriodella*, *Polyplacognathus*, *Rhodesognathus*, *Sagittodontina*, and *Scyphiodus*, differs from the first in that its genera show at least some similarity to other forms and an evolutionary origin can be proposed. With the exception of *Icriodella*, and possibly *Polyplacognathus*, these genera currently include only one species each. A third group of genera includes *Amorphognathus*, *Cahabagnathus*, *Eoplacognathus*, *Polonodus*, and *Pygodus*. Each of these genera contains more than one species, and these species are parts of rapidly evolving evolutionary lineages, the ancestor of which is known in at least some cases. Compared to the two other groups mentioned, this group has a far wider geographic distribution that in some cases is virtually cosmopolitan, and the genera of this group include many biostratigraphically important species. It is the evolution and evolutionary relationships of these taxa that are of major concern in the present study. For convenience, I will first discuss each of these lineages and will then deal with the other platform conodont genera in a separate section.

The AMORPHOGNATHUS lineage (Fig. 1)

The upper Middle and Upper Ordovician part of this lineage, which includes *A. tvaerensis* Bergström 1962, *A. superbus* Rhodes 1953, and *A. ordovicicus* Branson & Mehl 1933, was discussed in some detail by Bergström (1971a), who described the composition of the *Amorphognathus* apparatus and used the species mentioned as zonal indices in the North Atlantic Province conodont-zone succession. Subsequent work has confirmed the apparatus reconstruction and the biostratigraphic significance of these species, but it has also added important data on the early phylogeny of *Amorphognathus* (Sweet & Bergström 1972; Lindström 1977; Dzik 1976).

The idea that *Amorphognathus tvaerensis* evolved from *A. inaequalis* Rhodes 1953 (Bergström 1971a) is supported by new data from Poland (Dzik 1976) and Great Britain (Bergström coll.) and the ancestor of the latter species is now also clear. Forms with an array of elements similar to that of *A. ordovicicus* are known from rocks as old as Early Ordovician. Holodontiform (M) elements, which are very characteristic of the *Amorphognathus* apparatus, are represented by the types of *Lenodus* Sergeeva from the Arenigian of the Baltic region, and this genus has generally been interpreted as a junior synonym of *Amorphognathus* (Bergström 1971a, 1981b; Dzik 1976; Lindström 1977). The occurrence of this type of element in the Lower Ordovician is important because it gives a clue to the origin of *Amorphognathus*; that is, this evidence suggests (Lindström 1964, 1970, 1977; Bergström 1971a; Dzik 1976) that *Amorphognathus* evolved from the *Prioniodus* (*Baltoniodus*) stock in the middle Arenigian (Fig. 5). Lindström (1977) indicated that the non-holodontiform elements of primitive species of *Amorphognathus* such as *A. falodiformis* (Sergeeva) may be closely similar to those of *Prioniodus* (*Baltoniodus*) *navis* Lindström but the morphological differences, if any, have not yet been described. At any rate, only minor morphological modification is needed to derive the holodontiform (M) element of the former species from the oistodontiform element of the latter, and there is little doubt that the ancestor of *Amorphognathus* is in the *Prioniodus* (*Baltoniodus*) stock.

Although known in its main features, the pattern of

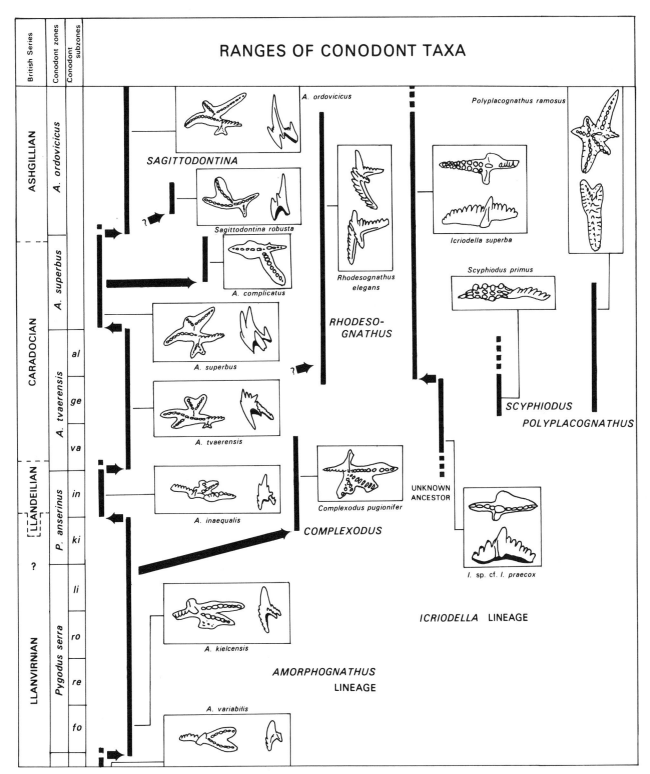

Fig. 1. Suggested evolutionary relationships and known stratigraphic ranges of taxa of the *Amorphognathus* and *Icriodella* lineages and of *Sagittodontina, Complexodus, Rhodesognathus, Scyphiodus,* and *Polyplacognathus.* Only key elements in the apparatuses are illustrated. The zonal framework is that of Bergström (1971a) and Löfgren (1978) with minor modifications. As in the other diagrams, horizontal lines from boxes mark the approximate stratigraphic position of the illustrated specimen(s). Abbreviations of subzonal designations are as follows: *va–fl, A. variabilis–M. flabellum; va–oz, A. variabilis–M. ozarkodella; su–gr, E. suecicus–S. gracilis; su–su, E. suecicus–P. sulcatus; fo, E. foliaceus; re, E. reclinatus; ro, E. robustus; li, E. lindstroemi; ki, A. kielcensis; in, A. inaequalis; va, P. (B.) variabilis; ge, P. (B.) gerdae; al, P. (B.) alobatus.*

evolution exhibited by *Amorphognathus* in the Llanvirnian and Llandeilian is still not safely established in all details. For instance, Löfgren (1978) records observations in support of the interpretation that a Swedish form identified as *A. variabilis* Sergeeva has an apparatus composed of only two types of platform elements and, accordingly, is closely similar to that of *Eoplacognathus* (cf. p. 41). A different opinion is expressed by Dzik (1976) and Lindström (1977) who include seven types of elements in the apparatus of *A. variabilis* and, with justification, regard this as a typical *Amorphognathus* apparatus. This view is supported by the fact that platform conodont elements strikingly similar to the types of Sergeeva's *A. variabilis* occur with holodontiform (M) elements in Hubei Province, The People's Republic of China (Bergström coll.). However, it appears that the only way to definitely establish the nature of the *A. variabilis* apparatus is to reinvestigate Sergeeva's type collection and, preferably, also topotype material. Nevertheless, it is of interest to note that regardless of the outcome of such a study, there is evidence that there was at least one platform conodont species, that referred to as *A. variabilis* by Dzik (1976), in the late Arenigian – early Llanvirnian that had an array of elements similar to that in advanced species of *Amorphognathus*.

From this form Dzik (1978, Fig. 1) derived a stratigraphically slightly younger species, *A. kielcensis* Dzik 1976, which differs from the former species in minor details in the pastiniscaphate (P) and holodontiform (M) elements. This species, which is widely distributed geographically in the Baltic area but somewhat erratic in its occurrence, ranges from strata at least as old as about the base of the *E. foliaceus* Subzone in Sweden. With its ancestor, it shares the character of having a posterolateral process nearly as long as the posterior process and also the tendency of developing, in mature amorphognathiform elements, lateral expansions of especially the posterolateral process, which bear an irregular ornamentation of small ridges, nodes, and denticles.

An extreme case of the latter tendency, in the same position on the pastiniscaphate element, is exhibited by *Complexodus pugionifer* (Drygant 1974), type species of *Complexodus* Dzik 1976. This species has a distinct anterior lobe on the posterolateral process that in adult specimens tends to have an irregular denticulation (Dzik 1978, Fig. 1:20, Pl. 13:6). In other respects, the pastiniscaphate element of *C. pugionifer*, the only type of element currently known in its apparatus, is so reminiscent of that of *A. kielcensis* that it seems justified to suggest that the latter species is the ancestor of *Complexodus* (Fig. 1).

A group of closely related species, including *Amorphognathus inaequalis*, *A. tvaerensis*, *A. superbus*, and *A. ordovicicus*, forms a lineage that extends from the upper Middle Ordovician to near the top of the Upper Ordovician. To this group belong also *A. complicatus* Rhodes 1953 and some other, but still incompletely known, forms such as *A. lindstroemi* (Serpagli 1967). As noted by Bergström (1971a) and Dzik (1976, 1978), the oldest species of this group, *A. inaequalis*, apparently evolved into the stratigraphically next younger species, *A. tvaerensis*. Morphological differences between these species are minor but recognizable in the platform as well as the holodontiform (M) elements (Lindström 1977; Dzik 1976). I share the opinion of Dzik (1976, 1978) that *A. inaequalis* is closely related to the *A. variabilis* of Dzik – *A. kielcensis* stock and there seems to be evidence supporting the view that it evolved directly from the latter species. Likewise, there is very little doubt that *A. tvaerensis* evolved from *A. inaequalis* as suggested by Bergström (1971a).

Representatives of *Amorphognathus tvaerensis*, *A. superbus*, and *A. ordovicicus* show considerable intraspecific variation and the currently used morphologic concept of these species is relatively broad. Both *A. tvaerensis* and *A. superbus*, and *A. superbus* and *A. ordovicicus*, are connected by morphological transitions,

and relations between these species appear to be rather clear. The characteristic species *A. complicatus* is interpreted as an offshoot from *A. superbus*; apart from the simple posterolateral process in the pastiniscaphate elements of the former species these species are closely similar to each other. Contrary to a recent statement by me (Bergström 1981b), *Amorphognathus* seems to be restricted to the Ordovician because a recent taxonomic and stratigraphic reevaluation of some reported Silurian occurrences of the genus failed to establish a single reliable record in that system. Also, among the Silurian platform-conodont genera, none has been identified as a likely descendant of *Amorphognathus* and the lineage might have become extinct in latest Ordovician time.

The ICRIODELLA lineage (Fig. 1)

The type, and probably most common species, of *Icriodella* is *I. superba* Rhodes 1953. Since this species, and the genus, were defined in terms of multielement taxonomy (Bergström & Sweet 1966; Webers 1966), only a rather limited amount of basic new data has been published on Ordovician forms whereas Silurian species have been discussed much more extensively. Although *I. superba* is widespread geographically in North America and Europe and shows a certain degree of morphological variation that has led to taxonomic splitting (Rhodes 1953; Orchard 1980), the collections studied have not permitted the recognition of distinct evolving morphological trends in this species.

The stratigraphically oldest representatives of *Icriodella superba* known are from the Harnagian of the Welsh Borderland (Bergström 1971a) and Rocklandian strata in New York State (Schopf 1966) and Vermont (Roscoe 1973). Slightly older (Costonian and possibly latest Llandeilian) strata in the Welsh Borderland contain a distinct species (Fig. 6 A–H) that may be conspecific with *I. praecox* Lindström, Racheboeuf & Henry 1974, which was originally described from the Middle Ordovician of Brittany. This form is characterized particularly by the fact that it has a single, rather than double row of denticles on the anterior process of the platform element although the other elements of its apparatus are quite similar to those of *I. superba*. The suggestion of Lindström, Racheboeuf & Henry (1974) that *I. praecox* is the ancestor of *I. superba* appears to be fully supported by the evidence at hand, and the latter species is likely to have evolved from the former in late Costonian or early Harnagian time in northwestern Europe (Fig. 1).

The ancestor of *Icriodella praecox* has not yet been identified. However, the platform element of this species shows some similarity in the development of its anterolateral process to *Amorphognathus* and *Prioniodus* (*Baltoniodus*) although its ramiform elements differ from those of the latter genera not only in the denticulation and other morphological details but also, and probably more significantly, in the absence of a tetraprioniodontiform element. Bearing in mind the very close similarity between the two *Icriodella* species, I reject Dzik's (1976:411) proposal that *I. praecox* should be referred to *Prioniodus* rather than to *Icriodella* because of its lack of a double denticle row. In my opinion, following that suggestion would lead to a highly artificial taxonomy. Nevertheless, the suggestion that *Icriodella* initially evolved from a prioniodontidean ancestor appears as likely today as when it was proposed more than ten years ago (Lindström 1970; Sweet & Bergström 1972, Fig. 3).

Because the youngest reported occurrence of *Icriodella superba* is in the Upper Ordovician (Richmondian) of the North American Midcontinent (Sweet & Bergström 1976), it is evident that the species has a rather long vertical range. However, as noted above, the species exhibits no very obvious morphological trends through its range although it cannot be ruled out that such trends may be masked by the considerable intraspecific variation and by the fact that the platform elements are susceptible to breakage and complete specimens are uncommon. Orchard (1980) described *I. prominens* from the lower to middle Ashgillian of Great Britain. Although the morphological features of that species are not entirely clear from the original description and illustrations, it may be conspecific with a previously unreported, but quite distinctive, species from strata of the same age in Sweden (Bergström coll.).

No representative of *Icriodella* is currently known from the uppermost Ordovician and it is unclear if the lowermost Silurian (Lower Llandoverian) species of the genus reported from North America (Pollock, Rexroad, & Nicoll 1970; McCracken & Barnes 1981; Nowlan 1981), as well as possibly slightly stratigraphically younger forms from Great Britain (Aldridge 1972), were derived from *I. superba*, *I. prominens*, or both. At any rate, the Early Silurian species are so similar to the Ordovician ones that there can be no doubt that they belong in the same genus. Aldridge (1972) described the vertical distribution and phylogenetic relationships of the *Icriodella* species in the British Llandoverian and a closely similar stratigraphic distribution pattern has recently been found in eastern Canada (Nowlan 1981).

As currently known, *Icriodella* does not range above the Wenlockian and the later history of the *Icriodella* lineage is somewhat unclear because of the poor record of this type of conodonts in much of the Wenlockian and the Ludlovian. However, the Upper Silurian *Pedavis* Klapper & Philip 1971, *Pelekysgnathus* Thomas 1949, and *Icriodus* Branson & Mehl 1938 exhibit so many morphological similarities to *Icriodella* that it seems justified to assume that these three genera are descendants of the *Icriodella* lineage. Particularly striking are the morphological changes in the ramiform elements, which by loss of the processes have attained a coniform shape. It is also of interest to note that *Icriodus* in the Devonian exhibits a taxonomic diversity far greater than that of any portion of the *Icriodella* lineage.

The CAHABAGNATHUS lineage (Fig. 2)

The platform conodonts herein referred to *Cahabagnathus* nom. nov. have been included in *Polyplacognathus* Stauffer 1935 by all previous authors except Drygant (1974), who proposed the preoccupied generic designation *Petalognathus* for this group of conodonts. *Petalognathus bergstroemi* Drygant 1974, type species of Drygant's genus, is clearly a junior synonym of *Cahabagnathus sweeti* (Bergström 1971), here designated type species of *Cahabagnathus*. The Middle Ordovician conodonts referred to this genus are interpreted to represent a lineage separate from those of other genera, and the introduction of a generic designation appears justified, especially as these conodont elements have a characteristic appearance and a wide distri-

bution. For a generic diagnosis of *Cahabagnathus*, see p 51. Although reported from Scotland (Bergström 1971a), Scandinavia (Hamar 1966; Bergström 1971a), the USSR (Drygant 1974; Moskalenko 1977), and The People's Republic of China (An 1981), representatives of *Cahabagnathus* are currently known primarily from North America, where they are widespread in the lower Middle Ordovician, especially in the Appalachians.

As shown by collections from eastern Tennessee, Virginia, and Oklahoma, five distinct species of *Cahabagnathus* occur in stratigraphic order and form an evolutionary lineage of intergrading successive morphotypes (Fig. 2). Illustrations of the four youngest of these species and their ranges were given by Bergström (1973, Fig. 4) and Bergström & Carnes (1976, Fig. 2) but the oldest species in the lineage, *Cahabagnathus* n. sp. A (Fig. 6 Q, R), which is currently known only from Oklahoma, has not been illustrated previously. Two of these species, *C. friendsvillensis* and *C. sweeti*, were named by Bergström (1971a), and for two others, *C. chazyensis* (=*Polyplacognathus friendsvillensis–sweeti* transition of Bergström 1978; Fig. 6 M–P) and *C. carnesi* (=*P. sweeti*, late form of Bergström & Carnes 1976; Fig. 6 K–L), specific names are formally proposed in the present paper.

The apparatus of *Cahabagnathus* n. sp. A and other species of the genus is composed of two types of paired platform elements which are stelliplanate and pastiniplanate, respectively. As described in the systematic part of the present paper, morphologic evolution in the *Cahabagnathus* lineage can be recognized in both of these element types, and is expressed primarily in the following features (Fig. 2) : (1) Gradually decreasing size, and ultimate disappearance, of the anterior lobe of the anterolateral process in the stelliplanate element; (2) widening of the posterior process of the pastiniplanate elements in adult specimens, which is well displayed in the transition of *C. friendsvillensis* into *C. chazyensis*; and (3) a progressive change in the appearance of the main denticle row at the junction of the posterior and anterior processes in the pastiniplanate element as illustrated in Fig. 2. It is of interest to note that three of the *Cahabagnathus* species exhibiting these changes have been found in successive order in the same formation at three localities, and the development of the most advanced species, *C. carnesi*, from its ancestor, *C. sweeti*, is documented in two sections. Furthermore, where representatives of the *Cahabagnathus* lineage have been found at other localities in North America and elsewhere, the order of succession of species is always the same.

When searching for an ancestor of the *Cahabagnathus* lineage, it is perhaps natural to turn first to the *Eoplacognathus* lineage, the other common and widespread group of early Middle Ordovician platform conodonts. In basic morphology, the two types of elements of the oldest known species of *Cahabagnathus*, *C. n. sp. A*, are not very unlike two of the four types of elements in the apparatus of species of *Eoplacognathus* such as *E. foliaceus* and *E. reclinatus* (Fig. 2). The *Cahabagnathus* elements differ, however, in occurring in mirror-image pairs and in having a distinct node and ridge ornamentation lateral to the central denticle row on the upper surface of the processes. It is my conclusion that these differences are more significant than they may appear at first sight and they occur in all growth stages. Although it cannot be ruled out completely

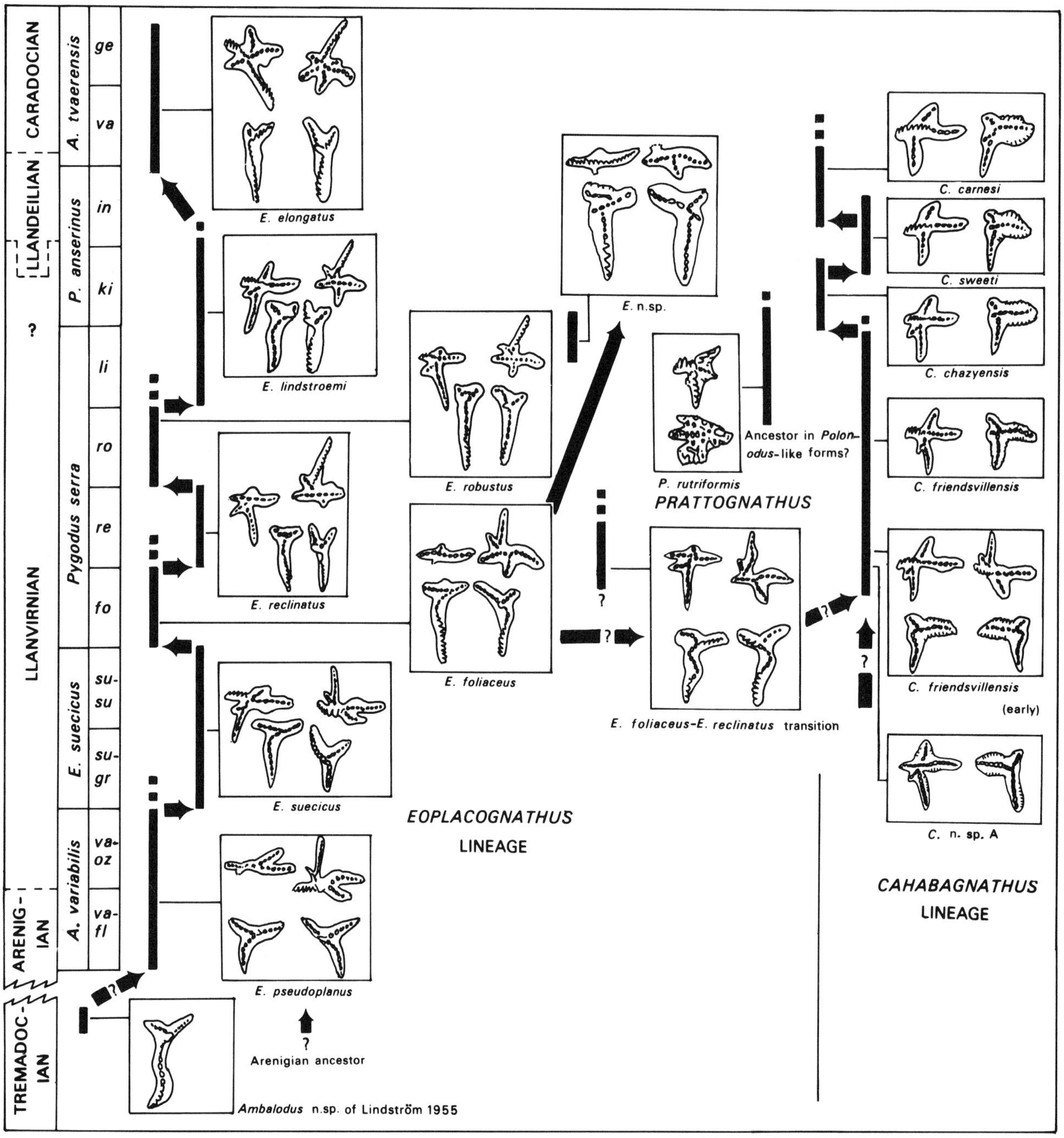

Fig. 2. Suggested evolutionary relationships and known stratigraphic ranges of taxa of the *Eoplacognathus* and *Cahabagnathus* lineages and of *Prattognathus*. In the case of *Prattognathus* and most species of *Cahaba-gnathus*, only one of the paired symmetry variants of each element is illustrated. For explanation of subzonal abbreviations, see Fig. 1.

in the present state of our knowledge that the ancestor of *Cahabagnathus* may belong to the *Eoplacognathus* lineage, no truly intermediate form has been found to support this idea. A potential ancestor is the *E. foliaceus* – *E. reclinatus* transition of Harris *et al.* (1979, Fig. 16), which occurs with, as well as stratigraphically below, primitive forms of the *Cahabagnathus* lineage in several North American sections. However, even though elements of this species are reminiscent in some respects of those of a species of *Cahabagnathus*, its apparatus has four types of platform elements that do not occur in mirror-images, and thus, this form is referable to *Eoplacognathus*. It should also be noted that early species of *Cahabagnathus* are unknown in Europe, which appears to have been a center of

morphological diversification of *Eoplacognathus*. Furthermore, there are undescribed, and still poorly known, platform conodont elements with a surface ornamentation quite similar to that of representatives of *Cahabagnathus* in the early Middle Ordovician of North America, for instance, Gen. et sp. nov. of Bergström (1978, Pl. 79:17). It is quite conceivable that the ancestor of *Cahabagnathus*, although not identified at the present time, may be found among that complex of species, which occurs lower stratigraphically than *Cahabagnathus* n. sp. A. Clearly, additional studies are needed to clarify the origin of the *Cahabagnathus* lineage.

In the successions studied so far, the youngest species of the *Cahabagnathus* lineage, *C. carnesi*, disappears without any obvious descendant. Representatives of *Polyplacognathus ramosus* Stauffer 1935 have two types of platform elements that occur in mirror-images, and have a surface ornamentation similar to that in species of *Cahabagnathus*. However, *P. ramosus* has a different, and more complex, process arrangement in the stelliplanate element, and the pastiniplanate element also differs conspicuously from that of *Cahabagnathus*. Further, in view of the fact that evolutionary trends in the *Cahabagnathus* lineage are in some respects opposite to what one would expect if it were the progenitor of *Polyplacognathus*, I consider it unlikely that the latter genus has its roots in the *Cahabagnathus* lineage. Other interpretations are discussed below (p. 45).

The EOPLACOGNATHUS *lineage (Fig. 2)*

This lineage, which includes more species than any other Ordovician platform conodont lineage, was recognized by Bergström (1971a), who described six of its species on a multielement basis, namely *Eoplacognathus suecicus* Bergström 1971, *E. foliaceus* (Fåhraeus 1966), *E. reclinatus* (Fåhraeus 1966), *E. robustus* Bergström 1971, *E. lindstroemi* (Hamar 1962), and *E. elongatus* (Bergström 1962). Viira (1972) discussed the symmetry of elements of this type, and Bergström (1973) illustrated the range, and the four types of elements in the apparatus, of each of five species in this lineage. Good schematic illustrations of some element types were given by Dzik (1976, 1978) who also discussed the evolutionary trends briefly. He included *E. pseudoplanus* (Viira 1974) and *E. zgierzensis* Dzik 1976, which occur in strata older than those dealt with by Bergström (1971a), in the lineage and agreed with Bergström (1971a) that the ancestor of the primitive *Eoplacognathus* species is likely to be *Amorphognathus variabilis* Sergeeva 1963 or a closely related form. Löfgren (1978), in a detailed discussion, considered *E. pseudoplanus* a questionable junior synonym of *E. suecicus*, and expressed the opinion that true *A. variabilis* is likely to have lacked ramiform elements in the apparatus, and also in other respects to have been more similar to *Eoplacognathus* than to *Amorphognathus*. In accordance with this, she referred the latter species with question to *Eoplacognathus* and considered it to be the oldest species in the *Eoplacognathus* lineage.

A review of collections at hand and of available data from the literature suggests that it is possible to recognize two groups of species within the *Eoplacognathus* lineage. The oldest of these, which includes *E. suecicus*, *E. pseudoplanus*, and related forms and which is here referred to as the *E. suecicus* – *E. pseudoplanus* species complex, is characterized by the fact that processes of pastiniplanate elements are of subequal length and the dextral and sinistral stelliplanate elements are in many cases closely similar to each other. The various species within this complex have been separated mainly on the basis of minor differences in the mutual length of the processes and their angles with each other.

As noted by Löfgren (1978), there is considerable intraspecific variation in both fully developed elements and between growth stages in this species complex, and there appears to be a virtually continuous series of transitional forms between its morphological end members. This makes separation into well defined taxa difficult, and the introduction of several species on the basis of very small collections with little, if any, consideration of intraspecific variation has led to taxonomic

problems, some of which have been discussed by Löfgren (1978).

Bergström (1971a) and Dzik (1976) suggested that the ancestor of early species of *Eoplacognathus* may be *Amorphognathus variabilis* or a closely similar form. As noted above, there is still some uncertainty regarding the apparatus of the latter species although now-available data appear to indicate that it has a full set of ramiform elements and is a true *Amorphognathus*. Evolution of a typical *Eoplacognathus* from *A. variabilis* would have involved loss of the ramiform elements but otherwise no drastic morphological changes, and this ancestry for *Eoplacognathus* remains a distinct possibility. Alternatively, *Eoplacognathus* might have evolved from an as-yet-unknown ancestor without ramiform elements in the apparatus.

In discussing the origin of *Eoplacognathus*, it is also appropriate to draw attention to the fact that Lindström (1955, 1964, Fig. 35D) described a platform conodont quite similar to the pastiniplanate element of *Eoplacognathus* from lowermost Ordovician (upper Tremadocian) strata in south-central Sweden (Fig. 2). Unfortunately, only a few specimens are known and most of them are incomplete. Although there are undescribed platform conodonts in early and middle Arenigian strata in Europe and North America (Bergström coll.), none of those collected thus far represents this particular type, and relations between the Tremadocian form and *Eoplacognathus* remain obscure. Lindström (1964) interpreted it as a homeomorph of *Eoplacognathus*, but in the present state of our knowledge it cannot be ruled out that it may be the evolutionary ancestor of the *Eoplacognathus* lineage.

The post-*Eoplacognathus suecicus* portion of the *Eoplacognathus* lineage includes a series of species connected by morphological intermediates. I refer to this group of species as the *E. foliaceus* – *E. elongatus* species complex. Morphological evolution within this complex is expressed mainly in changes in the mutual length, direction, and number of processes of the platform elements (for illustrations, see Bergström 1971a, 1973; Bergström & Carnes 1976; Dzik 1976, 1978). All of these species have four types of morphologically different platform elements that do not occur in mirror-images, and they are also characterized by the very long anterior process of the pastiniplanate elements (Fig. 2).

Typical specimens of *Eoplacognathus foliaceus* appear abruptly in my sections and I have not seen any actual specimens, or illustrations, of elements truly transitional between this species and *E. suecicus*. Although there is a distinct morphological gap between the species mentioned, *E. foliaceus* can be derived rather readily from *E. suecicus*, which is likely to be its ancestor. An interesting separate lineage in the *E. foliaceus* – *E. elongatus* species complex is represented by the still incompletely known *Eoplacognathus* n. sp. A of Bergström (1971a) which occurs in strata close to the *Pygodus serra* – *P. anserinus* zonal boundary in Baltoscandia (Bergström 1971a), Poland (Dzik 1978, Fig. 1: 13 – listed as *E. l. robustus*) and The People's Republic of China (Bergström coll.; Ni 1981). I interpret this form as a descendant of *E. foliaceus*, or a closely related species, and it seems to have considerable stratigraphic significance.

The youngest known species of the *Eoplacognathus foliaceus* – *E. elongatus* complex, *E. elongatus*, disappeared in the early Caradocian without leaving any obvious descendants. The idea that *Polyplacognathus* Stauffer evolved from *Eoplacognathus*,

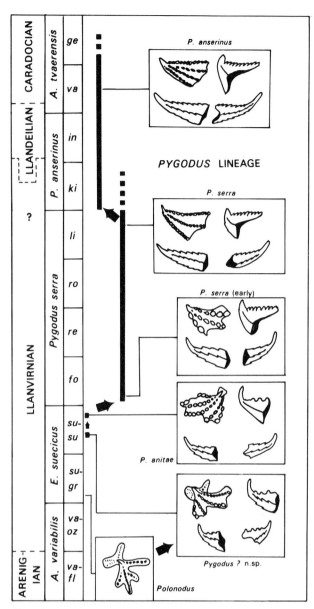

Fig. 3. Suggested evolutionary relationships and known stratigraphic ranges of taxa in the *Pygodus* lineage. For explanation of subzonal abbreviations, see Fig. 1. Only one of the paired symmetry variants of each element is illustrated.

that in each case, it differs markedly from that of *Eoplacognathus*. Accordingly, it seems very likely that the similarities mentioned are cases of homeomorphy rather than characters implying phylogenetic relationship, and that the *Eoplacognathus* lineage became extinct in Middle Ordovician time.

The Pygodus lineage (Fig. 3)

One of the morphologically most distinct, and biostratigraphically most useful, of the Ordovician platform genera is *Pygodus* Lamont & Lindström 1957. Representatives of *P. serra* (Hadding 1913) and *P. anserinus* Lamont & Lindström 1957 are present not only in an unusually wide range of facies but they are also among the most widespread geographically of all compound Ordovician conodonts.

Pygodus was discussed in some detail, and defined in terms of multielement taxonomy, by Bergström (1971a). Shortly afterward, at the *Marburg Symposium on Conodont Taxonomy* in 1971, I described the morphological evolution of the *Pygodus* lineage and discussed its origin but my conclusions and illustrations have not been published. On the basis of Swedish collections very similar to mine, Löfgren (1978) arrived at conclusions that agree closely with those I presented orally in Marburg. Some additional data supporting this interpretation have become available subsequently but, as noted below, some points are still unclear even if the general evolutionary scheme appears rather safely established.

Information now at hand strongly suggests that the origin of the stratigraphically oldest well-known species of *Pygodus*, *P. serra*, can be traced back via two primitive species, here referred to as *P. anitae* n. sp. and *P.?* n. sp. (Fig. 3), to forms that have been referred to *Polonodus* Dzik 1976 in the recent literature. The origin of *Polonodus* is still uncertain but it should be noted that fragments of primitive platform conodont elements similar in surface sculpture and in development of the basal cavity are not uncommon in Baltic strata as old as the *Megistaspis planilimbata* Limestone (early Arenigian) (Bergström coll.). Unfortunately, affinities of these fragments are obscure because no specimen complete enough to reveal the architecture of the platform elements of this species has been found thus far. At any rate, in cannot be ruled out that the *Polonodus* lineage might extend far down in the Lower Ordovician.

As presently understood, *Polonodus* includes a group of early Llanvirnian platform conodonts of complex architecture among which only one species, *P. clivosus* (Viira 1974), has been named formally. Although specimens of *Polonodus* have been collected in Estonia, Sweden, Poland, Newfoundland, New York State, and Nevada, the total number of elements recovered to date is small, and most of the specimens are fragmentary. This has made it difficult to evaluate the intraspecific variation in species of the genus, and the taxonomy of *Polonodus* is still poorly understood. Nevertheless, the evolution of *Polonodus* into *Pygodus* can readily be traced in the pygodontiform elements, although there are also rather striking morphological changes in the other elements of the apparatus. As noted by Löfgren (1978), perhaps the most characteristic feature in this evolution is the gradual reduction and, ultimately, the virtual disappearance of the posterior processes in the pygodontiform element. This morphological change is associated with a simplification in the branching of

put forward without supporting evidence by Barnes & Fåhraeus (1975), is not unreasonable on morphological grounds but as noted below (p. 45), there are also other possible ancestors of *Polyplacognathus*. It is interesting that elements of some post-Ordovician platform conodonts have the basic morphology of elements of *Eoplacognathus*. This applies to the Early Silurian *Astropentagnathus* Mostler 1967 and the Late Silurian *Ancoradella* Walliser 1964 which are similar to stelliplanate elements of *Eoplacognathus* in terms of process organization and other morphological features. Although different in surface ornamentation, the Devonian *Ancyrognathus* Branson & Mehl 1934 and the Mississippian *Doliognathus* Branson & Mehl 1941 are reminiscent in basic morphological plan to the pastiniplanate elements of *Eoplacognathus*. The appearance of the apparatus is still incompletely known in some of these genera but information currently available (Schönlaub 1971; Klapper & Philip 1971) suggests that the anterior processes, which ultimately merge into a subtriangular platform.

Early forms of *Polonodus*, such as those illustrated as *P. clivosus* (Viira) by Löfgren (1978, Pl. 16:12A, 12B), have two posterior processes that are platform-like and distinctly shorter than the lateral and anterior ones (Fig. 3). Stratigraphically slightly younger forms have relatively shorter posterior processes and two anterior processes with double denticle rows (Fig. 3). Such elements have been referred to as *Pygodus?* sp. B by Löfgren (1978, Pl. 16:2, 3) and they are herein referred to as *Pygodus?* n. sp. Still higher stratigraphically, there are specimens with the posterior processes reduced to just a short simple lobe posterior to the cusp. The anterior part of these elements consists of two confluent lobes that bear, in most specimens, double rows of denticles (Fig. 3). I regard these specimens as representatives of the earliest known typical species of *Pygodus*, herein named *Pygodus anitae* n. sp. (Fig. 6V–Z). Specimens of this type were illustrated as *Pygodus* sp. C by Löfgren (1978, Pl. 16:5, 6). Representatives of the next younger well-defined form, *P. serra*, have only three denticle rows on the anterior platform, and this species is connected with *P. anitae* via morphological intermediates [*P. serra* (early) in Fig. 3]. As noted by Bergström (1971a), *P. serra* evolved into *P. anserinus*. The latter two species, as well as *P. anitae*, and, probably, the *Polonodus* species have at least three types of elements in the apparatus in addition to the pygodontiform elements (Fig. 3). These elements also underwent a rather conspicuous evolution in the early Middle Ordovician but it was not as rapid, and is not as easily defined morphologically, as that of the pygodontiform elements. Evolutionary changes in elements of these types are illustrated schematically in Fig. 3.

Pygodus anserinus, the youngest known species of the *Pygodus* lineage, apparently became extinct in the early Caradocian in view of the fact that the stratigraphically youngest specimens found (Bergström 1971a) are from the *Prioniodus* (*B.*) *gerdae* Subzone of the *Amorphognathus tvaerensis* Zone. There are no other conodonts even remotely similar to *Pygodus* in younger Ordovician strata, and it appears that this lineage left no descendants. The two species from the Llandoverian of the Carnic Alps described by Walliser (1964) as *P.? lenticularis* and *P. lyra* in all probability are not related to *Pygodus* at even the family level. The latter species was based on elements now interpreted (Walliser 1972) as part of the apparatus of *Apsidognathus tuberculatus* Walliser 1964, the type species of *Apsidognathus*, the apparatus of which is very different from that of *Pygodus* (Klapper 1981). *Pygodus? lenticularis* is based on a single specimen that is morphologically rather different from *Pygodus* and it can clearly be rejected as a representative of the latter genus. Accordingly, it appears fully justified to consider these Silurian elements as, at the most, only homeomorphs of the pygodontiform element of *Pygodus*.

Other platform conodonts

Several Ordovician platform conodont genera are monospecific and show little, if any, phylogenetic evolution through the stratigraphical intervals where they are present. Such genera include *Prattognathus*, *Rhodesognathus*, and *Scyphiodus*. To this group should probably also be added *Nericodus* and *Sagittodontina* but, as indicated below, these genera are currently so poorly

known in terms of their morphology, and they have such a restricted distribution, that there is little basis for an evaluation of possible morphological trends. It is also possible that *Polyplacognathus* and *Serratognathus* should be included in this group although several species have been referred to each of these genera; however, the information at hand is insufficient to evaluate the evolution within these genera, and in the case of *Polyplacognathus*, the generic reference of some species is questionable (cf. p. 41).

Species of some of the genera mentioned have a relatively wide geographic and stratigraphic range; such species are *Rhodesognathus elegans* (Rhodes 1953) and *Polyplacognathus ramosus* Stauffer 1935. The other genera are represented by species with far more restricted ranges. Clearly, this decreases the possibilities of postulating evolutionary relationships at the generic level.

Nericodus is one of the oldest platform conodonts known in the geologic record but only a few specimens of its type species, *N. capillamentum* Lindström 1955, have been found thus far and the genus is very poorly understood. The specimens available are all fragmentary and even basic features of their morphology and morphological variation are uncertain. Miller (1980, 1981) has recently restudied the type and other specimens, and his suggestion that the early Arenigian (*Paroistodus proteus* Zone) *Nericodus* is related at the family level to the Tremadocian *Clavohamulus* Furnish 1938 and *Hirsutodontus* Miller 1969 is not unreasonable on morphological grounds. Likewise, it is possible that *Nericodus* may be related to the Tremadocian *Ambalodus* n. sp. Lindström 1955. Indeed, it cannot now be ruled out that one, or both, of the latter taxa may represent a lineage leading to *Polonodus*, but supporting evidence through much of the Arenigian is needed to prove the correctness of this idea.

By contrast, rich collections have been assembled of *Polyplacognathus ramosus* (Fig. 1), the type of *Polyplacognathus*, and this species is widespread in faunas of North American Midcontinent type through much of the upper Middle Ordovician [Faunas 6 through 10 of Sweet *et al.* (1971); cf. Sweet & Bergström 1976]. Compared to many other Ordovician platform conodonts, this species exhibits a great morphological stability throughout its range, and evolving morphological features suitable for further taxonomic discrimination have not been recognized. Characteristically, the apparatus includes one type of paired stelliplanate elements and dextral and sinistral pastiniplanate elements that are closely similar to each other. Apparently, the apparatus lacked ramiform elements.

In terms of element types and surface ornamentation, *Polyplacognathus* is similar to *Cahabagnathus*. However, data at hand do not support the idea that the former evolved from the latter. It is also possible to derive *Polyplacognathus ramosus* morphologically from advanced representatives of *Eoplacognathus* such as *E. elongatus* by modification of the processes and development of a complex and irregular surface ornamentation. Yet, no transitional forms are known and I consider a direct ancestor–descendant relationship between these genera as unlikely even if it cannot be completely ruled out at the present time (Fig. 5). Another, and perhaps more likely, ancestor of *Polyplacognathus ramosus* might be present in the complex of early Middle Ordovician platform conodonts from

Siberia described as *P. petaloides, P.* sp., and *P. lingualis* by Moskalenko (1970, Pl. 14:2, 3a, 3b, 4, 5). Although her illustrated specimens are all fragmentary and not easily interpreted in terms of, for instance, number and mutual length of the processes, it is clear that the type of basal cavity and the general appearance and arrangement of the processes are reminiscent of those of *P. ramosus*. Furthermore, Moskalenko's specimens are associated with typical North American Midcontinent Province conodonts and occur in strata that have been correlated with the Llandeilian (Chugaeva 1976:286–287), that is, an interval older than the earliest known occurrences of *P. ramosus*, which are likely to be of early Caradocian age. The stratigraphically youngest occurrences of *P. ramosus* reported thus far are in the lower part of the *Amorphognathus superbus* Zone (Sweet & Bergström 1971, 1976) and the species apparently became extinct in the late Middle Ordovician without leaving any descendants.

Representatives of *Prattognathus* are currently known only from Alabama, Tennessee, and Nevada, and all occurrences are in a rather narrow stratigraphic interval, namely the upper part of the *Pygodus serra* Zone and the lower part of the *P. anserinus* Zone. As interpreted herein, the apparatus of *Prattognathus* includes two types of paired platform elements (Figs. 2, 6S–U), and no additional elements are recognized as probable, or possible, parts of its apparatus. Although different in details, one of the two types of platform elements, up to now known as *Polyplacognathus stelliformis* Sweet & Bergström 1962 is so similar in overall appearance to the stelliplanate element in the apparatuses of *Eoplacognathus, Polyplacognathus* and, in particular, *Cahabagnathus* that it is quite natural to regard it as corresponding to that type of element in the apparatus. The other type of platform element in the apparatus of *Prattognathus*, originally described as *Polyplacognathus rutriformis* Sweet & Bergström 1962 is also a stelliplanate element but it is here interpreted as corresponding to the pastiniplanate element in species of *Cahabagnathus*. Although specimens of the multielement species *P. rutriformis* have been found in association with representatives of *Cahabagnathus friendsvillensis*, the type of stelliplanate element just mentioned is more similar to that of advanced species of *Cahabagnathus*, such as the specimen of *C. chazyensis* illustrated by Bergström (1978, Pl. 79:15), than to that of *C. friendsvillensis*.

Available specimens of *Prattognathus rutriformis* are all quite distinct morphologically, and no forms have been found that are transitional with a species of *Cahabagnathus* or any other early Middle Ordovician platform genus. Accordingly, the origin of *Prattognathus* is enigmatic. However, based on morphology, and its stratigraphic and geographic occurrence, one may perhaps suggest that the ancestor of *Prattognathus* may be present among some poorly known platform taxa in the Whiterockian of North America, for instance, '*Polyplacognathus*' n. sp. A of Fåhraeus (1970) (cf. Bergström 1980, Fig. 2M) and/or *Polonodus?* *newfoundlandensis* of Stouge (1980, Pl. 13:14–16). Both these species occur in an interval that is considerably older than that in which *P. rutriformis* occurs.

The peculiar appearance of the second type of stelliplanate element in *Prattognathus rutriformis* is not at all what one would expect in the ancestor of *Polyplacognathus ramosus*, and the relationships between *Prattognathus* and *Polyplacognathus* are unclear. However, regardless of the precise nature of their relationship, it seems both natural and justified to group these two genera, along with *Cahabagnathus* and *Eoplacognathus*, in Polyplacognathidae Bergström 1981.

The type of *Rhodesognathus, R. elegans,* has been recorded from upper Middle Ordovician strata at many localities in the North American Midcontinent and from a few in the Appalachians and northwestern Europe. As noted by Roscoe (1973) and Bergström (1981b), its apparatus contains a set of ramiform elements that are currently indistinguishable from corresponding elements of the apparatus of *Amorphognathus* species, which commonly occur together with representatives of *R. elegans*. The form described by Dzik (1976) as *R. elegans polonicus* apparently lacks ramiform elements and it differs also in other respects from typical *R. elegans* to the extent that it seems justified to regard it as a separate species; indeed, I hesitate to refer this form without question to *Rhodesognathus*. Likewise, the form described from the Middle Ordovician of Brittany by Lindström, Racheboef & Henry (1974) as '*Prioniodus*' (*Rhodesognathus?*) n. sp. aff. *Prioniodus variabilis* Bergström 1962 and *Prioniodus gerdae* Bergström 1971 seems more similar to *Prioniodus* (*Baltoniodus*) than to *Rhodesognathus*, particularly in that the edge of the lateral process continues into the anterior edge of the cusp rather than into the denticle anterior to the cusp as is characteristic of *Rhodesognathus* (Bergström & Sweet 1966:392).

As currently known, *Rhodesognathus* ranges from the lower Caradocian to a level well up in the Ashgillian (Sweet & Bergström 1976). The striking similarity between the ramiform elements of *Rhodesognathus* and *Amorphognathus* suggests close relationship, but no transitional specimens have been described. However, based on the admittedly incomplete data at hand it seems reasonable to suggest that *Rhodesognathus* evolved from *Amorphognathus* in the early Middle Ordovician (Fig. 1). The genus apparently became extinct in Late Ordovician time without leaving any descendants.

Up to now, the concept of *Sagittodontina* has been based solely on the fragmentary original specimens from the Upper Ordovician of Thuringia described by Knüpfer (1967). As shown by Bergström & Massa (in preparation), the type species of *Sagittodontina, S. bifurcata* Knüpfer 1967, has an apparatus of paired pastiniscaphate and pastinate elements and an array of ramiform elements of the same general types as in species of *Amorphognathus* (Fig. 4). In the appearance of both the platform and the ramiform elements, *Sagittodontina* is clearly different from *Amorphognathus* and other Ordovician platform genera, and its relationships are not clear. The architecture of its apparatus suggests that *Sagittodontina* might have its origin in the *Amorphognathus* lineage (Figs. 1, 5) but there is no direct evidence to support this idea. An alternative interpretation is that *Sagittodontina* represents a separate evolutionary lineage from a prioniodontacean ancestor parallel to that of *Amorphognathus*. Additional material is clearly needed to show, which, if either, of these interpretations is correct.

Representatives of *Scyphiodus* (Figs. 1, 5), the only Middle Ordovician platform conodont genus that appears to be indigenous to the North American Midcontinent, are relatively common in several formations there (see, for instance, Webers 1966 and Sweet 1982) but attempts to reconstruct the apparatus of *S. primus* Stauffer 1935, the type and only known

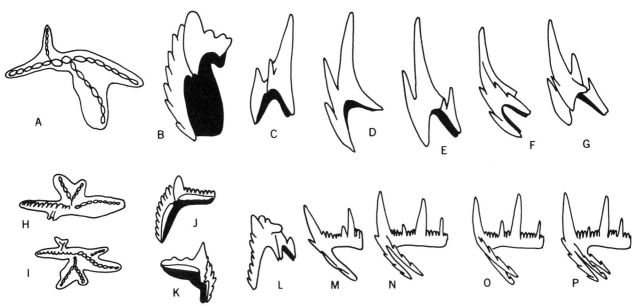

Fig. 4. Comparison between elements in the apparatus of *Sagittodontina bifurcata* Knüpfer 1967 (A-G) and those of *Amorphognathus tvaerensis* Bergström 1962 (H-P).

species of the genus, have thus far been unsuccessful. The fact that no ramiform elements have been identified as regular associates of the platform elements led Webers (1966), Sweet & Bergström (1972), Sweet (1982), and Bergström (1981b) to suggest that the apparatus might have been unimembrate and might have included only paired platform elements. These authors also noted that apart from the platform-development of the anterior process, *Scyphiodus* elements are remarkably similar to blade-like elements in the apparatus of *Bryantodina* and *Plectodina*, which are present in the same stratigraphic interval.

If *Scyphiodus* is closely related to one, or both, of these genera, it is, of course, to be expected that its ramiform elements, if present, would be similar to those of the genera mentioned, and there is a possibility that their true relationship has not been recognized. This possibility should be kept in mind in future studies of *Scyphiodus*-bearing samples as an alternative to the interpretation of *Scyphiodus* as having a unimembrate apparatus. Regardless of whether or not *Scyphiodus* had a more complex apparatus than is apparent now, it appears likely that this genus evolved in the Middle Ordovician from a stock such as *Plectodina* rather than from another platform conodont lineage such as *Amorphognathus*, *Icriodella*, or *Polyplacognathus* to which *Scyphiodus* is probably not closely related.

Patterns of evolution

It is beyond the scope of the present paper to discuss in detail the speciation processes in Ordovician platform conodonts. However, some of the taxa studied appear to show an interesting pattern of evolution that warrants some brief comments.

As shown by the summary diagram of my interpretation of the relations between Ordovician platform conodont taxa (Fig. 5), at least five lineages can be recognized, each apparently quite independent of the others, namely those of

Amorphognathus, Cahabagnathus, Eoplacognathus, Icriodella, and *Pygodus.* In terms of evolutionary processes, these long-ranging lineages are of particular interest. Each includes series of successive species showing little overlap in their individual vertical ranges (Fig. 1–3). In most cases, these species are distinguished from each other on characters in the platform elements showing rapid evolution. Also the nonplatform elements in the apparatuses of *Amorphognathus* and *Pygodus* exhibit evolutionary changes, and such changes in the holo-dontiform (M) elements of *Amorphognathus* have been used for separating taxa at the species level. The rate of evolution varied greatly between different types of elements; some morphologically conservative elements, such as the ramiform S elements, may be so similar in several species that they are virtually indistinguishable whereas the platform and M elements show rapid morphological change.

These conodont species lineages have a low species diversity at virtually every level in the Middle and Upper Ordovician; indeed, as shown in Fig. 5, each lineage is, as a rule, represented by only a single species at a particular time plane. It could perhaps be argued that this low diversity is a taxonomic artifact caused by the use of a very broad species concept. However, the concept employed does not differ appreciably from that used in the classification of other conodonts, and study of collections from areas widely separated geographically suggests that representatives of a particular genus show surprisingly little regional variation. For instance, platform elements of the genera listed above from East Asia and North America are, as a rule, indistinguishable from those of Baltoscandia.

In these lineages, the common type of speciation, as expressed by changes in morphology of the skeletal elements, is characterized by periods of rather slow to almost imperceptible morphological change, which are interrupted by relatively short episodes of accelerated evolution that is, in some cases, almost explosive. In a series of samples through a section representing such an episode, it is in many instances possible

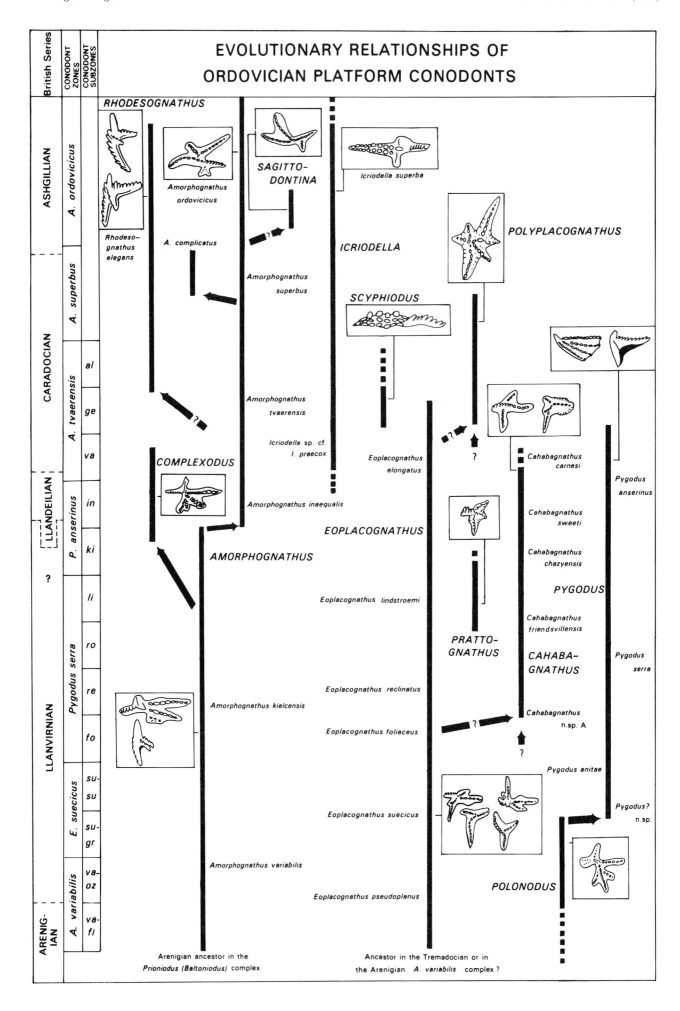

EVOLUTIONARY RELATIONSHIPS OF ORDOVICIAN PLATFORM CONODONTS

(for instance, the transition from *Pygodus serra* to *P. anserinus*) to trace the gradual, but rapid, development of one species into another through transitional morphotypes. Such a speciation event appears to be of the same type as those described recently in Permian and Triassic conodonts (Carr 1982; Wardlaw 1982). Because it seems to have occurred in the same time interval over a large region, and its geological record is preserved in a very narrow stratigraphic interval, such an event represents an excellent datum for regional biostratigraphic correlation. In fact, such levels have been used as zonal boundaries in the North Atlantic Province Middle and Upper Ordovician conodont zone succession since its introduction (Bergström 1971a).

Obviously, we have no way of knowing if the conodont 'species' we recognize in these lineages on the basis of combinations of certain morphological characters in the skeletal elements correspond to 'biologic' species but this is a problem conodonts share with other fossil groups. Assuming that our conodont taxonomy at the species level is indeed comparable to that in extant organisms, it appears that the much-publicized model involving a tree- or bushlike pattern of evolution does not describe the evolutionary scheme I envision for Ordovician platform conodonts very well. This scheme seems to combine features of both of the two contrasting concepts of speciation, phyletic gradualism and punctuated equilibria, as they were described by Eldredge & Gould (1972). However, further studies are needed to establish the detailed nature of this scheme as well as its occurrence among other types of conodonts. Because they show rapid evolution and can be obtained in large numbers from closely spaced samples, conodonts are well suited for studies of speciation processes, and further investigations of these matters are likely to yield results applicable not only to conodonts but also to other types of organisms.

Biostratigraphical remarks

The zonal scheme introduced by Bergström (1971a) for the Middle and Upper Ordovician of the North Atlantic Province was based largely on taxa of *Amorphognathus*, *Eoplacognathus* and *Pygodus*, and has proved to be applicable widely in Europe and North America as well as to successions in Asia and Australia. Since 1971, many new data have become available that bear on the biostratigraphic use of Ordovician platform conodonts, and some of this information is briefly discussed below.

Tremadocian and Arenigian

Although An (1981) recognized a *Serratognathus diversus* Zone and a *S. bilobatus* Zone in the upper Tremadocian–lower Arenigian of the People's Republic of China, platform conodonts have thus far been used very little in the Lower Ordovician conodont biostratigraphy. Some of the reasons for this are that they are uncommon, not very diversified morphologically, and still poorly known. The appearance in the upper

Fig. 5. Summary diagram showing suggested interpretation of evolutionary relationships of Middle and Upper Ordovician platform conodonts. Only key elements of some apparatuses are illustrated. For explanation of subzonal abbreviations, see Fig. 1.

Arenigian of the first major platform conodont lineage, that of *Amorphognathus*, is of major biostratigraphic significance. Taxa in the oldest part of this lineage (*A. falodiformis*, etc.) clearly have potential as zonal indices but they are still too poorly known both in terms of morphology and stratigraphic range to be useful biostratigraphically. Until recently, the oldest unit based on *Amorphognathus* was the *A. variabilis* Zone. However, on the basis of a new species said to be present in an interval just below that of *A. variabilis*, An (1981) introduced an *A. antivariabilis* Zone. An's zonal index is incompletely known and the characteristics he used to separate it from *A. variabilis* need further evaluation. However, if this species proves to be distinct, it may be the direct ancestor of *A. variabilis*. Although *A. antivariabilis* was said to be present below the interval of *A. variabilis*, An (1981, Fig. 3) correlated the base of his *A. antivariabilis* Zone with a level well above the base of the North Atlantic *A. variabilis* Zone. No reason was given for this somewhat surprising correlation.

Llanvirnian

Several different schemes of Llanvirnian conodont zones have been proposed (Sergeeva 1966 a, b; Lindström 1971; Bergström 1971a; Viira 1974; Dzik 1976, 1978; Löfgren 1978) but because of both taxonomic and biostratigraphic problems, the zonal classification has not yet been stabilized, especially in the lower Llanvirnian where the problems center round the *A. variabilis* Zone. Lindström (1971), Dzik (1978), and Löfgren (1978) have each given a different scope to this unit, and even identification of the zonal index remains problematic (for discussion, see Löfgren 1978:58). In Dzik's (1978) scheme, the *A. variabilis* Zone is overlain by the *Eoplacognathus pseudoplanus* Zone and the zonal boundary is placed at the level of transition of *E. zgierzensis* Dzik into *E. pseudoplanus*. Regrettably, there is some uncertainty regarding the morphological characteristics of *E. zgierzensis* because Dzik's one-sentence original description is too vague to permit definite assessment of the differences in morphology between it and *E. pseudoplanus*, and the illustrated holotype, which is from an erratic boulder of indeterminable stratigraphic origin, is just a fragment that is unworthy to serve as a morphological reference for the species (cf. Löfgren 1978). Potential confusion is also introduced by the fact that Dzik's (1976, Fig. 7) drawing of reconstructed complete specimens of *E. pseudoplanus* is mislabeled and should instead be *E. zgierzensis* (Dzik pers. comm. 1982). Löfgren (1978) discussed in some detail several of the problems related to the conodont biostratigraphy of this interval. She recognized an *A. variabilis* (*Eoplacognathus? variabilis in* Löfgren 1978) Zone and an *E. suecicus* Zone, each of which was subdivided into two subzones. My Swedish collections suggest that the top of Dzik's (1978) *E. pseudoplanus* Zone corresponds to a level in the middle part of the *E. suecicus–P. sulcatus* Subzone of Löfgren (1978). Because the top of the *E. suecicus* Zone was defined in the same way by Löfgren (1978) and Dzik (1978), the scope of Dzik's (1978) *E. suecicus* Zone is markedly smaller than that employed by Löfgren (1978). As noted by Dzik (1978:52–53), the precise scope of his *A. variabilis* and *E. pseudoplanus* zones is uncertain in terms of Baltoscandic standard units, and my attempts to pin down the boundary between these zones in the Swedish sections studied have been unsuccessful. Accordingly, for the purpose of the present paper, I prefer to follow Löfgren's (1978) zonal scheme, which is firmly based on the faunal succession in specific Swedish successions.

The upper Llanvirnian *E. foliaceus* and *E. reclinatus* subzones are well established and noncontroversial. As noted by Bergström (1971 b), a primitive form of *Pygodus serra* (Fig. 3) appears very near the base of the *E. foliaceus* Subzone in Baltoscandia, and it seems justified to follow Löfgren (1978) in defining the base of the *P. serra* Zone as coinciding with the base of the *E. foliaceus* Subzone. Interestingly, the stratigraphically oldest specimens of *Amorphognathus kielcensis* Dzik 1976 known to me are from strata very close to the base of the *P. serra* Zone, so the appearance of that characteristic species may also be a useful guide to the zonal boundary level in the absence of specimens of *Pygodus* or *Eoplacognathus*.

In a detailed study of the very condensed Middle Ordovician Mójcza sequence in the Holy Cross Mountains, Poland, Dzik (1978) dropped the *Pygodus serra* Zone and elevated its subzones to zones. He has subsequently used the same terminology when dealing with collections from other areas (Dzik 1980, 1981). Because the subzonal indices are far less widespread geographically than *P. serra*, abandonment of the *P. serra* Zone seems unjustified and may be considered a step backward in the process of developing a regionally useful Middle Ordovician conodont biostratigraphy.

Llandeilian

The base of the succeeding *Pygodus anserinus* Zone was originally defined (Bergström 1971a:97) as the level of appearance of *P. anserinus*. The latter species clearly developed from *P. serra*, which is present locally in the lowermost part of the *P. anserinus* Zone as pointed out long ago (Bergström 1971a:150). This fact was apparently overlooked or misunderstood by Fåhraeus & Hunter (1981) in their recent study of a Newfoundland succession that ranges across the zonal boundary. If those authors had followed the original definition of the *P. anserinus* Zone, they would have had no difficulty to pick its base and that level would have been very close to the one established by Bergström *et al.* (1974) in the same sections, but based on much smaller collections.

Bergström (1971 a) subdivided the *P. anserinus* Zone into a Lower and Upper Subzone with the subzonal boundary marked by the evolutionary transition of *Prioniodus (B.) prevariabilis* into *P. (B.) variabilis*. Dzik (1978) suggested that the evolutionary transition of *Amorphognathus kielcensis* into *A. inaequalis* also takes place near that level. I know of no confirmed record of the latter species in the Baltoscandic region, but support for Dzik's suggestion is given by the succession in the Smedsby gård drilling core, south-central Sweden (Bergström unpubl.) in which specimens of *A. kielcensis* occur up to a level less than 1 m below the base of the Upper Subzone. Also the distribution of other conodont taxa (expecially *Complexodus*) in that core and in other sections in Sweden suggests that the subzonal boundary and the level of the *A. kielcensis – A. inaequalis* transition are likely to be so close to each other stratigraphically that for all practical purposes, these levels can be considered to represent the same horizon. These subzones are in need of formal designations and I here propose that the Upper Subzone be renamed the *A. inaequalis* Subzone. It includes the interval of co-occurrence of *P.*

anserinus and *A. inaequalis* up to the level of appearance of *A. tvaerensis*, a direct descendant of *A. inaequalis*. Likewise, the Lower Subzone is renamed the *A. kielcensis* Subzone and defined as the interval of co-occurrence of *P. anserinus* and *A. kielcensis* up to the level of appearance of *A. inaequalis*, which is apparently a direct descendant of *A. kielcensis*. Because the latter species ranges through the *P. serra* Zone, presence of both *P. anserinus* and the subzonal index is needed for certain identification of the *A. kielcensis* Subzone. In the absence of these species, the subzonal boundary can be recognized, as previously, as the level of transition of *Prioniodus (B.) prevariabilis* into *P. (B.) variabilis*. Because Dzik (1978) elected to use the level of appearance of *A. inaequalis* as the top of his *P. anserinus* Zone, his scope of the latter zone is the same as the *A. kielcensis* Subzone as defined herein.

In this connection it is of interest to note that specimens of *A. inaequalis* are now known (Bergström coll.) from the conglomerate unit in the very uppermost part of the Ffairfach Group below the base of the key section of the Llandeilian near Afon Cennen at Llandilo. This proves that the base of the type Llandeilian is coeval with a level somewhere within the *A. inaequalis* Subzone and not with the base of the *P. anserinus* Zone as was tentatively suggested by Bergström (1971a). This new evidence is in excellent agreement with that from the megafossils (Addison *in* Williams *et al.* 1972) and confirms that the much cited correlation of the Llandeilian with the *Glyptograptus teretiusculus* Zone is erroneous. As noted by Bergström *et al.* (1974:1653) and Jaanusson (1979:A144–A145), if one follows the traditional practise of defining the top of the Llanvirnian as the top of the *Didymograptus murchisoni* Zone, and if one accepts the conodont evidence of the age of the strata immediately below the base of the Llandeilian in its type area, then one will have to conclude that there is a post-Llanvirnian, pre-Llandeilian interval which has no series designation in the British succession and which corresponds to most of, if not the entire, *Glyptograptus teretiusculus* Zone. Obviously, the new conodont data presented above increase the magnitude of this interval in terms of biostratigraphic units, and it would be most welcome if the proper authorities soon addressed the problems of the Llanvirnian–Llandeilian boundary and proposed a workable solution.

Although additional work is needed to establish their precise ranges in terms of the North Atlantic conodont zonal succession, species of *Cahabagnathus* are clearly useful index fossils in the upper Llanvirnian and the Llandeilian. As shown in Fig. 2, the known range of *C. friendsvillensis* approximates that of the *P. serra* Zone; *C. chazyensis* and *C. sweeti* occur in the *P. anserinus* Zone; and *C. carnesi* in the latter zone and in the lowermost *A. tvaerensis* Zone.

Caradocian

The next younger conodont zone, the *Amorphognathus tvaerensis* Zone, has been recognized widely in Europe and North America, as have the lower two of its three subzones, the *Prioniodus (B.) variabilis* and *P. (B.) gerdae* subzones. The uppermost subzone, the *P. (B.) alobatus* Subzone, although widely recognized in Baltoscandia, has been recorded from only a single locality in North America (Kennedy *et al.* 1979). Bergström (1971a), Dzik (1978), and Kennedy *et al.* (1979) have all commented on the similarity between *P. (B.) variabilis* and *P. (B.) alobatus*, and it may well be that these species have an ancestor–descendant relationship. This is supported by the fact that a few specimens approaching the appearance of *P. (B.) alobatus* have been found in the *P. (B.) variabilis* Subzone in Sweden (Bergström unpubl.). Nevertheless, the three sub-zones are readily distinguishable in the Baltoscandic sections studied.

The precise position of the Llandeilian–Caradocian boundary in the North Atlantic conodont zone succession remains somewhat uncertain. The occurrence of specimens resembling *A. tvaerensis* in the upper Llandeilian of the type area (Bergström 1971b:184), and of *A. tvaerensis* and *Eoplacognathus elongatus* in basalmost Caradocian (Costonian) strata in the Narberth area to the west of Llandilo (cf. Addison *in* Williams *et al.* 1972:36), together with the fact that the type Llandeilian is directly overlain by graptolitic shales of the *Nemagraptus gracilis* Zone (Williams 1953), suggest that the top of the type Llandeilian in all probability is coeval with a level in the *Prioniodus (B.) variabilis* Subzone (Fig. 5).

Also the next younger zone, the *Amorphognathus superbus* Zone, has been recognized at many localities in Europe and North America. The zonal index is a direct descendant of *A. tvaerensis*, and the principal difference between these species has been taken to be the appearance of the dextral pastini-scaphate element (Bergström 1962, 1971a:136). Dzik (1978:55) proposed a different circumscription of *A. tvaerensis* and *A. superbus* based on the holodontiform element and in accordance with this, a downward adjustment of the base of the *A. superbus* Zone. However, from my experience with large collections of *A. tvaerensis* and *A. superbus*, it is my definite impression that the gradual reduction and ultimate disappearance of the 'extra' posterolateral process in the dextral pastiniscaphate element is a more useful feature for separating these taxa within the interval of morphological intergradation than minor changes in the morphologically rather variable holodontiform element, which, furthermore, is only sparsely represented in most collections. Accordingly, I prefer to retain the traditional morphological scope of these species.

No formal subzones have been proposed within the *A. superbus* Zone. At least in the Baltoscandic region, the upper part of the zone is characterized by the appearance of *A. complicatus* and *Hamarodus europaeus*. It is still not known if this distribution pattern also pertains to other regions, so it is probably premature to introduce a subzonal subdivision for this zone at the present time.

Ashgillian

The base of the next younger zone, the *A. ordovicicus* Zone, is marked by the evolutionary transition of *A. superbus* into *A. ordovicicus* (Bergström 1971a), and this level has been recognized both in Europe and North America in strata of early Ashgillian age. Based on his investigation of several Ashgillian faunas from Great Britain, Orchard (1980) suggested that the zonal boundary is in the lower Cautleyan Stage of the Ashgillian. I have had the opportunity to examine his collections at Cambridge University and in my opinion, the conodont evidence for a Cautleyan age of this zonal boundary is inconclusive. It appears that this boundary could as well be in the (upper) Pusgillian. Clearly, additional work is needed in the British sections to solve this problem. No formally defined

subzones have as yet been proposed within the *A. ordovicicus* Zone but as noted by Bergström (1971a) and Orchard (1980), there are indications that subdivision of the zone may ultimately be possible although the appreciable differences in the local ranges of several potentially useful taxa have thus far prevented the establishment of regionally useful subzones.

Systematic paleontology

Lack of space prevents a full description of the many platform conodont taxa dealt with above. However, in order to validate new generic and specific names, I furnish descriptions of a few selected taxa. The morphologic terminology follows, where appropriate, that employed in the recent conodont volume of the *Treatise on Invertebrate Paleontology* (Clark *et al.* 1981).

CAHABAGNATHUS nom. nov.
Synonym. – *Petalognathus* Drygant 1974 (preoccupied, cf. Bergström 1981b:W129).

Derivation of name. – Referring to Cahaba Valley, Alabama, where representatives of the genus were first found in North America.

Type species. – *Polyplacognathus sweeti* Bergström 1971.

Diagnosis of the genus. – A genus of conodonts with an apparatus having one type of stelliplanate and one type of pastiniplanate elements, both occurring in mirror-images; no ramiform elements are known to belong to this apparatus. Stelliplanate elements with unbranched platformed posterior and postero-lateral processes, an unbranched or bifid platformed antero-lateral process, and a platformed or bladelike unbranched anterior process. Pastiniplanate elements with a broad, unbranched platformed posterior process, a short platformed lateral process, and a considerably longer, almost bladelike to platformed, anterior process that makes an angle of about 90° with the posterior process. All elements with a central denticle row on all processes, and lateral to this row a conspicuous but irregular ornamentation of transverse ridges and nodes; there is no distinct cusp in either type of element. Basal cavity shallow, restricted, with central pit, and wide recessive margin.

Remarks. – In order to conform with the orientation used in *Eoplacognathus*, and at the suggestion of Dr. J. Dzik, orientation of the pastiniplanate element is changed 90° from that adopted previously (Bergström 1971a, Fig. 14, 1973, Fig. 1, 1978, Pl. 79:13, 15; Bergström & Carnes 1976, Fig. 2) but the process designations are the same. In this new orientation, the anterior process is directed downward and the posterior one horizontally.

Only two other Ordovician conodont genera exhibit an even superficial similarity to *Cahabagnathus*, namely *Eoplacognathus* and *Polyplacognathus*. *Cahabagnathus* is distinguished from *Eoplacognathus* by having two types of elements, both of which are present in mirror-image pairs, rather than four types of elements that do not occur in mirror-image pairs, and by having a well-developed ornamentation over the entire upper platform surface rather than only a central denticle row. It differs from *Polyplacognathus* in the outline of both types of elements, especially the pastiniplanate one. As noted above (p. 41), it is possible that *Cahabagnathus* might have its ancestor in the *Eoplacognathus* lineage (Fig. 2) but other evolutionary relationships are also conceivable. It seems unlikely that *Polyplacognathus* evolved from the *Cahabagnathus* lineage.

Detailed descriptions of two (*C. friendsvillensis, C. sweeti*) of the five species currently included in the genus were given by Bergström (1971a:142–144) and there is no need to redescribe these species here, especially as I plan to deal with the morphology and ontogenetic development of these species in a separate study. Below, I describe two stratigraphically important, but previously unnamed, species.

CAHABAGNATHUS CARNESI n. sp.
Fig. 6K, L

Synonymy. – □1975. *Polyplacognathus sweeti* Bergström (late form) – Carnes, p. 197–202, Pl. 8:10, 13, 15 (only). □1976 *Polyplacognathus sweeti* Bergström (late form) – Bergström & Carnes, Fig. 2:6.

Derivation of name. – In honor of Dr. John B. Carnes, the discoverer of the present species.

Type locality. – Cuba, Hawkins County, Tennessee (Carnes 1975:62–69).

Type stratum. – Holston Formation, sample 73CC2-16 of Carnes (1975); *Prioniodus (B.) variabilis* Subzone.

Holotype. – OSU 37186 (Fig. 6K).

Diagnosis. – A *Cahabagnathus* species similar to *C. sweeti* (Bergström 1971) but distinguished from the latter by the fact that in the pastiniplanate element the main denticle row is not straight but makes a distinct bend toward the anterior process at the junction of the posterior and anterior processes.

Description. – In all essential features, the stelliplanate element of the present species is closely similar to that of *C. sweeti*. Although there may be some minor differences in the outline of the processes (Fig. 2), they are hardly distinctive enough to separate these elements from the corresponding ones of *C. sweeti*, particularly in view of the rather considerable variation of these features in the collections at hand. Also the pastiniplanate elements are similar to those of *C. sweeti* but they can be readily distinguished by the difference in the appearance of the main denticle row mentioned in the diagnosis.

Remarks. – The morphological features used to separate this species from *C. sweeti* may appear to be rather insignificant at first sight but they seem to be quite constant in collections of hundreds of specimens from a limited stratigraphic interval at several localities. As described by Carnes (1975), specimens of *C. carnesi* first appear rarely in large populations of *C. sweeti* but in stratigraphically slightly younger samples, the former species becomes completely dominant and only very rare elements of *C. sweeti* are present. Some morphological intermediates connect *C. carnesi* with *C. sweeti*, and there is no doubt that the latter species is its ancestor. Being characteristic of a narrow stratigraphic interval above that characterized by rich occurrences of *C. sweeti*, the present species has the potential to

be a useful guide fossil. The vertical ranges of these species show some overlap (Fig. 2) but also this overlap has potential correlative usefulness.

Known occurrences. – Tennessee, Hawkins County, Cuba, Holston Formation (Carnes 1975); and Grainger County, Thorn Hill, Holston Formation (Carnes 1975). In addition, according to Dr. T. W. Broadhead (personal communication,

---

Fig. 6. Elements of important platform conodont species discussed in the text. □A. *Icriodella* cf. *I. praecox* Lindström *et al.* 1974, pastinate pectiniform Pa element, OSU 37180, lateral view, ×84. 'Narberth Group', Llandeilian–Costonian transition beds, Bryn-banc quarries, 3 km E of Narberth, Wales, sample 79B40-1. □B. Same specimen as A, upper view (note single row of denticles on anterior process). □C. Same species as A, pastinate pectiniform Pa element, OSU 38181, lateral view, ×22. Costonian beds, Evenwood quarry (Bergström 1971a), 13 km SE of Shrewsbury, Welsh Borderland, sample W66–13. □D. Same specimen as C, upper view, ×22. □E. Same species, locality, and sample as A, one of two types of tertiopedate S elements, OSU 37182, lateral view, ×84. □F. Same species, locality, and sample as A, bipennate element, OSU 37183, lateral view, ×84. □G. Same species, locality, and sample as A, tertiopedate Pb element, OSU 37184, lateral view, ×70. □H. Same species, locality, and sample as A, one of two types of tertiopedate S elements, OSU 37185, posterior view, ×84. □I. *Cahabagnathus sweeti* (Bergström 1971), pastiniplanate element, VPIL 4597, upper view, ×35. 6 m above base of Effna Limestone in section along railroad at Montgomery Lime Production Company, Ellett Valley, Montgomery County, Virginia. Coll. J. M. Wilson. □J. Same species, locality, and sample as I, stelliplanate element, VPIL 4597, upper view, ×35. Coll. J. M. Wilson. □K. *Cahabagnathus carnesi* n. sp. (holotype), pastiniplanate element, OSU 37186, upper view, ×78. Holston Formation, sample 73CC2-16 of Carnes (1975), Cuba, Hawkins County, Tennessee. □L. Same species, locality, and sample as K, stelliplanate element, OSU 37187, upper view, ×78. □M. *Cahabagnathus chazyensis* n. sp. (syntype), pastiniplanate element, VPIL 4578A, upper view, ×68. 8.6 m above base of Lincolnshire Limestone, section at Montgomery Lime Production Company quarry, Ellett Valley, Montgomery County, Virginia. Coll. J. M. Wilson. □N. Same specimen as M, lower view, ×68. □O. Same species, locality, and sample as M (syntype), stelliplanate element, VPIL 4578A, upper view, ×62. Coll. J. M. Wilson. □P. Same specimen as O, upper–lateral view, ×62. □Q. *Cahabagnathus* n. sp. A, pastiniplanate element, OSU 37188, upper view, ×48. 106 m above base of McLish Formation in section on west side of U.S. highway 77, north of Ardmore, Carter County, Oklahoma, sample 72SE-416 of Sweet & Bergström. □R. Same species, locality, and sample as Q, stelliplanate element, OSU 37189, upper view, ×48. □S. *Prattognathus rutriformis* (Sweet & Bergström 1962), one of two types of stelliplanate element, OSU 36373, upper view, ×53. 36 m above base of section of Little Oak Limestone at quarry 3 km north of Pelham, Shelby County, Alabama (sample 80MS7-22 of Schmidt 1982). □T. Same specimen as S, under side (note appearance of basal cavity), ×53. □U. Same species, locality, and formation as S, other type of stelliplanate element, OSU 37190, upper view, ×53. 32 m above base of section (sample 80MS7-20 of Schmidt 1982). □V. *Pygodus anitae* n. sp., quadriramate element, LO 5621, lateral view, ×84. Top 0.5 m of Segerstad Limestone (sample J65–11), Lunne quarry, Brunflo, Jemtland, Sweden. □W. Same species, locality, and sample as V, tertiopedate element, LO 5622, lateral view, ×84. □X. Same species, locality, and sample as V, stelliscaphate element, LO 5623T (holotype), lateral view, ×84. □Y. Same specimen as X, upper view, ×84. □Z. Same species, locality and sample as V, tertiopedate element, LO 5624, lateral view, ×84. Abbreviations of repositories of figured specimens are as follows: OSU, Department of Geology & Mineralogy, Orton Geological Museum, The Ohio State University, Columbus, Ohio, USA; VPIL, Department of Geological Sciences, Virginia Polytechnic Institute and State University, Blacksburg, Virginia, USA; and LO, Institute of Palaeontology, University of Lund, Sweden.

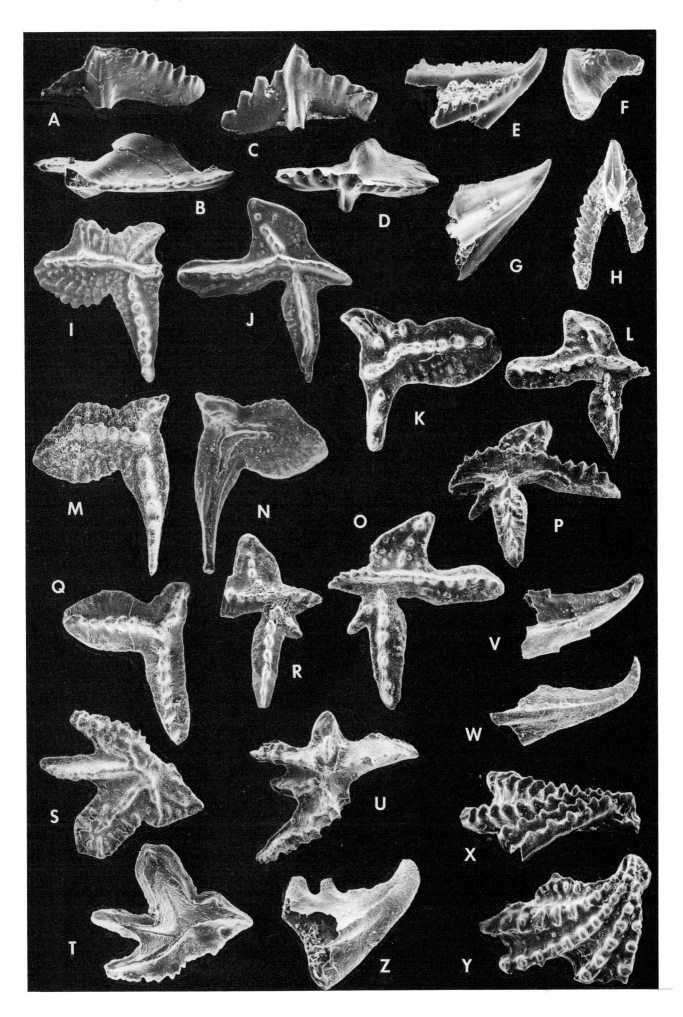

1982), elements of the present species have been found at other localities in the western thrust belts of eastern Tennessee.

Material. – More than 500 elements. It appears as if the ratio between pastiniplanate and stelliplanate elements is 1:1.

CAHABAGNATHUS CHAZYENSIS n. sp.
Fig. 6M-P

Synonymy. – □1972 *Polyplacognathus sweeti* Bergström – Raring, p. 116–117, Pl. 2:15, 18, 19. □1973 *Polyplacognathus friendsvillensis-P. sweeti* Bergström transition – Bergström, Fig. 2:12. □1976 *Polyplacognathus friendsvillensis-P. sweeti* transition – Bergström & Carnes, Fig. 2:4. □1977 *Polyplacognathus friendsvillensis* – Read & Tillman, p. 160, Fig. 4. □1978 *Polyplacognathus friendsvillensis-P. sweeti* transition – Bergström, Pl. 79:12, 13, Figs. 5, 6. □1979 *Polyplacognathus friendsvillensis*– Markello, Tillman, & Read, p. 64–67, Figs. 11, 18.

Comments to the synonymy list. – The references of Read & Tillman (1977) and Markello, Tillman, & Read (1979) include no formal descriptions or illustrations of the present species and the synonymy is based on my own study of their collections. Inclusion in the synonymy list was deemed appropriate because of the biostratigraphic importance of their papers, as well as the fact that the late C. G. Tillman assembled the finest collection of this species in existence.

Derivation of name. – From the Middle Ordovician Chazy Group, Champlain Valley, New York and Vermont in which the species is the dominant platform conodont and where specimens were first collected by Raring (1972).

Type locality. – Montgomery Lime Product Quarry, Ellett Valley, Montgomery County, Virginia (Wilson 1977).

Type stratum. – Lincolnshire Limestone, 26 feet above base, *Pygodus anserinus* Zone (sample VPIL 4578).

Syntypes. – VPIL 4578A (Fig. 6M–P).

Diagnosis. – A *Cahabagnathus* species characterized by the stelliplanate element with a bifid anterolateral process and a relatively broad, crudely square, posterolateral process; and pastiniplanate elements with a broad posterior process with an almost straight main denticle row extending to near the platform margin in the proximal part of the process.

Description. – A good description of both element types of this species was given by Raring (1972) on the basis of specimens from the Chazy Group, but because that description is not readily accessible, the critical features are dealt with herein.

    The stelliplanate elements show close similarity to those of *C. friendsvillensis* in most important respects, including the presence of a bifid anterolateral process, but in most instances, those of the present species may be distinguished by their broader and less pointed, in many cases crudely square, posterolateral process with markedly bent, rather than slightly evenly curved, anterior margin. The pastiniplanate element is closer morphologically to that of *C. sweeti* than to

that of *C. friendsvillensis* in that it has a much wider posterior process, and a different appearance of the denticle row proximally in the posterior process, compared to the latter species (Fig. 2).

Remarks. – Although elements of *C. chazyensis* are morphologically intermediate between those of *C. friendsvillensis* and *C. sweeti*, the present species is quite distinct and easily recognizable when both types of elements are present. It is of interest to note that Wilson (1977) found unmixed populations of this species through more than 80 m of strata in Ellett Valley, where there is a very minor overlap with the range of *C. sweeti* in the topmost part of the interval of *C. chazyensis*. Likewise, at its known occurrences in Texas, Tennessee, and Champlain Valley, *C. chazyensis* is the only *Cahabagnathus* species present through an interval ranging in thickness from several tens of meters to more than 100 m. There is no doubt that the present species is a descendant of *C. friendsvillensis*, and morphological intermediates connect it with *C. sweeti* at the top of its range.

Known occurrences. – Tennessee, Blount County, Friendsville, Lenoir Limestone (Bergström coll.) and Hawkins County, St. Clair, Lenoir Limestone (Bergström coll.); Virginia, Montgomery County, Ellett Valley, New Market and Lincolnshire formations (Wilson 1977); Texas, Brewster County, Marathon area, Woods Hollow Shale (Bergström 1978); New York and Vermont, Champlain Valley, Chazy Group, upper Crown Point and lower Valcour formations (Raring 1972). All known occurrences are apparently in the lower part of the *Pygodus anserinus* Zone but it is possible that the species might be present in the uppermost part of the *P. serra* Zone.

Material. – More than 200 elements. It appears as if the ratio between pastiniplanate and stelliplanate elements is 1:1 (cf. Wilson 1977).

PRATTOGNATHUS n. gen.

Derivation of name. – Named for the Pratt family, long-time residents at Pratt Ferry, Alabama, the type locality of *P. rutriformis*.

Type species. – *Polyplacognathus rutriformis* Sweet & Bergström 1962 (Fig. 6S-U). [The specific designation *rutriformis* has been used in a multielement sense for this species by Bergström (1973), Bergström & Carnes (1976) and Harris *et al.* (1979)].

Diagnosis of genus. – A genus of conodonts with apparatus having two types of asymmetrical albid platformed elements, both stelliplanate and occurring in mirror-image pairs; no ramiform elements known in the apparatus. One of these types with five, short, more or less bladelike processes that branch out from a common point in the middle of the unit. The other type with four processes, three of which are platformed and one bladelike, all branching from a central point. All processes in both types of elements with central row of denticles, and marginal to that an irregular ornamentation of low ridges and scattered nodes. Basal cavity broad but shallow beneath platformed processes, slitlike beneath bladelike processes.

Remarks. – In their study of the conodonts from the Pratt Ferry beds of Alabama, Sweet & Bergström (1962) described in some detail the elements herein referred to the multielement species *P. rutriformis* as *Polyplacognathus rutriformis* and *P. stelliformis*. Subsequently, these types of elements have been found at several other localities in Alabama (Schmidt 1982), eastern Tennessee (Bergström 1973; Bergström & Carnes 1976), and Nevada (Harris *et al.* 1979). Although no occurrence of abundant specimens of these elements is known, they are associated so constantly with each other that it seems very likely that they belong to one and the same apparatus. Furthermore, because this apparatus differs appreciably in several important respects from those of previously named conodont genera, it is appropriate to introduce a new generic name.

Prattognathus is distinguished from *Eoplacognathus* by having two types of elements, present in mirror-image pairs, rather than four types of elements that do not occur in mirror-image pairs. In addition to the central denticle row, these elements have an irregular pattern of ridges and nodes on the upper surface. The genus can be separated from *Cahabagnathus* and *Polyplacognathus* by the process arrangement and the outline of the elements, especially the one with dominantly platformed processes. It differs from *Polonodus* in the number and arrangement of the processes, and the appearance of the basal cavity, but, as noted above, it cannot be ruled out that *Prattognathus* may have its ancestor among forms similar to those referred to *Polonodus* by some authors.

Occurrence. – Specimens of *P. rutriformis*, the only known species of *Prattognathus*, have been found in the upper part of the *Pydodus serra* Zone and the lower part of the *P. anserinus* Zone at several localities in the Southern Appalachians and central Nevada (see references under Remarks).

Pygodus *Lamont & Lindström 1957*

Type species. – *Pygodus anserinus* Lamont & Lindström 1957

Pygodus anitae *n. sp.*
Fig. 6V-Z

Synonymy. – □1978 *Pygodus* sp. C. – Löfgren, p. 97, Pl. 19:4–6, Figs. 32A-C.

Derivation of name. – In honor of Dr. Anita Löfgren, who first described elements of this species.

Type locality. – Lunne, Jämtland, Sweden.

Type stratum. – Segerstad Limestone, 20 cm below top. Sample J65-11 of Bergström. 60 cm below top of the *Eoplacognathus suecius* Zone.

Holotype. – LO 5623T (Fig. 6X, Y).

Diagnosis. – A *Pygodus* species with at least two pairs of denticle rows on the anterior platform of the pygodontiform element and a short single-lobed posterior platform. Haddingodontiform element pastinate, with weakly denticulated anterior

and antero-lateral processes and a posterior process with a few equal-sized, short, suberect denticles. Quadriramate and alate elements subpyramidal with weakly denticulated processes.

Description. – Pygodontiform element of subtriangular shape with well-developed anterior platform, short, laterally compressed cusp, and a narrow posterior platform that is not appreciably longer than the cusp. Anterior platform with conspicuous central longitudinal depression, which is bordered on each side by a double row of short denticles; these four denticle rows extend from the anterior margin of the anterior platform to, or close to, the base of the cusp. In addition, a fifth denticle row may be developed along the anterior part of the lateral margin of the anterior platform. The basal cavity is very shallow, and extends over the entire lower surface of the element.

The haddingodontiform (pastinate) element has a relatively short, suberect cusp, weakly denticulated or undenticulated anterolateral and anterior processes, and a posterior process with a few distinct, short, suberect to slightly reclined, denticles that in most specimens tend to be separated from each other. The three processes are of about the same length and are connected with thin laminae to form a subpyramidal structure with a large basal cavity that extends to the base of the cusp.

In most respects, the subpyramidal quadriramate and alate elements are closely similar to corresponding elements in *Pygodus anserinus* and *P. serra* but they tend to be wider basally and less extended longitudinally, and have less strongly and less regularly denticulated processes.

Remarks. – The present species is a morphological intermediate between *Pygodus*? n. sp. and *P. serra*, and specimens transitional to the latter are present in the lower part of the *Eoplacognathus foliaceus* Subzone [*P. serra* (early) *in* Fig. 3; also cf. Löfgren 1978, Fig. 32D-F]. However, both the pygodontiform and haddingodontiform elements differ appreciably from those of typical specimens of *P. serra* and the two species are readily separable.

Known occurrence. – In Sweden, the species is characteristic of a narrow interval in the upper part of the *Eoplacognathus suecicus* Zone but well-preserved specimens are uncommon. The species is represented in my collections from Lunne, Kårgärde, and Vikarbyn (for the location of these sections, see Bergström 1971a), and Löfgren (1978) described elements from her Gusta section.

Material. – Several tens of specimens.

Acknowledgements. – I am most indebted to Dr. Walter C. Sweet for reading the manuscript and offering useful advice for its improvement, and for the many stimulating discussions on conodonts we have had through the years. I have also had the benefit to discuss many matters related to particularly Baltic conodonts with Dr. Jerzy Dzik during his stay at The Ohio State University in 1981–1982. Also, thanks are due to Mrs. Helen Jones, Mrs. Karen Tyler, and Mrs. Maureen Lorenz for skilled technical assistance, and to Drs. Anita Löfgren and Lennart Jeppsson for their editorial and other comments that substantially improved my manuscript.

References

Aldridge, R. J. 1972: Llandovery conodonts from the Welsh Border-land. *Bulletin of the British Museum (Nat. Hist.) Geology 22,* 125–231.

An Tai-xiang 1981: Recent progress in Cambrian and Ordovician conodont biostratigraphy of China. *Geological Society of America Special Paper 187,* 209–217.

Barnes, C. R. & Fåhraeus, L. E. 1975: Provinces, communities, and the proposed nectobenthic habit of Ordovician conodon-tophorids. *Lethaia 8,* 133–149.

Bergström. J., Bergström, S. M., & Laufeld, S. 1968: En ny skärning genom överkambrium och mellanordovicium i Rävatofta-området, Skåne. *Geologiska Föreningens i Stockholm Förhandlingar 89,* 460–465.

Bergström, S. M. 1962: Conodonts from the Ludibundus Limestone (Middle Ordovician) of the Tvären area (S. E. Sweden). *Arkiv för Mineralogi och Geologi 3:1,* 1–61.

Bergström, S. M. 1964: Remarks on some Ordovician conodont faunas from Wales. *Acta Universitatis Lundensis Sec. II, 3,* 1–66.

Bergström, S. M. 1971a: Conodont biostratigraphy of the Middle and Upper Ordovician of Europe and eastern North America. *Geological Society of America Memoir 127,* 83–157.

Bergström, S. M. 1971b: Correlation of the North Atlantic Middle and Upper Ordovician conodont zonation with the graptolite succession. *Colloque Ordovicien–Silurien, Brest, septembre 1971. Mémoires du Bureau de recherches géologiques et minières 73,* 177–187.

Bergström, S. M. 1971c: Taxonomy, phylogeny, and stratigraphic significance of Middle and Upper Ordovician platform conodonts from the North Atlantic area. *Symposium on Conodont Taxonomy, Marburg/Lahn, 1971. Abstracts of Lectures,* 2–3.

Bergström, S. M. 1973: Biostratigraphy and facies relations in the lower Middle Ordovician of easternmost Tennessee. *American Journal of Science 273-A,* 261–293.

Bergström, S. M. 1978: Middle and Upper Ordovician conodont and graptolite biostratigraphy of the Marathon, Texas graptolite zone reference standard. *Palaeontology 21,* 723–758.

Bergström, S. M. 1980: Whiterockian (Ordovician) conodonts from the Hølonda Limestone of the Trondheim region, Norwegian Caledonides. *Norsk Geologisk Tidsskrift 59,* 285–297.

Bergström, S. M. 1981a: Biostratigraphical and biogeographical significance of conodonts in two British Middle Ordovician olistostromes. *Geological Society of America Abstracts with Programs 13,* 271.

Bergström, S. M. 1981b: Family Balognathidae – Family Ulrichodinidae. *In* Robison, R. A. (ed.): *Treatise on Invertebrate Paleontology, Part W, Supplement 2 Conodonta,* W120–W135, W137–W140, W142–W148. Geological Society of America and the University of Kansas.

Bergström, S. M. 1983: Biostratigraphic integration of Ordovician graptolite and conodont zones – A regional review. *Geological Society of London Special Publication* (in press).

Bergström, S. M. & Carnes, J. B. 1976: Conodont biostratigraphy and paleoecology of the Holston Formation (Middle Ordovician) and associated strata in eastern Tennessee. *Geological Association of Canada Special Paper 15,* 27–57.

Bergström, S. M., Riva, J., & Kay, M. 1974: Significance of conodonts, graptolites, and shelly faunas from the Ordovician of Western and North-Central Newfoundland. *Canadian Journal of Earth Sciences 11,* 1625–1660.

Bergström, S. M. & Sweet, W. C. 1966: Conodonts from the Lexington Limestone (Middle Ordovician) of Kentucky and its lateral equivalents in Ohio and Indiana. *Bulletins of American Paleontology 50,* 271–441.

[Carnes, J. B. 1975: Conodont biostratigraphy in the lower Middle Ordovician of the western Appalachian thrust-belts in northeast-ern Tennessee. *Unpublished Ph. D. dissertation, Department of Geology and Mineralogy, The Ohio State University, Columbus, Ohio.*]

Carr, T. R. 1982: Conodont datum planes: Examples of conodont speciation events from the Permian and Triassic. *Geological Society of America Abstracts with Programs 14(5),* 256.

Chugaeva, M. N. 1976: Ordovician in the North-Eastern U.S.S.R. *in* Bassett, M. G. (ed.): *The Ordovician System: Proceedings of a Palaeontological Association Symposium, Birmingham, September 1974,*

283–292. University of Wales Press and National Museum of Wales, Cardiff.

Clark, D. L. *et al.* 1981: Conodonta. *In* Robison, R. A. (ed.): *Treatise on Invertebrate Paleontology, Part W Miscellanea.* The Geological Society of America and The University of Kansas.

Drygant, D. M. 1974: [New Middle Ordovician conodonts from North-western Volyn.] *Paleontologicheskii Sbornik 11,* 54–58 (in Ukrainian).

Dzik, J. 1976: Remarks on the evolution of Ordovician conodonts. *Acta Palaeontologica Polonica 21,* 395–455.

Dzik, J. 1978: Conodont biostratigraphy and paleogeographical relations of the Ordovician Mojcza Limestone (Holy Cross Mts., Poland). *Acta Palaeontologica Polonica 23,* 51–72.

Dzik, J. 1980: Ontogeny of *Bactrotheca* and related hyoliths. *Geologiska Föreningens i Stockholm Förhandlingar 102,* 223–233.

Dzik, J. 1981: Larval development, musculature, and relationships of *Sinuitopsis* and related Baltic bellerophonts. *Norsk Geologisk Tidsskrift 61,* 111–121.

Eldredge, N. & Gould S. J. 1972: Punctuated equilibria: An alterna-tive to phyletic gradualism. *In* Schopf, T. J. M. (ed.): *Models in Paleobiology,* 82–115. Freeman, Cooper & Company. San Francisco.

Fåhraeus, L. E. 1970: Conodont-based correlations of Lower and Middle Ordovician strata in Western Newfoundland. *Geological Society of America Bulletin 81,* 2061–2076.

Fåhraeus, L. E. & Hunter D. R. 1981: Paleoecology of selected conodontophorid species from the Cobbs Arm Formation (middle Ordovician), New World Island, north-central Newfoundland. *Canadian Journal of Earth Sciences 18,* 1653–1665.

Hamar, G. 1966: The Middle Ordovician of the Oslo region, Norway. 22. Preliminary report on conodonts from the Oslo–Asker and Ringerike districts. *Norsk Geologisk Tidsskrift 46,* 27–83.

Harris, A. G., Bergström, S. M., Ethington, R. L., & Ross, R. J., Jr. 1979: Aspects of Middle and Upper Ordovician conodont bio-stratigraphy of carbonate facies in Nevada and Southeast Califor-nia and comparison with some Appalachian successions. *Brigham Young University Geology Studies 26,* 7–43.

Jaanusson, V. & Bergström, S. M. 1980: Middle Ordovician faunal spatial differentiation in Baltoscandia and the Appalachians. *Alcheringa 4,* 89–110.

Kennedy, D. J., Barnes, C. R., & Uyeno, T. T. 1979: A Middle Ordovician conodont faunule from the Tetagouche Group, Camel Back Mountain, New Brunswick. *Canadian Journal of Earth Sciences 16,* 540–551.

Klapper, G. 1981: Family Pterospathodontidae. *In* Robison, R. A. (ed.): *Treatise on Invertebrate Paleontology, Part W, Supplement 2 Conodonta* W135–W136. Geological Society of America and the University of Kansas.

Klapper, G. & Philip, G. M. 1971: Devonian conodont apparatuses and their vicarious skeletal elements. *Lethaia 4,* 429–452.

Knüpfer, J. 1967: Zur Fauna und Biostratigraphie des Ordoviciums (Gräfenthaler Schichten) in Thüringen. *Freiberger Forschungshefte C220,* 1–119.

Lindström, M. 1955: Conodonts from the lowermost Ordovician strata of south-central Sweden. *Geologiska Föreningens i Stockholm Förhandlingar 76,* 517–614.

Lindström, M. 1964: *Conodonts.* 196 pp. Elsevier Publishing Com-pany, Amsterdam.

Lindström, M. 1970: A suprageneric taxonomy of the conodonts. *Lethaia 3,* 427–445.

Lindström, M. 1971: Lower Ordovician conodonts of Europe. *Geological Society of America Memoir 127,* 21–61.

Lindström, M. 1977: Genus *Amorphognathus* Branson & Mehl. *In* Ziegler, W. (ed.): *Catalogue of Conodonts III,* 21–52. E. Schweizer-bart' sche Verlagsbuchhandlung, Stuttgart.

Lindström, M., Racheboeuf, P. R., & Henry, J.-L. 1974: Ordovician conodonts from the Postolonnec Formation (Crozon Peninsula, Massif Armoricain) and their stratigraphic significance. *Geologica et Palaeontologica 8,* 15–28.

Löfgren, A. 1978: Arenigian and Llanvirnian conodonts from Jämt-land, northern Sweden. *Fossils and Strata 13,* 1–129.

Markello, J. R., Tillman C. G., & Read, J. F. 1979: Field Trip No. 2. Lithofacies and biostratigraphy of Cambrian and Ordovician platform and basin facies carbonates and clastics, southwestern Virginia. *Geological Society of America, Southeastern Section meeting, Blacksburg, Virginia, Field Trip Guidebook, April 27–29, 1979*, 41–85.

McCracken, A. D. & Barnes, C. R. 1981: Conodont biostratigraphy and paleoecology of the Ellis Bay Formation, Anticosti Island, Quebec with special reference to late Ordovician – Early Silurian chronostratigraphy and the systemic boundary. *Geological Survey of Canada Bulletin 329*, 51–134.

McTavish, R. A. & Legg, D. P. 1976: The Ordovician of the Canning Basin. *In* Bassett, M. G. (ed.): *The Ordovician System: Proceedings of a Palaeontological Association Symposium, Birmingham, September 1974*, 447–478. University of Wales Press and National Museum of Wales, Cardiff.

Miller, J. F. 1980: Taxonomic revisions of some Upper Cambrian and Lower Ordovician conodonts with comments on their evolution. *University of Kansas Paleontological Contributions, Paper 99*, 1–43.

Miller, J. F. 1981: Family Clavohamulidae Lindström, 1970. *In* Robison, R. A. (ed.): *Treatise on Invertebrate Paleontology, Part W, Supplement 2 Conodonta*, W115–W116. Geological Society of America and the University of Kansas.

Moskalenko, T. A. (Москаленко, Т. А.) 1970: Конодонты криволуцкого яруса (средний ордовик) Сибирской платформы [Conodonts of the Krivaya Luka Stage (Middle Ordovician) of the Siberian platform]. *Академия наук СССР, Сибирское отделение, Труды Института геологии и геофизики 61*, 1–131.

Moskalenko, T. A. (Москаленко Т. А.) 1977: Отряд Conodontophorida Eichenberg. *In* Каныгин, А. В., Москаленко, Т. А., Ядренкина, А. Г. & Семенова В. С.: О стратиграфическом расчленении и корреляции среднего ордовика Сибирской платформы, 32–43. *Академия наук СССР, Сибирское отделение, Труды Института геологии и геофизики 372*.

Nasedkina, V. A. & Puchkov, V. N. (Наседкина, В. А. & Пучков В. Н.) 1979: Среднеордовикские конодонты севера Урала и их стратиграфическое значение [Middle Ordovician conodonts of the northern Urals and their stratigraphic significance]. *In* Г. Н. Папулов и В. Н. Пучков (eds.): Конодонты Урала и их стратиграфическое значение [Conodonts from the Urals and their stratigraphic significance], 5–24. *Академия наук СССР, Уральский научный центр, Труды Института геологии и геохимии 145*.

Ni Shizhao 1981: [Discussion on some problems of Ordovician stratigraphy by means of conodonts in eastern part of Yangtze Gorges region.] *Selected papers on the 1st Convention of Micropaleontological Society of China (1979)*, 127–134 (in Chinese).

Nicoll, R. S. 1980: Middle Ordovician conodonts from the Pittman Formation, Canberra, ACT. *BMR Journal of Australia Geology & Geophysics 5*, 150–153.

Nowlan, G. S. 1981: Some Ordovician conodont faunules from the Miramichi anticlinorium, New Brunswick. *Geological Survey of Canada Bulletin 345*, 1–35.

Orchard, M. J. 1980: Upper Ordovician conodonts from England and Wales. *Geologica et Palaeontologica 14*, 9–44.

Pollock, C. A., Rexroad, C. B., & Nicoll, R. S. 1970: Lower Silurian conodonts from northern Michigan and Ontario. *Journal of Paleontology 44*, 743–764.

[Raring, A. M. 1972: Conodont biostratigraphy of the Chazy Group (lower Middle Ordovician), Champlain Valley, New York and Vermont. *Unpublished Ph. D. dissertation, Lehigh University, Bethlehem, Pennsylvania.*]

Read, J. F. & Tillman, C. G. 1977: Field trip guide to lower Middle Ordovician platform and basin facies rocks, southwestern Virginia. *In* Ruppel, S. C. & Walker, K. R. (eds.): The ecostratigraphy of the Middle Ordovician of the Southern Appalachians (Kentucky, Tennessee, and Virginia), USA: A field excursion, 141–171. *University of Tennessee Department of Geological Sciences Studies in Geology 77-1.*

Repetski, J. E. & Ethington, R. L. 1977: Conodonts from graptolite facies in the Ouachita Mountains, Arkansas and Oklahoma. *Arkansas Geological Commission. Symposium on the geology of the Ouachita Mountains 1*, 92–106.

Rhodes, F. H. T. 1953: Some British Lower Palaeozoic conodont faunas. *Philosophical transactions of the Royal Society of London B:237*, 261–334.

[Roscoe, M. S. 1973: Conodont biostratigraphy and facies relationships of the lower Middle Ordovician strata in the Upper Champlain Valley. *Unpublished M. Sc. thesis, The Ohio State University, Colombus, Ohio.*]

[Schmidt, M. A. 1982: Conodont biostratigraphy and facies relations of the Chickamauga Limestone (Middle Ordovician) of the Southern Appalachians, Alabama and Georgia. *Unpublished M. S. thesis, The Ohio State University.*]

Schönlaub, H. P. 1971: Zur Problematik der Conodonten-Chronologie an der Wende Ordoviz/Silur mit besonderer Berücksichtigung der Verhältnisse im Llandovery. *Geologica et Palaeontologica 5*, 35–57.

Schopf, T. J. M. 1966: Conodonts of the Trenton Group (Ordovician) in New York, Southern Ontario, and Quebec. *New York State Museum and Science Service Bulletin 405*, 1–105.

Sergeeva, S. P. (Сергеева С. П.) 1966a: Биостратиграфическое распространение конодонтов в тремадокском ярусе (ордовик) Ленинградской области [The biostratigraphic distribution of conodonts in the Tremadocian stage (Ordovician) of the Leningrad region]. *Докл. АН СССР 167:3.*

Sergeeva, S. P. (Сергеева С. П.) 1966b: Распространение конодонтов в нижнеордовикских отложениях Ленинградской области [The distribution of conodonts in the Lower Ordovician deposits of the Leningrad region]. *Уч. зап. Лен. гос. пед. ин-та им. А. Е. Герцена, 290.*

Sheng Shen-fu 1980: The Ordovician System in China. Correlation chart and explanatory notes. *International Union of Geological Sciences Publication No.1*, 1–7.

Simes, J. E. 1980: Age of the Arthur Marble: conodont evidence from Mount Owen, northwest Nelson. *New Zealand Journal of Geology and Geophysics 23*, 529–532.

[Stouge, S. S. 1980: Conodonts of the Table Head Formation (Middle Ordovician), western Newfoundland. *Unpublished Ph. D. dissertation, Memorial University of Newfoundland, St. John's, Newfoundland.* 413 pp.]

Sweet, W. C. 1981: Morphology and composition of elements. *In* Robison, R. A. (ed.): *Treatise on Invertebrate Paleontology, Part W, Miscellanea, Supplement 2 Conodonta*, W5-W20. Geological Society of America and the University of Kansas.

Sweet, W. C. 1982: Conodonts from the Winnipeg Formation (Middle Ordovician) of the northern Black Hills, South Dakota. *Journal of Paleontology 56*, 1029–1049.

Sweet, W. C. & Bergström, S. M. 1962: Conodonts from the Pratt Ferry Formation (Middle Ordovician) of Alabama. *Journal of Paleontology 36*, 1214–1252.

Sweet, W. C. & Bergström, S. M. 1971: The American Upper Ordovician Standard. XIII. A revised time-stratigraphic classification of North American upper Middle and Upper Ordovician rocks. *Geological Society of America Bulletin 82*, 613–628.

Sweet, W. C. & Bergström, S. M. 1972: Multielement taxonomy and Ordovician conodonts. *Geologica et Palaeontologica SB 1*, 29–42.

Sweet, W. C. & Bergström, S. M. 1973: Biostratigraphic potential of the Arbuckle Mountains sequence as a reference standard for the Midcontinent Middle and Upper Ordovician. *Geological Society of America Abstracts with Programs 5*, 355.

Sweet, W. C. & Bergström, S. M. 1974: Provincialism exhibited by Ordovician conodont faunas. *Society of Economic Paleontologists and Mineralogists Special Publication No. 21*, 189–202.

Sweet, W. C. & Bergström, S. M. 1976: Conodont biostratigraphy of the Middle and Upper Ordovician of the United States Midcontinent. *In* Bassett, M. G. (ed.): *The Ordovician System: Proceedings of a Palaeontological Association Symposium, Birmingham, September 1974*, 121–151. University of Wales Press and National Museum of Wales.

Sweet, W. C., Ethington, R. L., & Barnes, C. R. 1971: North American Middle and Upper Ordovician conodont faunas. *Geological Society of America Memoir 127*, 161–193.

Tipnis, R. S., Chatterton, B. D. E., & Ludvigsen, R. 1978: Ordovician conodont biostratigraphy of the Southern District of Mackenzie, Canada. *Geological Association of Canada Special Paper 18*, 39–91.

Viira, V. 1972: On symmetry of some Middle Ordovician conodonts. *Geologica et Palaeontologica 6*, 45–49.

Viira, V. (Вийра, В.) 1974: Конодонты ордовика Прибалтики [*Ordovician conodonts of the East Baltic*]. 1–142. 'Валгус.' Tallinn.

Walliser, O. H. 1964: Conodonten des Silurs. *Abhandlungen des hessischen Landesamtes für Bodenforschung 41*, 1–106.

Walliser, O. H. 1972: Conodont apparatuses in the Silurian. *Geologica et Palaeontologica SB 1*, 75–80.

Wardlaw, B. R. 1982: A conodont speciation event in the Early Triassic of western North America. *Geological Society of America Abstracts with Programs 14(7)*, 642.

Webers, G. F. 1966: The Middle and Upper Ordovician conodont faunas of Minnesota. *Minnesota Geological Survey Special Publication SP-4*, 1–123.

Williams, A. 1953: The geology of the Llandeilo district, Carmarthenshire. *Quarterly Journal of the Geological Society of London 108*, 177–208.

Williams, A., Strachan I., Bassett, D. A., Dean, W. T., Ingham, J. K., Wright, A. D., & Whittington, H. B. 1972: A correlation of Ordovician rocks in the British Isles. *Geological Society of London Special Report No. 3*, 1–74.

[Wilson, J. M. 1977: Conodont biostratigraphy and paleoecology of the Middle Ordovician (Chazyan) sequence in Ellett Valley, Virginia. *Unpublished M. S. thesis, Virginia Polytechnic Institute and State University, Blacksburg, Virginia.*]

Relationships between Ordovician Baltic and North American Midcontinent conodont faunas

JERZY DZIK

FOSSILS AND STRATA

Dzik, Jerzy 1983 12 15: Relationships between Ordovician Baltic and North American Midcontinent conodont faunas. *Fossils and Strata*. No. 15, pp. 59–85. Oslo. ISSN 0300-9491. ISBN 82-0006737-8.

During most of the Early and Middle Ordovician, epicontinental seas of Laurentia, Baltica, and islands in the Iapetus Ocean between them were centers of diversification of different groups of conodonts. The most important were Phragmodontidae in Laurentia, Periodontidae in the Iapetus area, and Balognathidae in Baltica. Spectra of fossil assemblages (analyzed in terms of conodont lineages) were very stable through time in each of these biogeographic provinces. Relatively few lineages passed the biogeographic boundaries and successful immigrations were rare. An especially stable composition characterizes the Baltic faunas. A profound change in the composition of faunas in Baltica took place in the Oanduan (late Caradocian) through influx of several lineages previously confined to islands in the Iapetus. Conodont faunas of Laurentia, at times enriched by Iapetus-born lineages, evolved rather gradually until the end of the Ordovician. Diversity of Ordovician conodont assemblages seems to depend more on local ecologic factors and bathymetry than on climate. □*Conodonta, biostratigraphy, biogeography, evolution, Ordovician.*

ECOS III

A contribution to the Third European Conodont Symposium, Lund, 1982

Jerzy Dzik, Department of Geology and Mineralogy, The Ohio State University, Columbus, Ohio 43210, U.S.A. (Present address: Zakład Paleobiologii PAN, Aleja Żwirki i Wigury 93, PL-02-089 Warszawa, Poland); 17th August, 1982.

Ordovician bioprovincialism has received much attention from conodont workers and several important contributions to its recognition have been published (for review, see Jaanusson 1979; Sweet & Bergström 1974; Jaanusson & Bergström 1980; Lindström 1976 b). The profound difference between conodont faunas of the North American Midcontinent and those of the southeastern Appalachians and Europe has been recognized since the Sweet *el al.* (1959) paper. A boundary separating these two distinct biogeographic units has been traced along the Helena–Saltville fault in the southern Appalachians and corresponding structural features to the north (Bergström 1971; Jaanusson & Bergström 1980). Although changes in the distribution of particular Late Ordovician conodont taxa have been discussed in detail by Sweet & Bergström (1974) and faunal changes across the Appalachians in the Middle Ordovician have been presented by Bergström & Carnes (1976), much remains to be done regarding the pattern of distribution and shifts of Early and Middle Ordovician conodont lineages in Europe as well as precisely tracing relationships of those lineages that are thought to be typical of provinces. It must be stated, however, that data are still far from complete, and numerous key faunas (for example those from the Middle Ordovician of Great Britain and the Sudeten Mountains) have not yet been described in terms of multielement taxonomy. Accordingly, this paper is just a preliminary report on a study that remains to be completed.

Terminology

Choice of the best element-notation system begins to be a difficult task for conodontologists. Among several proposed systems of terminology for elements in the conodont apparatus (for review, see Sweet 1981b; also Barnes *et al.* 1979 and Dzik & Trammer 1980) two seem to be applicable for most apparatuses. Those are Jeppsson's (1971) and Sweet's (1981b) systems. When applied to sextimembrate apparatuses they are easily transferable to each other.

For the purpose of the present paper Jeppsson's notation has been chosen. It was originally introduced as a tool for expressing homology of elements. Symbols of all types of elements are derived by abbreviations of former form-taxonomic generic names, which seems to be in good agreement with tradition in biological terminology. In the same way terms for, say, larval stages (echinospira, pilidium, calyptopis, zoëa, etc.), or organs (stigmaria, helens, aptychi, etc.) have been introduced, all being based on names of former taxa. This does not leave a place for uncertainty regarding reference for homologization. Particular types of elements are identified in any apparatus by homologization with particular elements of the *Ozarkodina* apparatus. There is no assumed *a priori* order in arrangement of the types of elements in the apparatus, therefore no trouble appears when insertion of a new type of element into the transition series appears necessary.

The apparatus of *Amorphognathus* is proposed here as a homologization standard for septimembrate apparatuses. The complete set of elements in such apparatuses contains, according to this notation (in supposed order), sp, oz, tr, pl, ke, hi, and ne types. Sweet (1981b) and Bergström (1981) proposed to consider the keislognathiform element of *Amorphognathus* as pl (Sb) and consequently created 'location' Sd for tetraprioniodontiform elements. There are, however, analogous tetraramous elements in the pl (Sb) location in apparatuses other than those of the Balognathidae (for example

Microzarkodina and *Paraprioniodus*), while it has been asserted that the keislognathiform element is a homologue of one of the Sc elements in those Balognathidae that do not have it (Dzik 1976). In the typical apparatus of the Balognathidae, Sa, tetraprioniodontiform (Sb), keislognathiform, and Sc elements form a single symmetry transition series, and a keislognathiform element should be inserted between Sb and Sc elements. This could be done either by introducing a third letter to the notation (like Sbb) or by replacing Sc elements *sensu* Sweet (1981b) to Sd location and reserving the Sc location only for homologues of the keislognathiform elements of *Amorphognathus* (which I would prefer). Kuwano (1982) proposed symbols Sa-b and Sb for homologues of pl and ke elements in the *Ozarkodina excavata* (Branson & Mehl) apparatus, respectively.

Meanings of other terms used in the text are explained in the *Treatise on Invertebrate Paleontology* (Sweet 1981 b). There also all taxonomic, auctor references that are not listed in this paper can be found.

Comments on methodology

One basic trouble in establishing a good scientific framework for a discussion of conodont provincialism is the lack of unequivocal meaning for units in paleobiogeography. Concepts of high-rank biogeographic units are usually based on more or less objectively determined differences in the composition of faunistic assemblages from different areas. This does not create special difficulties unless problems of the boundary between particular units and their change in geologic time are involved. Then any biogeographic unit appears to have a particular meaning that depends on methods used in recognizing it and the particular group of fossils used, rather than on objective factors (Jaanusson 1979). For these reasons Lindström (1976 a) has proposed to separate the territorial and faunistic aspects of provincialism, and to abandon the first one and shift the discussion toward aspects of the evolution of particular faunas. These may completely change their distribution in time. This means that he has proposed to study high-rank communities (faunas) rather than high-rank ecosystems (provinces and realms in common understanding). Data discussed below show, however, that there is no visible integration among species, either in their distribution within faunas or within particular conodont communities. This fits well with data concerning the distribution of other groups of fossils as well as with theoretical considerations (Hoffman 1979). There is also little evidence of invasion of other areas by entire groups of conodonts that form the core of particular faunas. Rather, each species seems to change the area of its distribution separately, and when all the 'fauna' is replaced by another assemblage the replacing assemblage is rarely the 'fauna' that occurred in another area before. At least some conodont species may occur alone. The concept of a conodont 'fauna' thus appears quite foggy and there is little chance to make it more objective.

For all these reasons I prefer to define biogeographic units not on the basis of their supposed internal integration but on external factors. A boundary between two such units should be drawn on the supposed discontinuity in spatial distribution of environmental factors, which may be expressed in an abrupt change in the distribution of some organisms. The concepts of island biogeography (MacArthur & Wilson 1967) are applicable to such biogeographic units. It is not difficult to find boundaries in the terrestrial environment with the features required by this definition. The marine environment is much more continuous, but discontinuities in the distribution of particular marine environments such as margins of the continental shelf, areas of convergence of warm and cold water, thermoclines, a range of uplifts of oceanic currents carrying biogenes, etc., can be used to delineate boundaries between marine ecosystems. All these factors may separate oceanic water into discrete cells that are different in the composition of their faunal and floral assemblages. Features of these cells may change with the evolution of climate and even more profoundly with the tectonic evolution of continents (Williams 1976). In the present paper the Ordovician epicontinental seas of Laurentia (the Midcontinent province), Baltica (the Baltic province), and islands in the Iapetus Ocean between them (the North Atlantic province) are considered to be paleobiogeographic units of this kind (Fig. 1).

Four main kinds of faunal processes may be connected with the evolution of biogeographic units defined above: (1) Phyletic evolution of particular species. (2) Change in the distribution of particular species within boundaries of the unit. (3) Immigrations of species from other ecosystems, which displace local species from their niches, but do not change the state of faunal equilibrium. (4) Immigrations of species that occupy previously uninhabited niches, or destruction of the previous distribution of niches, in both cases changing faunal equilibrium.

How can one recognize these basically different processes from the limited fossil record? It would be especially important to have a tool that permits distinction between rebuilding of the fossil assemblage caused by a local environmental change and immigration of extrinsic faunal elements. Solution of this problem seems to be the first step to the serious considerations of bioprovincialism (see discussion in Sweet & Bergström 1974; Barnes *et al.* 1973; Bergström & Carnes 1976). A strict distinction between species of different lineages coexisting at the same time and chronospecies being part of the same (monospecific at any time) lineage must be made in any analysis of this kind. Because of objective limitations of the fossil record, particular fossil assemblages must be analyzed in terms of lineages rather than particular chronospecies. Similarly, a direct comparison of the same paleoenvironments in different provinces seems to be outside the possibilities provided by sedimentological analysis. Fossils alone seem to be the most sensitive environmental indicators. A consideration of the entire spectrum of environments within two provinces would be a way to reach these objectives; but rarely, if ever, are enough data available to reconstruct such spectra for particular time units. Some support is fortunately given by Walther's law. One may assume that vertical changes in the composition of conodont assemblages, which reflect the influence of changing environments through time, are characteristic for each biogeographic unit, for in such units different parts of the ecospace are inhabited by organisms that represent lineages specific to it. Different provinces are thus expected to have different dynamics of fossil assemblages in relation to environmental changes.

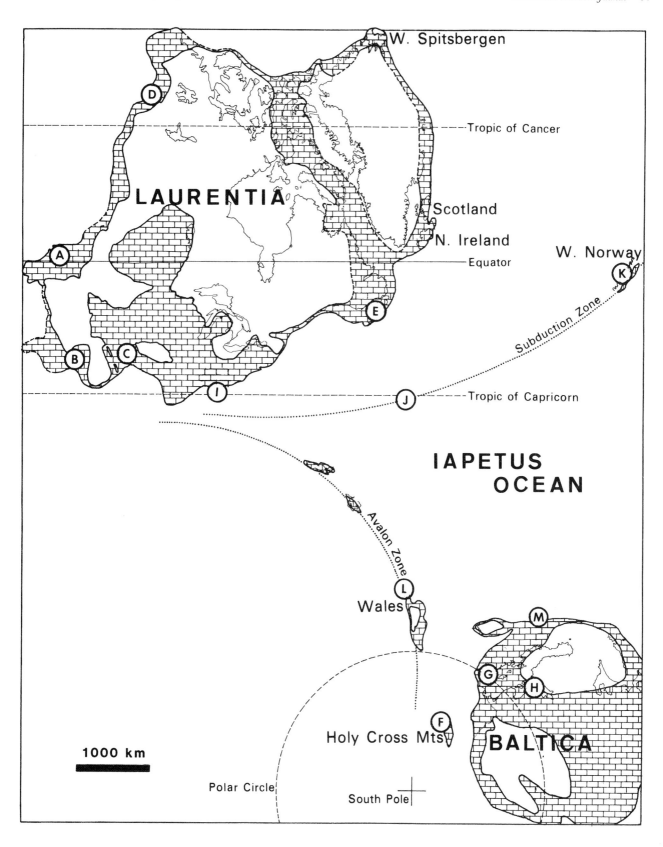

Fig. 1. Presumed geographic position of main localities discussed in the text in the late Early and early Middle Ordovician according to Bergström (1979) and Bruton & Bockelie (1980). □A. Ibex area of Utah and Basin Ranges of Nevada (Ethington & Clark 1981; Harris *et al.* 1979). □B. Marathon area of Texas (Bergström 1978). □C. Arbuckle Mountains of Oklahoma. □D. Sunblood Range of the District of Mackenzie (Tipnis *et al.* 1979). □E. Western shore of Newfoundland (Fåhraeus & Nowlan 1978; Stouge in press). □F. Holy Cross Mountains (Dzik 1978). □G. Island of Öland (van Wamel 1974). □H. Estonia (Viira 1974). □I. Southeastern Appalachians (Bergström & Carnes 1976). □J. Central Newfoundland (Fåhraeus & Hunter 1981). □K. Trondheim area of Norway (Bergström 1979). □L. Type section of Llandeilo, Wales (Bergström 1971). □M. Kalkberget and Gammalbodberget sections, Jämtland (Löfgren 1978).

Logs of contribution of particular conodont lineages (species) in samples plotted against time (rock thickness) are used here as a practical tool for recognition of the dynamics of assemblages (Fig. 11). This method of presentation of quantitative data on conodont distribution has been extensively used for biostratigraphic correlation and environmental analysis (Bergström & Sweet 1966; Sweet 1979 a, b; Jeppsson 1979). This kind of logs may also show whether a new species that appears in the assemblage is invading an unoccupied niche or displacing some earlier species from its niche. In the first case the appearance of a new species is accompanied by a proportionate reduction of the contribution of all other species, without a change in the relations among them. The same effect may also be produced by a relative increase in productivity by some species, however. Ecologic competition between new and old species has an effect only on that part of the assemblage affected by it (for examples see Fig. 11).

Sources of data

The present article owes its origin to the fortunate opportunity to compare data on the Baltic and Polish Ordovician conodonts I collected as a member of the Zakład Paleobiologii PAN in Warsaw, Poland, with extensive collections of Midcontinent conodonts gathered during many years by Walter C. Sweet, Stig M. Bergström, and their students at The Ohio State University in Columbus, Ohio. It should be noted that I have had free access at Ohio State University to all available collections, including several that have not been described in the literature. Many of these collections, as noted below, were carefully sorted before I saw them, and components of the individual species were arranged in different parts of the slides to reflect the taxonomic judgment of the sorter. These have been of great help to me. Other collections were not sorted when I examined them and are currently under study by others. My judgement as to the assemblages of species represented in the unsorted samples or as to the quantitative relations between particular species is thus no more than an approximation and may well be different from the one expressed when the collections have been completely prepared. The following sources of data have been particularly useful:

Midcontinent conodonts. – The main source of data concerning Midcontinent conodonts used here is a collection of samples from the Lower and Middle Ordovician of the Arbuckle Mountains, Oklahoma (see Sweet & Bergström 1973). Two long sections were sampled by Walter C. Sweet, Valdar Jaanusson, and Stig M. Bergström in 1972 and subsequently processed in the Department of Geology and Mineralogy of The Ohio State University. These include samples from the upper 234 feet (71.3 m) of the West Spring Creek Formation, the entire Joins, Oil Creek, McLish, Tulip Creek, and Mountain Lake formations at a locality along Interstate Highway 35 north of Ardmore, Carter County; and samples from the McLish, Tulip Creek, and Bromide formations from a section along Oklahoma Highway 99, south of Fittstown. The major part of the collection has apparatuses of particular species separated on the slides. This means that most of the

apparatuses have been reconstructed by Walter C. Sweet and Stig M. Bergström.

I have also examined samples from the Tumbez and Elway-Eidson formations of the Lay School section, Grainger County, Tennessee, collected and arranged by species on his slides by Carnes (1975) and the collection of Blackriveran conodonts from the Pan American Davidson core from a borehole in Richland County, Ohio, collected, arranged and tabulated by Votaw (1971). Neither Carnes nor Votaw has yet published his ideas about these samples. During a short stay in Washington, D. C., I saw samples from the Whiterockian of Nevada that form the basis of the Harris et al. (1979) paper. Some previously unpublished data concerning Nevada conodonts have also been supplied to me by Anita G. Harris (U. S. Geological Survey) and Stig M. Bergström, who also made available to me some of his samples from Marathon, Texas. John E. Repetski (U. S. Geological Survey) allowed me to examine his collection of conodonts from the Dutchtown Formation of Missouri. Published data used here are from sections in Nevada (Ethington & Schumacher 1969), Utah (Ethington & Clark 1981), the District of Mackenzie (Tipnis et al. 1979), the Melville Peninsula (Barnes 1977), Quebec (Barnes & Poplawski 1973), Scotland (Higgins 1967) and several other localities.

North Atlantic conodonts. – The section of the Mójcza Limestone at Mójcza, Holy Cross Mountains, Poland, which has been preliminarily described elsewhere (Dzik 1978), was fully sampled in 1979 and is currently under study at the Zakład Paleobiologii PAN. Szaniawski (1980) has published a description of Tremadocian conodonts from the same area. I saw the collection of Dr. Zdzisława Urbanek (University of Wrocław) from the Sudeten Mountains (see Baranowski & Urbanek 1972) in 1978. A description of some Bohemian conodonts is currently in press (Dzik 1983). Welsh conodont faunas have been partially described and reviewed by Bergström (1964, 1971) and I have seen some of his samples. Additional published data concern Newfoundland (Fåhraeus & Nowlan 1978; Fåhraeus & Hunter 1981; Stouge 1982 and in press), the Appalachians (Sweet & Bergström 1962; Bergström et al. 1972; Bergström & Carnes 1976; Landing 1976), the Armorican Massif (Lindström et al. 1974), and western Norway (Bergström 1979).

Baltic conodonts. – Baltic conodonts have been described in great detail in several papers summarized by Lindström (1971), Bergström (1971), van Wamel (1974), Viira (1974), Dzik (1976), and Löfgren (1978). Along with samples from erratic boulders of Baltic origin from northern Poland, I base my knowledge of Baltic faunas on samples from some Estonian and Swedish sections. Several samples were taken in 1977 from the Sukhrumägi section near Tallin. In 1980 I also sampled the upper part of the Langevoja and Hunderum substages at Hälludden, Öland, as well as the Ceratopyge Beds and Latorpian at the Ottenby cliff, Öland; the section partially described by Fåhraeus (1966) at Gullhögen quarry, Skövde, Västergötland, and a few other localities in Västergötland. Stig M. Bergström has shown me his collection of conodonts from the Gullhögen quarry, which covers the upper part that was not sampled by me. During a few days stay in

Tallinn I had an opportunity to study collections of conodonts from several boreholes described by Viive Viira (Geological Institute, Tallin; see Viira 1974). Especially important were cores from the Ohesaare and Kaagvere boreholes, which present a good record of the evolution of *Amorphognathus*. A detailed description of the conodont succession within the latter borehole has not been published.

Correlation

Although Baltic and Midcontinent Ordovician conodont faunas are basically different, several influxes of populations belonging to quickly evolving lineages took place between them, thus allowing some time correlation between them. These correlation horizons are discussed briefly below in stratigraphic order. The zonation proposed by Lindström (1971) and Bergström (1971), with a few additions by Dzik (1978), based on the evolution of Baltic lineages, is used here as a reference standard. A few zones previously proposed are omitted, namely the *Baltoniodus triangularis* and *Microzarkodina parva* Zones. The first is based on a species that is hard to identify (see Dzik 1984); the second is defined as an assemblage zone. In addition, the *Prioniodus alobatus* Zone, which may partially represent the *P. gerdae* Zone, is omitted because it is based on a species of unknown time of origin. No comparable zonation of the Midcontinent Ordovician has been proposed yet (see, however, Ethington & Clark 1981 and Stouge in press). Several Midcontinent conodont chronospecies have well-recognized phylogeny and their appearances supply several useful correlation horizons. These horizons are used here provisionally only for the purpose of having reference to some Midcontinent evolutionary events, not exactly correlated with the Baltic zones (see Figs. 11, 12). They are not intended to be boundaries of an established zonal scheme, which remains to be defined.

(1) *Glyptoconus*(?) *asymmetricus* (Barnes & Poplawski 1973) probably evolved from *G.* (?) *striatus* (Graves & Ellison 1941) at the time of deposition of the upper Fillmore Formation of the Ibex area, Utah (see Ethington & Clark 1981). The Joins Formation of Oklahoma, the Hølonda Limestone of Norway, and the *A. raniceps* Limestone of Öland are thus younger than, or at least contemporary with, the upper Fillmore.

(2) *Microzarkodina flabellum* (Lindström 1955) occurs in the highest Wah Wah and the lower half of the Juab Formation in the Ibex area (Ethington & Clark 1981). Because periodontids presumably originated in *Prioniodus elegans* Zone time and evolutionarily advanced *Microzarkodina* cannot be older than the top of the *Oepikodus evae* Zone, the uppermost Wah Wah is not older than upper *O. evae* Zone.

(3) *Histiodella holodentata* Ethington & Clark 1981 originated from *H. sinuosa* (Graves & Ellison 1941) in the upper part of the Lower Table Head Formation of Newfoundland (Stouge in press), in the middle of the Lehman Formation of the Ibex area, Utah (Ethington & Clark 1981), in the uppermost Antelope Valley Limestone at Steptoe, Nevada, and in the middle of the same formation in the Toquima Range, Nevada (Harris *et al.* 1979) It appeared in the Baltic area (Viira 1974)

and the Holy Cross Mountains, Poland (Dzik 1976, 1978), in the early Kundan, where it occurs together with *Amorphognathus variabilis* Sergeeva 1963. This means that the Baltic *A. variabilis* Zone is not older than the upper Lehman Formation of Utah and other occurrences of *H. holodentata*.

(4) *Eoplacognathus suecicus* Bergström 1971 occurs with the biostratigraphically probably very diagnostic *Phragmodus* sp. n. of Harris *et al.* (1979) in Nevada (Harris *et al.* 1979), below the first occurrences of *Phragmodus flexuosus* Moskalenko 1973.

(5) Early *P. flexuosus* is known to occur together with *Eoplacognathus reclinatus* (Fåhraeus 1966) (R. L. Ethington, personal communication).

(6) Advanced *P. flexuosus* (having a more prominent gradient in the size distribution of denticles on the pl element than earlier forms) occurs in the Kukrusean of Estonia with *Baltoniodus variabilis* (Bergström 1962) (Bergström 1971). Several co-occurrences of *Baltoniodus*, *Phragmodus*, and the *Polyplacognathus friendsvillensis–sweeti* lineage are known from the Appalachians (Bergström & Carnes 1976; Fåhraeus & Hunter 1981).

(7) *P. sweeti* Bergström 1971 appears in the Uhakuan of Volhynia (Drygant 1974).

(8) *Baltoniodus gerdae* (Bergström 1971) occurs in the Mountain Lake Formation of Oklahoma with typical *Phragmodus inflexus* Stauffer 1935 (Sweet & Bergström 1973).

(9) *Amorphognathus tvaerensis* Bergström 1962 occurs with *Phragmodus undatus* Branson & Mehl 1933 in the Bromide Formation of Oklahoma in the section south of Fittstown as well as in several other localities (Bergström & Sweet 1966). *P. undatus* is known also in the Oanduan of Estonia (Viira 1974) and the Mjøsa Limestone of Norway (Bergström 1971).

Patterns in the distribution of conodont lineages

Presumed relationships between Midcontinent and Baltic lineages and their connections with conodonts of other areas are discussed below. The groups have been arranged in family-rank taxa according to the supposed pattern of relationships among them. They may be arranged into four larger groups, which differ in the ground plan of the apparatus and of particular elements. The first such group is formed by conodonts included here in the family Chirognathidae, which are characterized by hyaline elements with strongly developed denticulation and with morphologically rather continuous transition between all elements of the apparatus. They are supposed to have evolved directly from the Cambrian Westergaardodinidae. The second group comprises the families Fryxellodontidae, Panderodontidae, Protopanderodontidae, and Ulrichodinidae, which have apparatuses of coniform elements but do not have a geniculate ne element; the Distacodontidae, with a geniculate ne element but a tr element without lateral processes; and possibly the Cordylodontidae and Multioistodontidae, whose relationships are inadequately known. the third and fourth groups include conodonts with rather highly differentiated apparatuses, originally with nondenticulated elements, ne being geniculate and tr with lateral processes. They differ in the number of processes developed in particular elements. The third group, consisting of the Prioniodontidae, Phragmodontidae, Balognathidae, Icriodontidae, and perhaps the Distomodontidae, has apparatuses with triramous sp, oz, and tr elements and a tetraramous (rarely triramous) pl element. The fourth one, including Oistodontidae, Periodontidae, and most of the post-Middle Ordovician conodonts, originally had biramous sp and oz elements and triramous tr and pl elements. Dzik (1976) proposed to include the first group in the suborder Westergaardodinina Lindström 1970, the second and third in the Prioniodontina Dzik 1976, and the fourth in the Ozarkodinina Dzik 1976.

Chirognathidae Branson & Mehl 1944

Chosonodina Müller 1964, which is widespread but never numerous in the North American Midcontinent (Mound 1965, 1968; Ethington & Clark 1981; Harris *et al.* 1979), seems to have evolved during deposition of the Joins and Oil Creek formations of Oklahoma from its original *Westergaardodina*-like shape toward a morphology similar to that of *Chirognathus* Branson & Mehl 1933. Typical *Chirognathus* occurs much higher, above the *B. gerdae* Zone (Webers 1966) and may belong to the same lineage, which seems to be confined to the Midcontinent. *Bergstroemognathus* Serpagli 1974 and *Appalachignathus* Bergström, Carnes, Ethington, Votaw & Wigley 1974 may be related to this group. They occur in the American part of the North Atlantic province and the Midcontinent, but, although having almost worldwide distribution (Serpagli 1974; Bergström & Carnes 1976; Cooper 1981), have not been recorded from the Baltic area. Juvenile specimens of Whiterockian *Leptochirognathus* Branson & Mehl 1943, different from older *Chosonodina* only in very robust denticulation, have a base with an acute outer side, which may be a remnant of the small lateral process of *Tripodus laevis* Bradshaw 1969 from the West Spring Creek Formation of Oklahoma. A similar process occurs in *Jumudontus gananda* Cooper 1981 with an almost worldwide occurrence (except the Baltic region), and in unnamed species of the same genus from the Joins Formation of Oklahoma (see McHargue 1975). All these conodonts are supposed to be indicative of warm, shallow-water environments (Barnes *et al.* 1973; Bergström & Carnes 1976).

Multioistodontidae Bergström 1981

I provisionally assemble here, following Bergström (1981), genera with apparatuses composed of hyaline, gently curved ramiform elements that bear some similarity to those of the Tremadocian *Cordylodus* Pander 1856. Among them, *Multioistodus* Cullison 1938 s.s. (restricted to species that lack a geniculate ne element and with rather little distinction between elements of the apparatus) and *Erismodus* Branson & Mehl 1933 (together with related forms) are almost exclusively Midcontinent lineages (Sweet 1982). *Erraticodon* Dzik 1978, which appeared briefly two times in the Baltic area (Dzik 1978), is more common on the American side of the North

Atlantic province (Sweet & Bergström 1962; Stouge in press), and in the Midcontinent (Harris *et al.* 1979; Ethington & Clark 1981) but it is a lineage that is rather typical of the Australian province (Bergström 1971; Cooper 1981). *Spinodus* Dzik 1976 may have originated from *Erraticodon* through some transitional forms that occur in the Midcontinent (see Bradshaw 1969, Pl. 137:7–11) and North Atlantic provinces (Landing 1976; Stouge in press). It occurs subordinately in the North Atlantic part of the Appalachians (Sweet & Bergström 1962; Bergström & Carnes 1976), central Newfoundland (Fåhraeus & Hunter 1981), Scotland and Wales (Bergström 1971), the Holy Cross Mountains, Poland, and the Baltic area (Dzik 1976).

Fryxellodontidae Miller 1981 (Fig. 2)

Discovery of two more specimens of *Nericodus* Lindström 1955 in the *P. proteus* Zone of Ottenby, Öland (Fig. 2 B, C), which appear to be morphologically similar to *Polonodus* Dzik 1976 (Fig. 2 A, D), suggests to me that these genera, together with *Fryxellodontus* Miller 1969, which shares a widely conical shape and lacks a well-defined cusp, represent a group of closely related lineages. Even if the Öland specimens represent a species intermediate between Tremadocian *Nericodus* and Late Arenigian *Polonodus*, this does not necessarily mean that *Polonodus* originated in the Baltic area. This genus, before its brief appearance in the Kundan of the Holy Cross Mountains and the Baltic area (Dzik 1976), seems to have been rather widespread in Texas, Nevada (Harris *et al.* 1979; Bergström 1978, 1979), and western Newfoundland where it is well represented in the Table Head Formation (Stouge in press). Although very variable, and exhibiting morphologic change during ontogeny, *Polonodus* does not seem to have had a differentiated apparatus, or to have been represented by many species.

Panderodontidae Lindström 1970 (Fig. 3)

This group (taken to include the Belodellidae Khodalevich & Tschernich 1973) is characterized by species whose skeletal elements have a very deep basal cavity and whose apparatuses have low morphologic diversification. Little is known about its early evolution. The most generalized morphology of elements is shown by *Scalpellodus* Dzik 1976, which is known to occur in the Baltic area from the end of the early Volkhovian (Löfgren 1978) but does not have any known Baltic ancestry. There is little morphologic difference between *S. latus* (van Wamel 1974), the oldest, and *S. cavus* (Webers 1966), the youngest species of this genus (Fig 3: 22, 27) and originally they were considered to belong to the same chronospecies (Dzik 1976). Löfgren (1978) has shown, however, that during the Early Ordovician some evolutionary transformation occurred in this lineage in the Baltic area, which was expressed mostly in the smoothing of the surface of elements and simplification of the apparatus. Elements of the *Scalpellodus* apparatus are difficult to distinguish from particular elements of Gen. n. B ('Ordovician *Belodella*') and the '*Paltodus*' *jemtlandicus* Löfgren 1978 group, therefore it is not easy to trace distribution of this genus on the basis of data in the literature. *Macerodus dianae* Fåhraeus & Nowlan 1978 from the Cow Head and St. George groups of Newfoundland (Stouge 1982), may represent the 'scandodontiform' element of *S. latus* or a related species (see

Löfgren 1978; Pl. 5:6). On the other hand elements of *Scalpellodus striatus* Ethington & Clark 1981 from the Fillmore Formation of Utah (Ethington & Clark 1981) may well belong to other genera with generalized morphology of elements of the apparatus. Certain *Scalpellodus* occurred in the Midcontinent from the time of the *E. suecicus* Zone (Harris *et al.* 1979), was also not uncommon in the Appalachians (Bergström & Carnes 1976) and lasted there even after its disappearance in the Baltic area. Possibly this lineage reappeared briefly in the Baltic area in the Ashgillian (Viira 1974).

Panderodus Ethington 1959, which differs from *Scalpellodus* in possessing a fissure on the lateral side of the elements, is known to occur in the *O. evae* Zone of Argentina (Serpagli 1974; Fig. 3:11 and possibly in Australia ('*Protopanderodus primitus* Druce' of Cooper 1981). It did not appear in the Baltic area before the middle Kundan (Dzik 1976) and appeared in the Midcontinent even later, the oldest ones there being known from the Crystal Peak Dolomite of Utah (Ethington & Clark 1981). It shows a scattered but wide distribution in both these provinces. Several other conodont lineages with elements bearing a lateral fissure may be derived from *Panderodus*, all of them being typical representantives of shallow-water communities. Among them *Dapsilodus* Cooper 1976 (Fig. 3:16), has very wide distribution, but *Scabbardella* Orchard 1980 (Fig. 3:19) seems to be confined mostly to Wales and the Holy Cross Mountains before the Ashgillian.

The most typical of the North American Midcontinent group of conodonts with coniform elements are robustly denticulated panderodontids, which were represented in the late Middle Ordovician by at least two lineages of the genera *Pseudobelodina* Sweet 1979 (Fig. 3:21), and *Belodina* Ethington 1959 (Fig. 3:17, 18), which in the Late Ordovician were supplemented by several lineages of *Culumbodina* Moskalenko 1973, *Plegagnathus* Ethington & Furnish 1959, and *Parabelodina* Sweet 1979 (Sweet 1979 b). The oldest known species of this group, *B. monitorensis* Ethington & Schumacher 1969, appears in the middle part of the McLish Formation of Oklahoma and initiated a lineage that became widespread in the Midcontinent (Sweet 1981 a; see also Nowlan 1979) but did not enter the Baltic area until the latest Caradocian (Viira 1974; Dzik 1976; S. M. Bergström, personal communication). Precise place of origin and ancestry of *Belodina* is unknown, although the ancestor was almost certainly some species of *Panderodus* (Sweet 1981a). The oldest known serrated *Panderodus*-like conodonts have been described from the Late Llanvirnian of the Baltic area (Dzik 1976) as *Belodella serrata* Dzik 1976 (Fig. 3:20). Its direct relationship to the Silurian *Belodella* is questioned (Ethington & Clark 1981) and it may rather be related to *Pseudobelodina*, which has more elaborated denticulation but still a low degree of diversification of elements in the apparatus (Sweet 1979 b). Because of the rarity of *B. serrata* in Baltic assemblages it can hardly represent a Baltic lineage. Roots of the *Belodina* group are to be looked for elsewhere.

Protopanderodontidae Lindström 1970 (Fig. 3)

This highly diversified group of conodonts (taken to include the Scolopodontidae Bergström 1981, Oneotodontidae Miller 1981, and Teridontidae Miller 1981) includes species having robust elements with a relatively shallow basal cavity and usually a distinct coniform tr element in the apparatus.

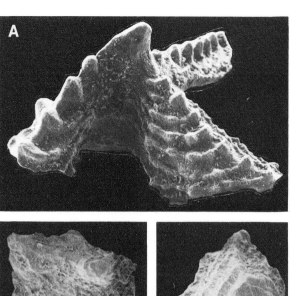

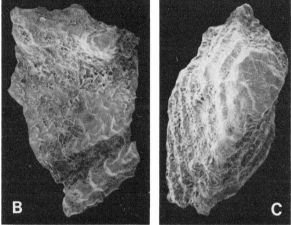

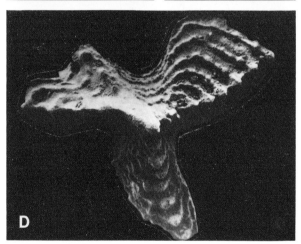

Fig. 2. ☐A, D. *Polonodus clivosus* (Viira 1974), erratic boulder of glauconitic limestone E-080, early *E. pseudoplanus* Zone as indicated by associated *Baltoniodus*, Mochty near Warsaw, Poland; specimen ZPAL CVI/217, oblique lateral and oral views to compare with *Nericodus*(?) sp.; ×100 and 95, respectively. ☐B, C. *Nericodus*(?) sp., Ottenby, southern Öland, sample Ot-7, 95 cm above the base of limestone sequence, *P. proteus*(?) Zone, Early Latorpian. ☐B. Specimen ZPAL CVI/381, oral view, ×48. ☐C. Specimen ZPAL CVI/382, ×88.

Although no species intermediate in age between the Tremadocian *Semiacontiodus nogamii* (Miller 1969) from the Notch Peak Formation of Utah and *S. cornuformis* (Sergeeva 1963) from the Baltic late Volkhovian and Kundan has been described (however, '*Scalpellodus latus*' of Cooper 1981 may belong here) there is no significant difference in morphology and composition of the apparatus that would substantiate

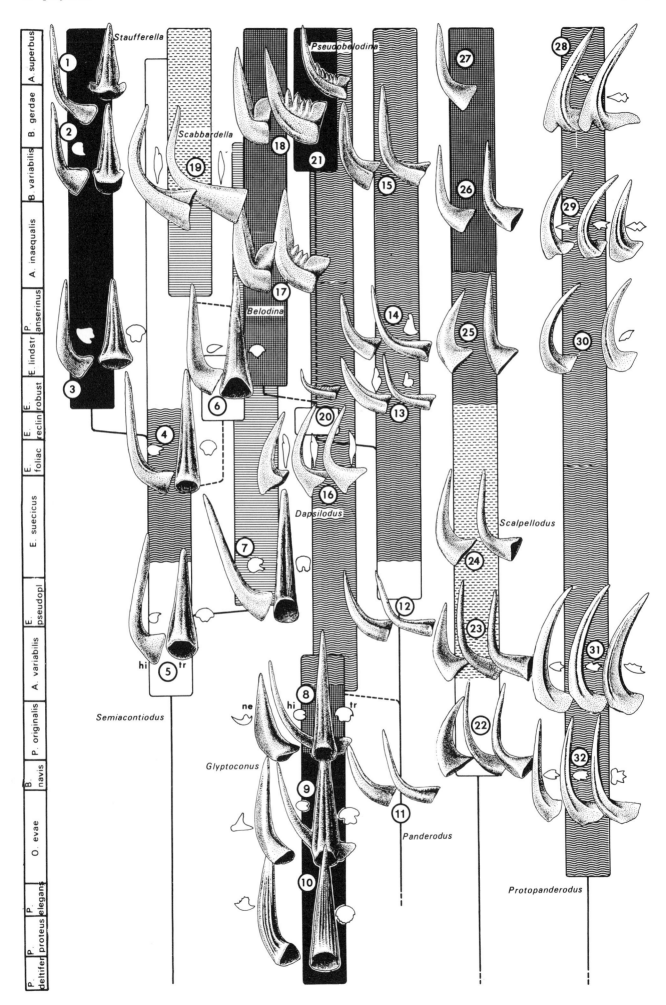

their generic distinction (Dzik 1976). *S. cornuformis* occurred in the Baltic area until the end of the Caradocian without significant change in morphology. A similar form occurs in the Antelope Valley Limestone of Nevada together with *Eoplacognathus suecicus*, but it is unclear whether it represents an influx of the Baltic lineages or is a continuation of a Tremadocian Midcontinent lineage of the genus. Certainly the Midcontinent population of *Semiacontiodus* evolved separately from the Baltic one later, in the time span between the deposition of the Tulip Creek and Mountain Lake formations of Oklahoma (Fig. 3:1–3), and developed a very characteristic incision of the lateral carinae close to the base (*Staufferella* Sweet, Thompson & Satterfield 1975). This feature seems to develop further (Webers 1966; Bergström & Sweet 1966; Sweet *et al.* 1975) and has some biostratigraphic potential. An opposite direction in evolution is represented by *S. longicostatus* (Drygant 1974) from the Holy Cross Mountains and Volhynia (Fig. 3:7), which lost lateral carinae on the tr element (Dzik 1976).

Another lineage of conodonts similar to *Semiacontiodus*, but having two types of symmetrical elements with *Panderodus*-like fissures, helps to link Midcontinent and Baltic conodont faunas (Fig. 3:8–10). Bergström (1979:303) has noted the common occurrence of asymmetrical, wide elements originally described as *Protopanderodus asymmetricus* Barnes & Poplawski 1973, with symmetrical prominently striated and costate elements in several localities, and suggested that they belong to the same apparatus. This species seems to have a very wide, but time–restricted occurrence, being known from the lower part of the Joins Formation of Oklahoma, the upper Fillmore to Juab formations of Utah (Ethington & Clark 1981), the Antelope Valley Limestone of Nevada (Harris *et al.* 1979), the Mystic Conglomerate of Quebec (Barnes &

Poplawski 1973), the Catoche, Port au Choix, and Table Head Formations of Newfoundland (Stouge 1982, and in press), the Hølonda Limestone of Norway (Bergström 1979), and the Kundan of Öland and Jämtland, Sweden (Löfgren 1978). Gracile striated elements of the same type occur throughout the Ibex section of Utah; however, below the range of *P. asymmetricus* they are associated with more elongate asymmetrical elements, which together may represent another part of the same assemblage for which the name *Glyptoconus(?) striatus* (Graves & Ellison 1941) is available (see Ethington & Clark 1981). Some hyaline symmetrical elements from the lowermost Fillmore Formation of Utah ('*Scolopodus*' *cornutiformis* Branson & Mehl 1933 and. *S. paracornutiformis* Ethington & Clark 1981), the Jefferson City Formation of Missouri (Kennedy 1980) and the West Spring Creek Formation of Oklahoma may represent roots of this lineage, which then appears to be very typical of the Midcontinent Early Ordovician faunas.

Until now the evolution of *Protopanderodus* Lindström, 1971 has been recognized in only very general terms (see Dzik 1976; Löfgren 1978; Bergström 1978; Harris *et al.* 1979; Kennedy *et al.* 1979) and connections between Midcontinent and Baltic populations cannot be traced exactly. It is not clear if Whiterockian *Protopanderodus* from Oklahoma (Fig. 3:31–32), is conspecific with contemporaneous Baltic populations. Some differences can be found between samples from the Oil Creek and McLish formations where a few *Protopanderodus* elements with a flat anterior side (Fig. 3:32) have been found.

Ulrichodinidae Bergström 1981 (Fig. 4)
For a long time *Ulrichodina* Furnish 1938 has been considered to have had a monoelemental apparatus (Sweet & Bergström 1972; Kennedy 1980; Ethington & Clark 1981), and to be an exclusively Midcontinent genus without known ancestry. In most of its known occurrences (never in great number) its symmetrical elements occur with asymmetrical elements of identical coloration and somewhat similar shape (Fig. 4:2). Such an association is also recorded in a sample from the top of the West Spring Creek Formation of Oklahoma, which includes asymmetrical elements that are similar to, or identical with, *Eucharodus parallelus* (Branson & Mehl 1933) (see Kennedy 1980). This is exactly the type of association typical of *Scandodus furnishi* Lindström 1955 (see Bergström 1981), occurring, among other localities, in the Latorpian of Ottenby cliff, Öland (Fig. 4:1). Below the range of typical *Ulrichodina*, in the House Formation of Utah, Ethington & Clark (1981) have found several elements which they identified as '*Scandodus*' sp. n. 5, that are very similar to *S. furnishi* and may represent a Midcontinent population of this species before its divergent evolution into *U. abnormalis* (Branson & Mehl 1933). This species has a tr element with a characteristic, but not unique, undulation of the base, which is known to occur also in *Paltodus* (see Szaniawski 1980). *Ulrichodina* (incl. *Scandodus*) did not develop a geniculate ne element, which separates it from *Paltodus*, and it may be a successor of *Utahconus* Miller 1980, which has a similar apparatus ·composition and element morphology (Miller 1980).

Distacodontidae Bassler 1925 (Fig. 4)
The oldest known representative of this group (taken to include the Drepanoistodontidae Bergström 1981), *Paltodus*

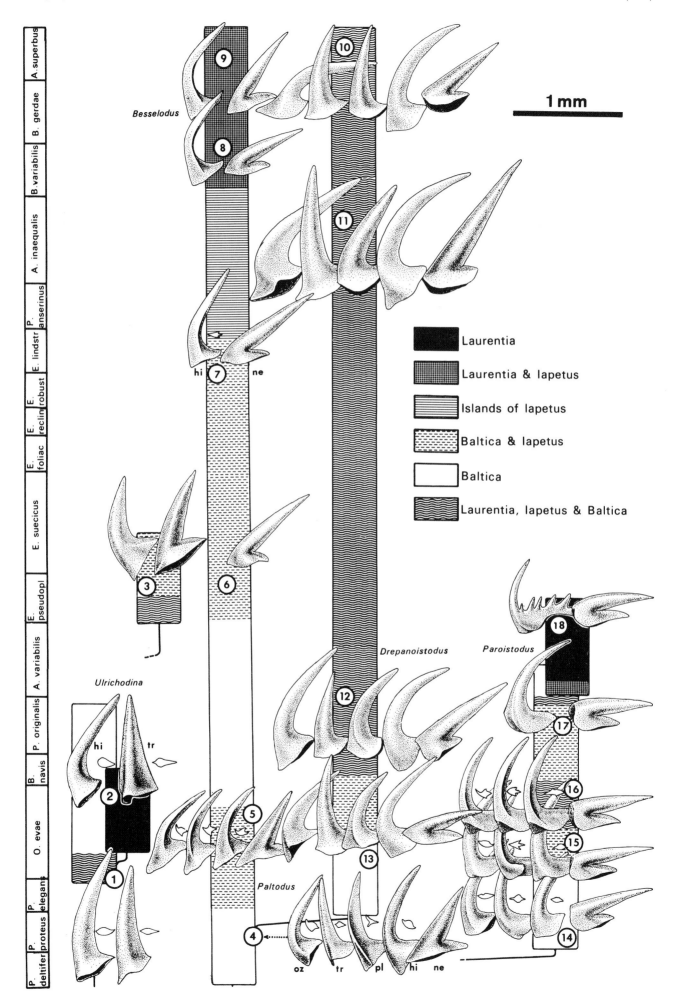

1 mm

Laurentia

Laurentia & Iapetus

Islands of Iapetus

Baltica & Iapetus

Baltica

Laurentia, Iapetus & Baltica

Besselodus

Ulrichodina

Paltodus

Drepanoistodus

Paroistodus

hi   tr

hi   ne

oz   tr   pl   hi   ne

deltifer Lindström 1955, from the *Ceratopyge* Limestone of Sweden and time-equivalent strata in Estonia and the Holy Cross Mountains (see Szaniawski 1980) still has ne elements of a very primitive shape, with indistinct geniculation, and tr elements of *Ulrichodina* shape (Fig. 4:4). Subsequent evolution of *Paltodus, Drepanoistodus,* and *Paroistodus* has been discussed by Lindström (1971) and Dzik (1976). With the possible exception of *Drepanoistodus* these lineages are confined in their phyletic evolution to the Baltic area, with several brief appearances in the Midcontinent (see Barnes & Poplawski 1973; Ethington & Clark 1981). It has been suggested by Stouge (in press) that element-species '*Cordylodus*' *horridus* Barnes & Poplawski 1973 was associated in the same apparatus with an ne element of *Paroistodus originalis* (Sergeeva 1963) shape. My study of material from Nevada described by Harris *et al.* (1979) supports this idea (Fig. 4:18). It seems that *P. horridus* is a Midcontinent offshoot of the Baltic *Paroistodus* lineage, which developed denticulation on the posterior process of non-ne elements in the same way as in *Protopanderodus insculptus* (Branson & Mehl 1933) (see Bergström 1978).

Lindström (1971) has distinguished two chronospecies of Baltic *Drepanoistodus,* which represent subsequent steps in relative elongation of the cusp of the ne element. Although such an evolutionary transformation obviously took place, it is obscured by high variability of this element within Baltic populations (van Wamel 1974; Dzik 1976; Löfgren 1978). There is no problem finding a distinction between the older of these chronospecies, *D. forceps* (Lindström 1955) (Fig. 4:13), and Midcontinent *D. angulensis* (Harris 1962) from the Joins and overlying formations of Oklahoma (Fig. 4:12) and contemporaneous strata of Utah (Ethington & Clark 1981). *D. angulensis* seems to be more advanced in morphology of the ne element than the second Baltic chronospecies, *D. basiovalis* (Sergeeva 1963), and there is a problem of distinction between *D. angulensis* and the younger *D. suberectus* (Branson & Mehl 1933). It remains also to be determined to what degree evolution of *Drepanoistodus* was independent in the Midcontinent and Baltic area.

Fig. 4. Proposed interrelationships between Baltic and Midcontinent lineages of Ulrichodinidae and Distacodontidae. (1) *Ulrichodina* (*Scandodus*) *furnishi* (Lindström 1955); after Bergström (1981). (2) *Ulrichodina abnormalis* (Branson & Mehl 1933), West Spring Creek Formation, Oklahoma. (3) '*Paltodus*' *jemtlandicus* Löfgren 1978; after Löfgren (1978). (4) *Paltodus deltifer* (Lindström 1955), Ceratopyge Beds of Öland. (5) *P. subaequalis* Pander 1856, Latorpian of Öland. (6) Gen. n. of Löfgren (1978). (7) *Besselodus semisymmetricus* (Hamar 1966); after Dzik (1976). (8) *B.* sp., Mountain Lake Formation, Oklahoma. (9) *B. variabilis* (Webers 1966); data from Webers (1966); see also Aldridge (1982). (10) *Drepanoistodus suberectus* (Branson & Mehl 1933); after Bergström & Sweet (1966) and Webers (1966). (11) Same species, McLish Formation, Oklahoma. (12) *D. angulensis* (Harris 1962), Oil Creek Formation, Oklahoma. (13) *D. forceps* (Lindström 1955); after Dzik (1976). (14) *Paroistodus numarcuatus* (Lindström 1955), Ceratopyge Beds of Öland. (15) *P. proteus* (Lindström 1955), Latorpian of Öland. (16) *P. parallelus* (Pander 1856), Latorpian of Öland. (17) *P. originalis* (Sergeeva 1963); after Dzik (1976). (18) *P.*(?) *horridus* (Barnes & Poplawski 1963), Antelope Valley Limestone, Nevada.

Oistodontidae Lindström 1970 (Fig. 5)

The oldest well-known species of this group (tentatively taken to include the Juanognathidae Bergström 1981) is '*Triangulodus*' *subtilis* van Wamel 1974, which appeared in the Baltic area in the *P. proteus* Zone (van Wamel 1974; Fig. 5:13). Having very primitive morphology of elements, the apparatus of this species differs from similar apparatuses of the Distacodontidae in triramous tr and pl elements and from *Eoneoprioniodus* (= *Triangulodus*) in biramous oz and triramous pl, as well as in albid, instead of hyaline, elements. It may be considered an ancestor of several lineages of *Protoprioniodus* McTavish 1973, which developed mostly in Australia (McTavish 1973; Cooper 1981) but occasionally invaded the Midcontinent (Ethington & Clark 1981) and the Baltic area (van Wamel 1974; Cooper 1981). To this genus may be assigned a lineage typical of the Midcontinent represented by '*Gothodus*' *marathonensis* Bradshaw 1969 (Fig. 5:11), which may be a direct successor of Australian '*Microzarkodina*' *adentata* McTavish 1973. It has a highly diversified apparatus, with denticulated ramiform elements, and was widely distributed from the District of Mackenzie (Tipnis *et al.* 1979), Utah (Ethington & Clark 1981), through Texas (Bradshaw 1969), Oklahoma (Mound 1965; McHargue 1975), to Scotland (Higgins 1967).

Histiodella Harris 1962 developed from a *Juanognathus*-like ancestor (Ethington & Clark 1981) that probably differed from '*T*'. *subtilis* only in the reduced posterior processes of tr and pl elements. The *Histiodella* lineage (Fig. 5:1–5), was typical of marginal areas of Laurentia, and is known to have occurred during the Early Whiterockian in the American part of the North Atlantic province (including the Hølonda Limestone of Norway; Bergström 1979) and in the Midcontinent (Barnes & Poplawski 1973; Landing 1976; Harris *et al.* 1979; Ethington & Clark 1981). Its advanced species, *H. holodentata* Ethington & Clark 1981, briefly invaded the Baltic area and the Holy Cross Mountains in the Kundan (Viira 1974; Dzik 1976, 1978). Evolution of denticulation in the *Histiodella* lineage provides an excellent tool for time correlation of the Midcontinent Early Whiterockian (McHargue 1982; Stouge in press).

The Midcontinent and Baltic lineages of *Oistodus* Pander 1856 differ from each other in the width of lateral processes of tr elements and they may have had an independent origin from *Protoprioniodus* (Fig. 5:14, 15).

Periodontidae Lindström 1970 (Fig. 5)

The group is tentatively taken to include the Cyrtoniodontidae Hass 1959 (see confusion with *Cyrtoniodus* in Bergström 1981). The oldest, still unnamed species of *Periodon* appears in the *P. elegans* Zone or slightly earlier (van Wamel 1974). It differs from contemporaneous Oistodontidae in the ramiform appearance of all elements except ne ones (Fig. 5:16). Even in the *O. evae* Zone some sp elements can be found with undenticulated blades. Tr and pl elements have well-developed lateral processes (Serpagli 1974; Landing 1976; Dzik 1976), which puts the species close to *Microzarkodina flabellum* (Lindström 1955), its supposed derivative (see Löfgren 1978). *Microzarkodina* has reduced posterior processes in these types of elements and in the Kundan developed denticulation on the anterior process of the sp element, being

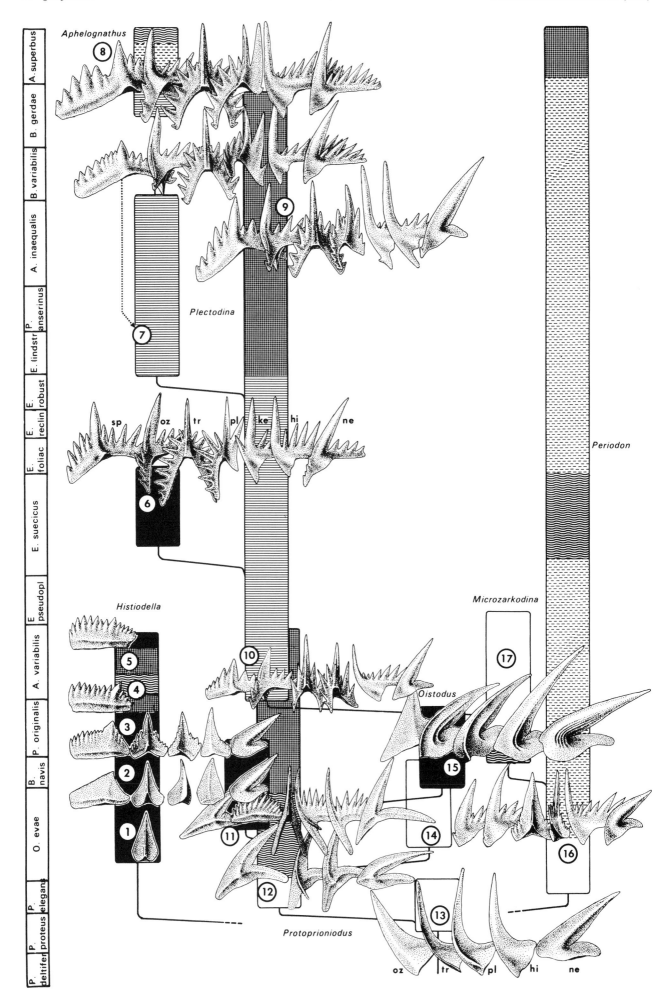

in these respects similar to the oldest species of *Plectodina* (see Sweet 1981a:246), *'Phragmodus' polonicus* Dzik 1978, which, however, has an oz element with an anterolateral process (Dzik 1978; Stouge in press; Fig. 5:10). Evolution of all these lineages seems to be concentrated in the North Atlantic province, but they differ in their distribution, and *Microzarkodina* very rarely invaded the Midcontinent (Ethington & Clark 1981) while *Plectodina* is unknown in the Baltic area.

The evolution of *Plectodina* has great importance for the later history of conodont faunas but it is not possible to reconstruct it now in detail. Presumably several independent lineages evolved in the seas of the Welsh Massif, the Appalachians, and the western margin of the Midcontinent. These are distinguishable mostly in the development of denticulation on the ne element. Welsh *P. flexa* (Rhodes 1952) (Fig. 5:7) may be an ancestor of younger *Aphelognathus rhodesi* (Lindström 1959) from the Crug Limestone (Orchard 1980; Sweet 1981a) which differs from it only in the complete reduction of the anterior ('lateral') process of the oz element (Fig. 5:8). This species was the most characteristic among non-Baltic forms that appeared in the Oanduan of the Baltic area (Fig. 10A–G; Viira 1974) and destroyed the climactic Baltic conodont community evolving there since the Arenigian (Fig. 11). The *Aphelognathus* lineages appeared in the Midcontinent somewhat later, the oldest species being *A. kimmswickensis* Sweet, Thompson & Satterfield 1975 of Kirkfieldian age (Sweet 1981a), which may even be an independent derivative of *P. aculeata* (Stauffer 1935) (Sweet 1981a).

Prioniodontidae Bassler 1925 (Fig. 6)

The group is taken to include the Oepikodontidae Bergström 1981 and Pygodontidae Bergström 1981. Still inadequately known species of *Acodus* Pander 1856 appearing in the Tremadocian of the Baltic area and the Holy Cross Mountains are supposed to be derivatives of *Paltodus*, and are the oldest representantives of this group. They differ from *Paltodus* in having much more prominent lateral ribs or processes of the oz, tr, and pl elements (Lindström 1971; Szaniawski 1980). Among Baltic *Acodus* s.l. are almost certainly the ancestors of both the Oistodontidae and the Prioniodontidae. Origin of the latter is marked by development of a fourth rib on the pl element. *Tripodus distortus* (Branson & Mehl 1933) (= *Diaphorodus delicatus*; see Lindström 1977; Kennedy 1980;

Ethington & Clark 1981), which is closely similar to Baltic *Acodus* and possibly conspecific with one of its species, is widespread in the Midcontinent, among other localities in the upper West Spring Creek Formation of Oklahoma (Fig. 8:22). Diversification of apparatuses with ramiform elements probably took place in Australia during the deposition of the Emanuel Group (McTavish 1973) but still little is known about the origin of the most characteristic and widespread species of this family: *Prioniodus elegans* Pander 1856; *Oepikodus evae* (Lindström 1955), and *Prioniodus(?) communis* (Ethington & Clark 1964). In the case of the first two species this significantly undermines concepts of zones based on their occurrences.

Prioniodus(?) communis (Fig. 6:15), which may be conspecific with *'Gothodus' microdentatus* van Wamel 1974, that appears in the Baltic area before *P. elegans* (see van Wamel 1974), and evolved in the Midcontinent into a form with rather robust lateral process of the oz (incl. sp) element (Fig. 6:16). I am inclined to identify this species, which occurs in the basal part of the Joins Formation of Oklahoma, with the Argentinian *P.(?) intermedius* Serpagli 1974 rather than with Australian *P.(?) minutus* McTavish 1973, as has been done by Ethington & Clark (1981). I would consider identity of *P.(?) intermedius* with *P.(?) communis* unlikely. *O. evae* deeply differs from *P.(?) communis* in having tetraramous rather than triramous tr elements (Fig. 6:1; Bergström & Cooper 1973).

In the Oklahoma section supposed *P.(?) intermedius* is followed, with some overlap, by a prioniodontid species with a *Scalpellodus*-shaped oz(sp) element in its apparatus, and probably conspecific or closely related to *'Belodella' robusta* Ethington & Clark (Fig. 6:17). This lineage occurs without significant changes throughout almost the entire Whiterockian (Bergström 1978; Harris *et al.* 1979; Stouge in press) with very few short-lived occurrences in the Baltic area (Löfgren 1978). The Oklahoma material is too meager to test suspicion of possible *P.(?) intermedius* – *'B'. robusta* relationships but because of the general plan of the apparatus, a prioniodontid relationship of the discussed lineage seems indisputable. It has nothing to do with Siluro-Devonian *Belodella* Ethington 1959, which is rather an offshoot of serrated panderodontids, and there is urgent need for a generic name for the Ordovician species.

An unnamed prioniodontid that occurs in the Fillmore Formation of Utah ('?*Ruetterodus* sp.' of Ethington & Clark 1981) and in the Joins Formation of Oklahoma ('*Haddingodus*' of Mound 1965; Fig. 6:7 herein) may help in understanding the apparatus of the bizarre North Atlantic *Pygodus* Lamont & Lindström 1957. Its apparatus has a prioniodontid plan, but differs in the rather peculiar denticulation and reduction of the posterior process of sp element while the anterior and lateral processes are connected by the expanded base (Fig. 6:7). Composition of this apparatus is reconstructed very provisionally and more detailed studies may change many details; however, it seems that typical *Pygodus* apparatuses can be derived from it through intermediate forms described by Landing (1976) as *Fryxellodontus? ruedemanni* Landing 1976 (incl. *Stolodus* sp. cf. *S. stola*) from the *O. evae* Zone of New York and by Löfgren (1978) from the Kundan of Sweden. The evolution of *Pygodus* has been used as ι basis for zonation of the Llandeilian of the North Atlantiι and Baltic provinces

Fig. 5. Proposed interrelationships among Baltic and Midcontinent lineages of Oistodontidae and Periodontidae. (1) *Histiodella* sp. n.; after Ethington & Clark (1981). (2) *H. altifrons* Harris 1962, Joins Formation, Oklahoma. (3) *H. sinuosa* (Graves & Ellison 1941), Oil Creek Formation, Oklahoma. (4) *H. holodentata* Ethington & Clark 1981; after Dzik (1978). (5) *H.* sp. n.; after Harris *et al.* (1979). (6) *Plectodina* cf. *joachimensis* Andrews 1967, Dutchtown Formation, Missouri. (7) *P. flexa* (Rhodes 1952), Llandeilo Limestone, Wales. (8) *Aphelognathus(?) rhodesi* (Lindström 1959), Oanduan of Baltic area. (9) *Plectodina* sp. n., 'Bromide' Formation, Oklahoma. (10) *P. polonica* (Dzik 1978); data from Dzik (1978) and Stouge (1980). (11) *Protoprioniodus(?) marathonensis* (Bradshaw 1969), Joins Formation, Oklahoma. (12) *P. elongatus* (Lindström 1955); after van Wamel (1974). (13) *P.(?) subtilis* (van Wamel 1974); after van Wamel (1974). (14) *Oistodus lanceolatus* Pander 1856 (not illustrated). (15) *O.(?) multicorrugatus* Harris 1962, Joins Formation, Oklahoma. (16) *Periodon* sp. n., uppermost Latorpian, Öland. (17) *Microzarkodina ozarkodella* Lindström 1971 (not illustrated).

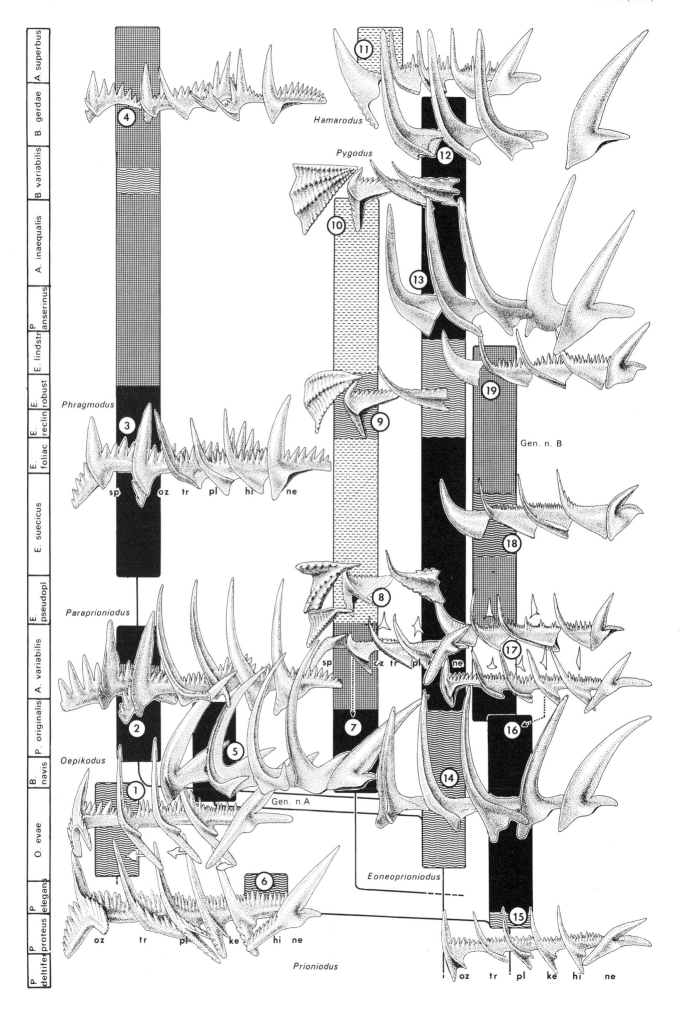

(Bergström 1971); therefore the mode of its evolutionary transformations is of particular interest to geochronology. Fåhraeus & Hunter (1981) have opposed the earlier assumption that *P. serra* (Hadding 1913) and *P. anserinus* Lamont & Lindström 1957, were parts of the same evolutionary lineage, and stated that *P. anserinus* evolved allopatrically from *P. serra* with significant overlap in their time distribution. Having *a priori* nothing against such an interpretation I must note that not enough evidence for it has been presented by Fåhraeus & Hunter (1981). Their data show an increase in the contribution of sp elements with four rows and a decrease of three-row ones. Although each type of sp elements is typical of the particular species, their common occurrence in samples intermediate in age between occurrences of typical populations of both species, does not necessarily mean that two genetically isolated populations are represented by each of these samples. Biometrical evidence for a morphologic gap between *P. serra* and *P. anserinus* morphotypes occurring in the same samples is necessary to show this. Also, it must be kept in mind that the fourth row on the sp element of *Pygodus* appears in ontogeny with some delay in respect to the other rows. A simple change in the population dynamics of intermediate populations may involve changes in the numerical contribution of particular morphotypes to the samples.

Hamarodus Viira 1974, which appears in the Baltic area (Bergström 1971; Viira 1974) and the Holy Cross Mountains (Dzik 1976, 1978) in the upper part of the *A. superbus* Zone (Fig. 6:11), also present in Wales (Orchard 1980) and the Carnic Alps (Serpagli 1967), is not known from the Midcontinent. Its apparatus, which has sp and oz elements with a very deep basal cavity, as well as tr and pl elements with reduced but still denticulated lateral processes, suggests a relationship to Gen. n. B ('Ordovician *Belodella*') rather than to *Periodon* (see Bergström 1981), which differs in the pattern of denticulation. Both interpretations would lead, to the same conclusion regarding its supposed North Atlantic origin.

Fig. 6. Proposed interrelationships among selected Baltic and Midcontinent lineages of Phragmodontidae and Prioniodontidae. (1) *Oepikodus evae* (Lindström 1955), Latorpian of Öland. (2) *Paraprioniodus costatus* (Mound 1965), Joins Formation, Oklahoma. (3) *Phragmodus flexuosus* Moskalenko 1973, Antelope Valley Limestone, Nevada. (4) *P. inflexus* Stauffer 1935; after Sweet (1981 a). (5) '*Multioistodus' compressus* Harris & Harris 1962?, Oil Creek Formation, Oklahoma. (6) *Prioniodus elegans* Pander 1856; after van Wamel (1974) and Bergström (1981). (7) *Pygodus*(?) aff. *ruedemanni* (Landing 1976), Joins Formation, Oklahoma. (8) *Pygodus* sp. n.; after Löfgren (1978). (9) *P. serra* (Hadding 1913); after Dzik (1976) and Löfgren (1978). (10) *P. anserinus* Lamont & Lindström 1957; after Bergström (1971) and Sweet & Bergström (1962). (11) *Hamarodus europaeus* (Serpagli 1967); after Dzik (1976) and Orchard (1980). (12) *Eoneoprioniodus alatus* (Dzik 1976), Mountain Lake Formation, Oklahoma. (13) Same species, McLish Formation, Oklahoma. (14) *E. cryptodens* Mound 1965, Joins Formation, Oklahoma. (15) '*Gothodus' microdentatus* van Wamel 1974 and *Prioniodus*(?) *communis* (Ethington & Clark 1964); after van Wamel (1974) and Ethington & Clark (1981). (16) *P.*(?) *intermedius* Serpagli 1974?, Joins Formation, Oklahoma. (17) Aff. '*Belodella' robusta* Ethington & Clark 1981, Oil Creek Formation, Oklahoma. (18) '*Belodella' jemtlandica* Löfgren 1978; after Löfgren (1978). (19) '*B.' nevadensis* (Ethington & Schumacher 1969), McLish Formation, Oklahoma.

Phragmodontidae Bergström 1981 (Fig. 6)

The oldest species of *Eoneoprioniodus* Mound 1965, which occurs in the *O. evae* Zone of Argentina (Serpagli 1974) and Australia (Cooper 1981), but invaded the Midcontinent somewhat later, during deposition of the Kanosh shale of Utah and the Joins Formation of Oklahoma (Mound 1965; Ethington & Clark 1981), and appearing even later briefly in the Baltic area (Lindström 1971; van Wamel 1974; Dzik 1976; Löfgren 1978), differs from *Acodus* and *Tripodus* only in having hyaline and generally much larger elements. Baltic *E. brevibasis* (Sergeeva 1963) is a direct successor of primitive Australian *E. larapintinensis* (Crespin 1941) rather than a relative of contemporaneous, alate Midcontinent *E. cryptodens* Mound 1965 (Fig. 6:14). *E. alatus* (Dzik 1976) has a Midcontinent ancestry and is common in the McLish and the Mountain Lake formations of Oklahoma (Fig. 6:13).

Conodonts that are similar in the apparatus organization pattern, but different in having processes of all except ne elements expanded into sharp-pointed blades, are known from the most of the early Whiterockian Midcontinent localities and were described by Ethington & Clark (1981) as *Multioistodus compressus* Harris & Harris 1962. Although the specific name may be appropriate (this is hard to decide without revision of the types of *M. compressus*), this apparatus (Fig. 6:5) is completely different from the apparatus of the type species of *Multioistodus, M. subdentatus* Cullison 1938, from the Dutchtown Formation of Missouri, which also occurs in the Joins Formation of Oklahoma. *M. subdentatus* has a cusp with a rounded cross-section, has weakly diversified and very variable elements of the apparatus, and lacks a geniculate ne element. It may rather be related to *Cordylodus*. In the Joins Formation a third *Multioistodus*-like species occurs, which however, has non-hyaline elements and differs from both species discussed above also in having rather short, straight and flat cusps. It seems to have no direct relation to '*M*'. *compressus* because its tr and pl elements have reduced posterior processes. Possible relationships to Oistodontidae seem to be unlikely because of the triramous oz(sp) element (Fig. 8:1). These two *Multioistodus*-like lineages require new generic names. The common occurrence of so many and such morphologically similar forms, which are unknown elsewhere, remains to be explained.

Another strange Midcontinent lineage is represented by '*Scandodus' sinuosus* Mound 1965, which in Utah and Oklahoma constitutes the core of most pre-*Phragmodus* assemblages (Fig. 11). Elements of its apparatus are almost homeomorphic with primitive *Eoneoprioniodus* species but, contrary to the reconstruction by Ethington & Clark (1981), the apparatus lacks geniculate ne elements. An origin from *Eoneoprioniodus* by reduction of the ne element, or from *Scolopodus*, seems equally probable.

Paraprioniodus costatus (Mound 1965) (Fig. 6:2) may have evolved from *Eoneoprioniodus cryptodens* and may have given rise to early *Phragmodus* Branson & Mehl 1933 (Sweet & Dzik in preparation). Until the end of the Ordovician *Phragmodus* was the most abundant conodont in the eastern Midcontinent. Before the appearance of *P. undatus* Branson & Mehl 1933 in the Oanduan of the Baltic area, the only Baltic record of this genus is two specimens of an advanced *P. flexuosus* Moskalenko 1973 found by Bergström (1971) in the Kukrusean of Estonia.

Balognathidae Hass 1959 (Figs. 7, 8)

The group is taken to include the Polyplacognathidae Bergström 1981. *Baltoniodus* Lindström 1971 developed its apparatus independently of and later than *Prioniodus* (Lindström 1971; Dzik 1983), and is a derivative of the younger part of possibly the same lineage of *Acodus* (Fig. 8:21). Until the invasion of the American part of the North Atlantic province, and subsequently the Midcontinent, by *B. variabilis* (Bergström 1962) and its successors, the evolution of this genus was almost completely restricted to the Baltic and adjacent areas (Dzik 1976). Some very rare species of *Baltoniodus* with incompletely known apparatuses occurred before the appearance of *B. variabilis* in the Midcontinent and the North Atlantic provinces, however (Fig. 8:18).

The origin from *Baltoniodus*, and the early evolution of *Amorphognathus* Branson & Mehl 1933, is well marked by changes in morphology of the ne element and development of a platform on the sp element (Fig. 8:8, 9). In the Late Volkhovian of the Baltic area there was a species of *Amorphognathus* similar to *Baltoniodus navis* (Lindström 1955) in morphology of the ramiform elements and with an ne element that differed from the corresponding element of *Baltoniodus* only in a much larger angle between the cusp and posterior process – *A. falodiformis* (Sergeeva 1963) (Lindström 1977; Fig. 7A–E herein). In some populations of *B. navis* with robust elements, the sp element has in some cases a lateral ridge on the posterior process (Dzik 1976, Fig. 22 a). Fragments of similar appearance with a somewhat better developed platform have been identified in a sample 30 cm below the top of the *Asaphus lepidurus* Zone (Volkhovian) at Hälludden, Öland (Fig. 7A) associated with typical ne elements of *A. falodiformis* (Fig. 7E). A sample from the overlying, about 10 cm thick bed contains sp elements with a somewhat more prominent posterolateral process and with an initial crista on the posterolateral side of the base of the ne element that is connected with a crista on the posterior process in its proximal part. This V-shaped arrangement of cristae appears to be well developed in a sample from the top layer of the Volkhovian at Hälludden. All available specimens of the sp element from several samples of the overlying *Asaphus expansus* Zone have a very weakly developed bifurcation of the anterolateral process, with its widened posterior part of the base lacking denticulation (Fig. 8:9). This is the typical *A. variabilis* Sergeeva 1963. Subsequent samples show, however, gradual development of the platform and posterolateral process, which is always shorter than the posterior one. Profound changes in the morphology of the anterolateral process occur between samples bordering the discontinuity surface between the *A. expansus* and *A. raniceps* Zones (Bohlin 1949; Jaanusson 1957). The basal sample from the *A. raniceps* Zone contains sp specimens with well-developed denticulation on both rami of the anterolateral process, which are of approximately equal length. Later evolution of *Amorphognathus* has been discussed by Dzik (1976, 1978). It may be worth adding that the transition from *B. navis* to *B. prevariabilis parvidentatus* (Sergeeva 1963) occurs in the uppermost Volkhovian and the transition from the latter chronosubspecies into *B. p. medius* (Dzik 1976) is in the middle of the *A. expansus* Zone, while the typical population of *B. p. medius* has been recognized 190 cm above the base of the *A. raniceps* Zone.

Elongation of the posterior ramus of the anterolateral process in the sp elements marks the origin and early development of *Eoplacognathus* Hamar 1966. The oldest species known, *E. zgierzensis* Dzik 1976, which has been found 260 cm below the top of the Kundan in the Gullhögen quarry, Västergötland (the best specimen being unfortunately lost during manipulations), and in the Ohesaare borehole, Estonia, has a posterior ramus of the process only about two times longer than anterior one, while those of *Amorphognathus* are of almost equal length and all younger species of *Eoplacognathus* have a posterior ramus several times longer than the anterior one in comparable ontogenetic stages of development. *E. zgierzensis* has an almost symmetrical pair of oz elements, which, as in *E. pseudoplanus* (Viira 1974), have all processes of similar length. Asymmetry is better visible in oz elements of *E. pseudoplanus* and even more distinct in *E. suecicus* Bergström 1971, which has longer anterolateral processes of these elements. Subsequent evolution of the Baltic *Eoplacognathus* lineage has been discussed by Bergström (1971). It seems that *Eoplacognathus* was confined in its phyletic evolution to the central part of the Baltic area while *Amorphognathus* evolved during the same time in areas close to the Holy Cross Mountains, Armorican, and Welsh massifs (Dzik 1978).

The lineage typical of the North American part of the North Atlantic province and the Midcontinent, represented by *Polyplacognathus friendsvillensis* Bergström 1971 and *P. sweeti* Bergström 1971, has been considered unrelated to Baltic *Eoplacognathus* and close to later *P. ramosus* Stauffer 1935 on the basis of the presence of marginal crenulation of the platform (Bergström 1971). I propose here another orientation of oz elements of *Polyplacognathus* than that proposed by Bergström (1971) with his posterior process being anterolateral (Fig. 8:3–5). When compared in this way with *E. suecicus*, which invaded the Midcontinent before the first appearance of *P. friendsvillensis*, not many differences can be found. Early populations of *P. friendsvillensis* from the lower part of the McLish Formation of Oklahoma contain juvenile oz elements that are very similar to those of *E. suecicus*. The typical feature of the American lineage, whose later evolution has been described by Bergström & Carnes (1976), seems to be a twisted row of denticles on the anterolateral process in its proximal part in the oz elements. At least two species of *Eoplacognathus* show this feature: *E. n. sp. A* of Bergström 1971 from the Furudal Limestone of Sweden and '*E. foliaceus–reclinatus* transition' of Harris *et al.* (1979) from the Antelope Valley Limestone of Nevada. There is no way to derive *P. ramosus* from *P. sweeti*. It can quite easily be derived from *E. elongatus* (Bergström 1971), however, which invaded the Midcontinent before the first appearance of *P. ramosus*. Some specimens of *E. elongatus* from the Mountain Lake Formation of Oklahoma have crenulation of the platform margin. It appears thus that *Polyplacognathus* in its current meaning consists of two lineages that are not directly related to each other. To avoid this obviously polyphyletic grouping I would suggest including *Eoplacognathus* in synonymy with *Polyplacognathus*, as no difference in apparatus composition or element construction has been indicated. *P. sweeti* has a very wide distribution and also invaded the Baltic area (Bergström 1971; Drygant 1974).

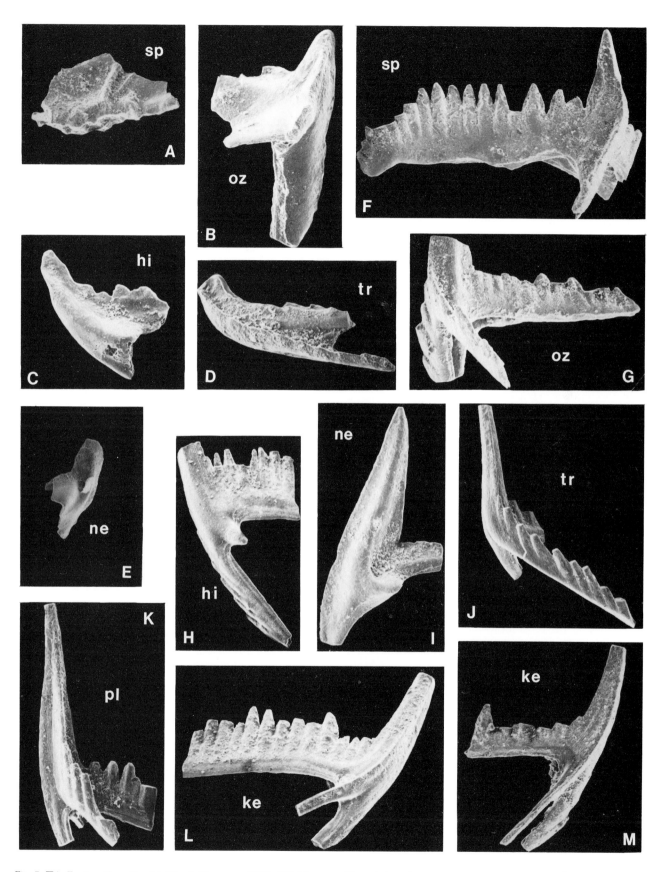

Fig. 7. □A–E. *Amorphognathus falodiformis* (Sergeeva 1963). □F–M. Associated with it *Baltoniodus navis* (Lindström 1955). Sample Ha 15, 30 cm below the top of Volkhovian, Hälludden, Öland. Specimens ZPAL CVI/383–396, respectively; all ×80.

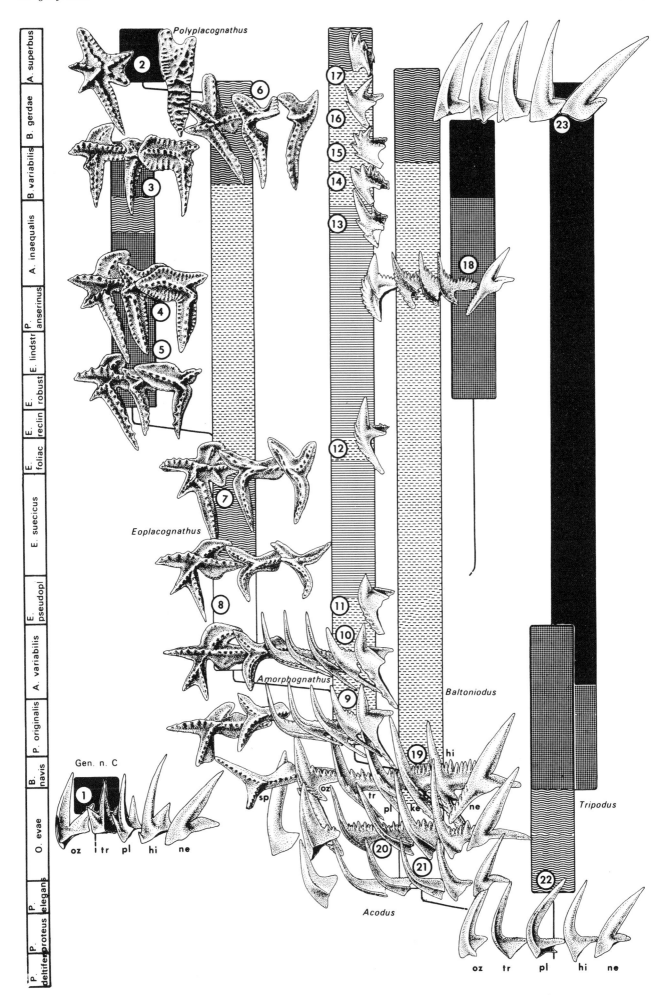

Complexodus Dzik 1976, which occurs in the Holy Cross Mountains in great numbers has an unknown origin, although it has a wide distribution and is also known from China (An 1982). Similarly unknown is the origin of *Rhodesognathus* Bergström & Sweet 1966, which is represented in the Midcontinent and Wales by species distinct from that of the Holy Cross Mountains, *R. polonicus* (Dzik 1976), which has a much better-developed platform. According to Sweet (1979 a) the apparatus of *Rhodesognathus* may include ramiform elements similar to those of *Amorphognathus*.

Icriodontidae Müller & Müller 1957

The oldest well-known species of this group, *Icriodella superba* Rhodes 1953, appears rather suddenly and without known direct ancestry in the Oanduan of the Baltic region (Viira 1974; Fig. 10H–J herein) and slightly earlier in the Midcontinent (Bergström & Sweet 1966; Webers 1966). Its apparatus composition suggests a direct relationship to the Balognathidae. The occurrence of its supposed ancestor in the Llanvirnian of the Armorican Massif (Lindström *et al.* 1974) as well as earlier appearance in Wales (Bergström 1971), suggest North Atlantic provenance of the lineage.

Ordovician paleobiogeography of the Baltic, Midcontinent, and adjacent areas

According to paleomagnetic data from the early Middle Ordovician of the Midcontinent area, the latter was part of the continental block of Laurentia, which was located close to the equator, while the latitude of the Baltic area at that time was approximately 60° S (Fig. 1; Bergström 1979). Few paleomagnetic data are available regarding the Early Paleozoic massifs in between, which were incorporated into larger continental blocks during the Caledonian and Hercynian orogenies. Probably, most of them originated as parts of island arcs bordering subduction zones of diverse ages. Their presumed position in the Ordovician may be reconstructed

with some degree of confidence on the basis of knowledge of their geological history (e.g., Bergström 1979; Dzik 1978, 1984; Bruton & Bockelie 1980). Epicontinental seas associated with each of these continental blocks certainly formed their own ecosystems, which were separated from other shelf-water ecosystems by extended areas of oceanic environment (Fig. 1). Therefore, they can be treated as distinct biogeographic units in terms of island biogeography (see MacArthur & Wilson 1967). Available quantitative data on the distribution of conodont assemblages in particular areas support such a view. It is well shown on logs of the contribution of the common conodonts that form the cores of particular assemblages, when high-rank taxonomic assignment of a particular conodont lineage is indicated (Fig. 11). Relative stability in the composition of conodont assemblages in time has been documented in many cases previously (Sweet 1979 a, b; Jeppsson 1979) and it is not especially surprising to find it also

---

Fig. 9 (p. 78). Oanduan conodonts with possibly Baltic or cosmopolitan provenance. All specimens from the erratic boulder E-305, Mochty near Warsaw, Poland. □A–G. *Amorphognathus superbus* (Rhodes 1952), specimens ZPAL CVI/361–367, respectively; A, B: ×67, C–E: ×90, F, G: ×117. □H.*Pseudooneotodus* sp., ZPAL CVI/368; ×180. □I, J, L. *Panderodus panderi* (Stauffer 1935), ZPAL CVI/369, 371; I, L: ×120, J: ×540. □K. *Panderodus* sp., ZPAL CVI/370, ×120.

Fig. 10 (p. 79). Oanduan conodonts with North Atlantic provenance. All specimens from the erratic boulder E-305, Mochty near Warsaw, Poland. □A–G. *Aphelognathus rhodesi* (Lindström 1959), ZPAL CVI/372–378, respectively; A: ×60, B, D, E, G: ×80, C: ×120, F: ×96. □H–J. *Icriodella superba* Rhodes 1952, ZPAL CVI/379, 380; H: ×80, I: ×240, J: ×60.

Fig. 11 (pp. 80–81). Logs of relative percent contribution of the most important lineages in the Early and Middle Ordovician of North America (left) and Europe (right). Family assignments of particular lineages indicated by patterns. Presented localities: □A. Woods Hollow Shale, Marathon, Texas, data from Bergström (1978). □B–H. Crystal Peak, Watson Ranch, Lehman, Kanosh, Juab, Wah Wah, and Fillmore Formations from selected localities in Ibex area, Utah (Ethington & Clark 1981). □I–L. Mountain Lake, Tulip Creek, McLish, and continuous section of Oil Creek and Joins Formations from Arbuckle Mountains, Oklahoma. □M. Jefferson City Formation, Missouri (Kennedy 1980). □N. Lenoir Limestone, Tennessee (Bergström & Carnes 1976). □O. Table Head Formation, Newfoundland (Stouge 1980). □P. Cow Head Group, Newfoundland (Fåhraeus & Nowlan 1978). □Q. Mójcza Limestone, Holy Cross Mountains, Poland (Dzik 1978). □R. Wysoczki chalcedonite, Holy Cross Mountains, Poland (Szaniawski 1980). □S–U. Gullhögen Formation (courtesy of Stig M. Bergström), Vikarby Limestone, and Kundan of Gullhögen quarry, Skövde, Sweden. □V. Uppermost Volkhovian and Lower Kundan of Hälludden, Öland. □W. Ceratopyge beds and Latorpian of Ottenby cliff, Öland. □X–Y. Gammalbodberget and Kalkberget sections of Jämtland, Sweden (Löfgren 1978). Conodont lineages: (1) *Erismodus*. (2) *Erraticodon*. (3) 'Scandodus' *sinuosus*. (4) *Eoneoprioniodus*. (5) Gen. n. A. ('*Multioistodus*'). (6) *Paraprioniodus*. (7) *Phragmodus*. (8) *Histiodella*. (9) *Periodon*. (10) *Microzarkodina*. (11) *Plectodina*. (12) *Prioniodus*. (13) *Oepikodus*. (14) *Protoprioniodus*. (15) *Pygodus*. (16) Gen. n. B ('*Belodella*'). (17) *Protopanderodus*. (18) *Semicontiodus*. (19) *Glyptoconus*. (20) *Belodina*. (21) *Panderodus*. (22) *Dapsilodus*. (23) *Scabbardella*. (24) *Paltodus*. (25) *Drepanoistodus*. (26) *Paroistodus*. (27) *Ulrichodina*. (28) *Acodus* and *Tripodus*. (29) *Baltoniodus*. (30) *Amorphognathus*. (31) *Eoplacognathus*. (32) *Polyplacognathus* sensu Bergström 1971. (33) *Complexodus*. (34) *Rhodesognathus*. (35) *Drepanodus*. (36) *Scalpellodus*. (37) *Cornuodus*. (38) *Multioistodus*. (39) *Hamarodus*.

Fig. 8. Proposed interrelationships among Baltic and Midcontinent lineages of Balognathidae and possibly related forms. (1) '*Acodus*' *auritus* Harris & Harris 1962?, Joins Formation, Oklahoma. (2) *Polyplacognathus ramosus* Stauffer 1935; after Bergström (1981). (3) *P. sweeti* Bergström 1971, early form, Mountain Lake Formation, Oklahoma. (4) *P. friendsvillensis* Bergström 1971, late form, Tulip Creek Formation, Oklahoma. (5) Same species, early form, McLish Formation, Oklahoma. (6) *Eoplacognathus elongatus* (Bergström 1962), Mountain Lake Formation, Oklahoma. (7) *E. suecicus* Bergström 1971; after Harris *et al.* (1979). (8) *E. zgierzensis* Dzik 1976, Ohesaare borehole, depth 509.78–510.35 m, Estonia. (9) *Amorphognathus falodiformis* (Sergeeva 1963), uppermost Volkhovian of Öland. (10) *A. variabilis* Sergeeva 1963, Lower Kundan of Öland. (11) Same species, Upper Kundan of Västergötland. (12) *A. kielcensis* Dzik 1976, Vikarby Limestone, Västergötland. (13) *A. inaequalis* Rhodes 1952?, Kaagvere borehole, depth 312.8 m, Estonia. (14–16) *A. tvaerensis* (Bergström 1962), Kaagvere borehole, depth 293.8–303.4 m, Estonia. (17) Same species, Bromide Formation, Oklahoma. (18) *Baltoniodus* sp. n., Mountain Lake Formation, Oklahoma. (19) *B. navis* (Lindström 1955); after Dzik (1976). (20) *B. crassulus* (Lindström 1955), uppermost Latorpian of Öland. (21) *Acodus deltatus* Lindström 1955, Latorpian of Öland. (22) *Tripodus distortus* (Branson & Mehl 1933), West Spring Creek Formation, Oklahoma. (23) *T.* sp. n., Mountain Lake Formation, Oklahoma.

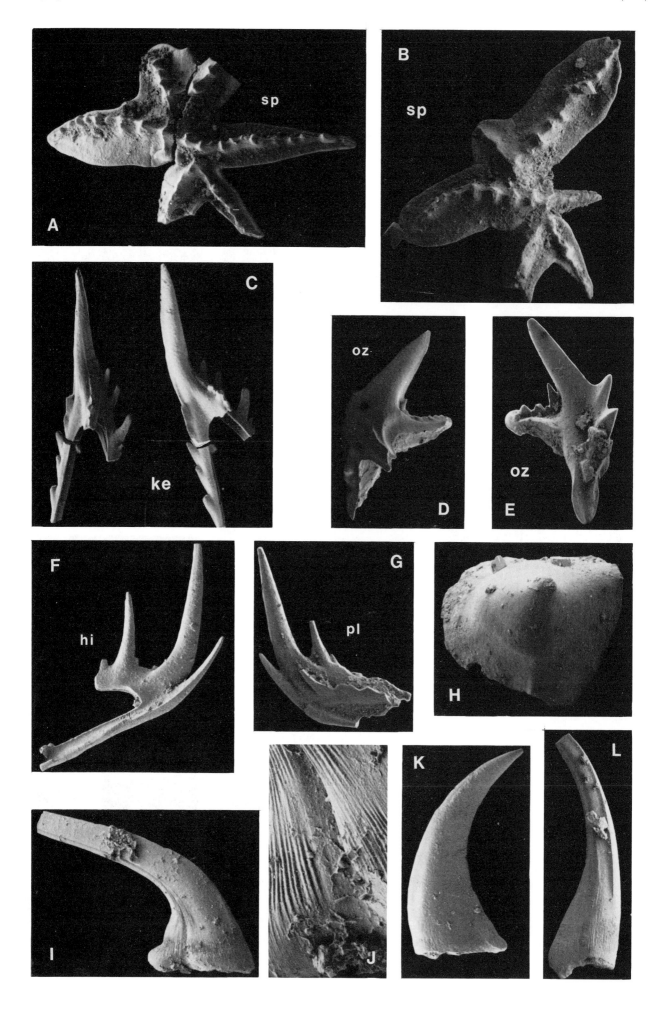

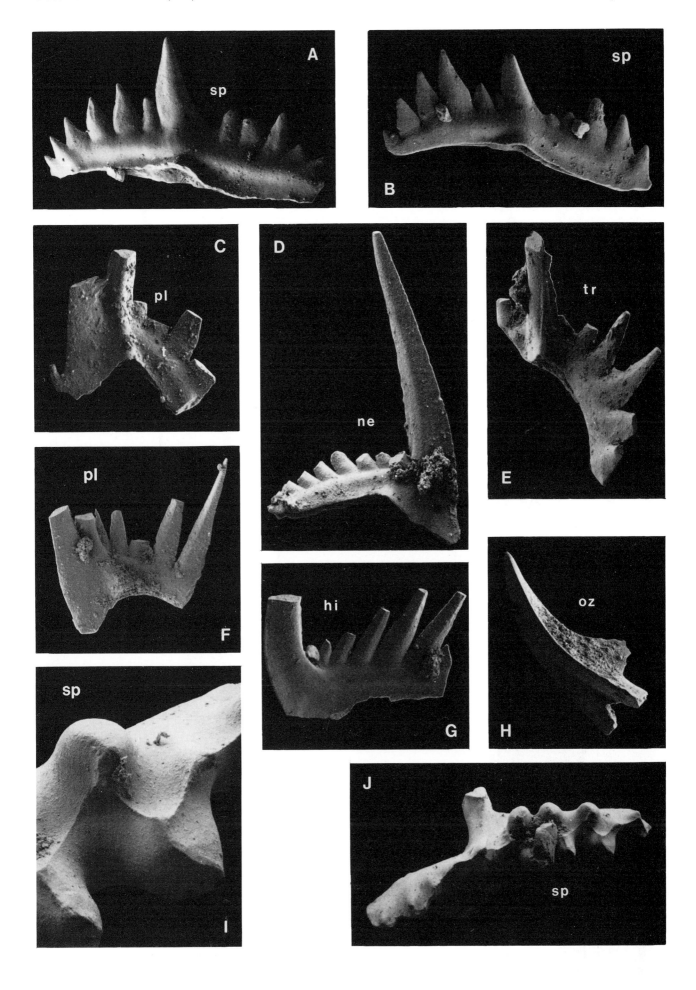

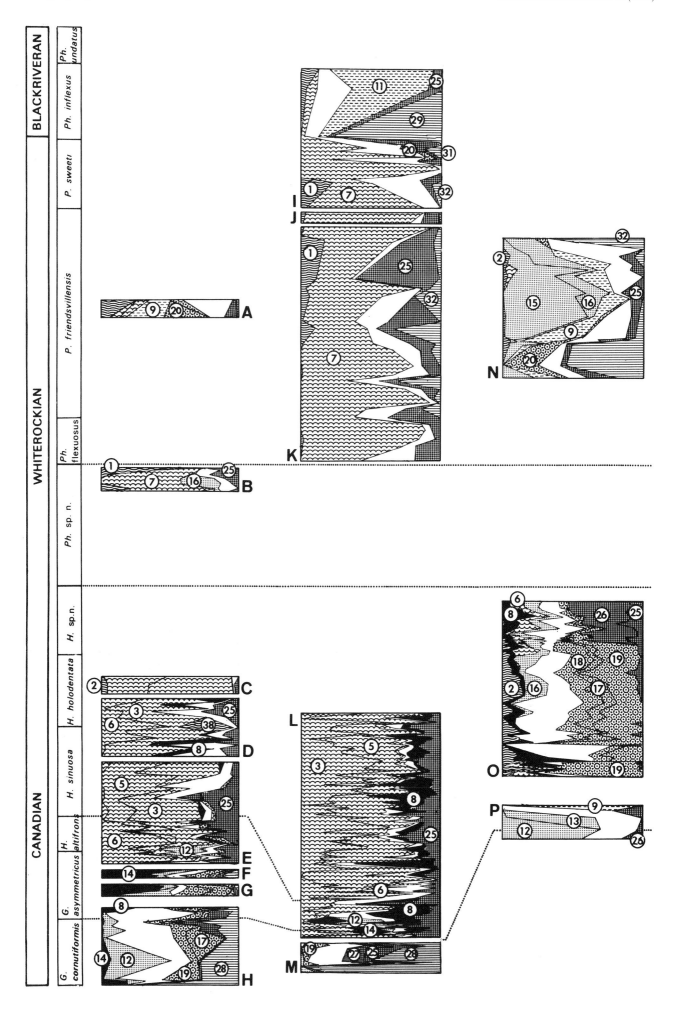

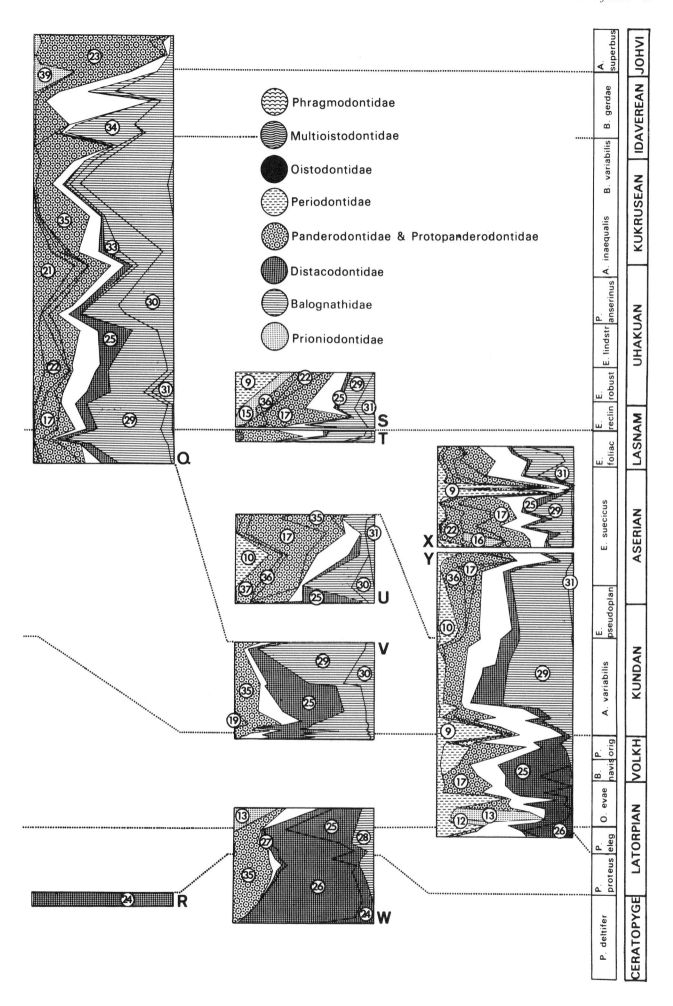

in the sections studied here. It appears, however, that during long time spans the cores of assemblages in particular areas continue to be formed by the same high-rank taxonomic units despite changes in contribution of particular evolutionary lineages belonging to these units, and that almost all areas have their own typical taxonomic units, which diversified mostly in only that area.

For instance, the evolution of the Balognathidae, with the exception of a single lineage, is confined exclusively to Baltica and adjacent islands in the Iapetus Ocean. Some lower-rank compartmentalization is visible in the case of the evolution of the *Amorphognathus* and *Eoplacognathus* lineages. The first evolved phyletically outside the central part of Baltica, with the best record on the island of the Małopolska Massif (Holy Cross Mountains; see Bergström 1971; Dzik 1976, 1978). The *P. friendsvillensis–sweeti* lineage, a supposed continuation of the *E. suecicus* lineage, underwent rapid phyletic evolution in the marginal seas of Laurentia and carbonate platforms possibly separated from them which were introduced later into the Appalachians (Bergström & Carnes 1976). This lineage did not appear in the Baltic area until the Uhakuan (Bergström 1971; Drygant 1974). The Phragmodontidae, a possibly Gondwana-born group, since its introduction occurred almost exclusively in the seas of Laurentia, where they underwent significant diversification. Some of their populations had areas of distribution that extended to islands in Iapetus and rarely expanded even to Baltica. The most successful branch of the Periodontidae had its evolution concentrated around the Iapetus islands. Each of its several lineages had somewhat different areas of distribution, some entering into the seas of Baltica (early *Periodon, Microzarkodina*), others tending toward Laurentia (*Plectodina–Aphelognathus* lineages). Similar patterns can be observed in the evolution of the Distacodontidae and the Panderodontidae, the latter developing to the originally(?) Midcontinent branch of *Belodina* and related genera. A few successfully developing lineages had extrinsic provenance, like *Protoprioniodus*(?) *marathonensis* of Laurentia with a probably Australian origin and *Complexodus pugionifer* of the Małopolska Island, known elsewhere only from China and Wales (S. M. Bergström, personal communication).

An independent evolution of conodont lineages confined to particular Middle Ordovician marine ecosystems involved also relative stability in the composition of particular ecologic groups of conodonts in each of the areas discussed. The pattern of vertical changes in the contribution of large groups of conodonts (Fig. 11), as well as other available data, suggest that different couples of high-rank taxa contributed to the total ecologic spectrum in each of the provinces. In Baltica (Fig. 11S–Y) this spectrum was represented mostly by shallow-water Panderodontidae and rather open-sea Balognathidae; in Laurentia (Fig. 11A–M) by Multioistodontidae and Phragmodontidae; in island areas of Iapetus (Fig. 11N–Q) by Protopanderodontidae and Periodontidae (with a significant contribution by *Amorphognathus*), and in equatorial Gondwana (Australia) probably by Multioistodontidae (*Erraticodon*) and Oistodontidae, respectively.

Among appearances of extrinsic lineages and interchanges of lineages developing in discussed areas two distinct classes can be recognized:

(1) Brief and numerically low contributions to the assemblage by appearances of species known in great numbers of specimens from other areas. Such are occurrences of *Histiodella holodentata* in the Małopolska Island and Baltica (Dzik 1978), *Glyptoconus*(?) *asymmetricus, Polonodus clivosus, Eoneoprioniodus alatus,* and *P. sweeti* in Baltica, all genuine lineages of Laurentia, as well as *Microzarkodina flabellum* and *Paroistodus parallelus,* both Baltic or Iapetus lineages, which appeared in the Midcontinent. Appearances of *Erraticodon* in discussed regions are of the same nature. No significant rebuilding of the assemblage is associated with appearances of this kind. They can be interpreted as effects of regional changes in the area of distribution of particular lineages observed close to their margins. Such populations, close to the margin of distribution of a species, must be more sensitive environmentally than highly productive populations in the center of the distribution area. Frequency-distribution data suggest that each conodont species had its own area of distribution with high productivity in the center (several species are known to contribute more than 90 % to samples from some areas) which decreases toward the margins. Rarely several species have the same center of distribution. No evident positive correlation in the distribution of different species has been observed, despite attempts to recognize persistent conodont associations (Bergström & Carnes 1976).

(2) Another class of 'migration' phenomena is represented by introductions of lineages that thereafter start to contribute significantly to the assemblages of some areas and to develop local phyletically evolving lineages, This is well exemplified by the introduction of *Eoneoprioniodus, Protoprioniodus* (?) *marathonensis, Paroistodus*(?) *horridus, Belodina, Eoplacognathus suecicus, E. elongatus,* and several other lineages into the Midcontinent. This is usually associated with the replacement of some previously occurring lineage or at least a significant rebuilding of the assemblage. Few events of this kind are observed in the Baltic area.

Generally Midcontinent and Baltic assemblages behaved in different ways. While Midcontinent faunas underwent rather gradual rebuilding during the Ordovician with several new lineages becoming permanent parts of communities (like Iapetus-born *Plectodina,* possibly Australian *Erismodus,* Baltic *Polyplacognathus* and *Amorphognathus*) the Baltic communities were much more conservative in their composition. Profound remodelling of them occurred in the Oanduan, when several lineages of the Iapetus provenance, represented by *Aphelognathus, Icriodella,* and *Phragmodus* (originally a Midcontinent form), were introduced. Although this assemblage has superficially a Midcontinent appearance, that is because of the approximately synchronous introduction of similar forms to the Midcontinent area, where they are not so distinct from local elements as they are in Baltica (see Bergström & Sweet 1966; Kennedy *et al.* 1979). Probably no direct Midcontinent to Baltic 'migration' took place at that time. Subsequent evolution of the Baltic assemblage with introduction of the numerically dominant conodonts *Hamarodus* and *Scabbardella* did not make it similar to that of Laurentia.

Sweet & Bergström (1974) suggested that rebuilding of the Baltic conodont fauna in *B. gerdae* Zone time was caused by a shift of the warm-water zone toward the pole accompanied by an expansion of the warm-water Midcontinent fauna. In terms of the continental drift concept this effect may be reached by moving the Baltic continental plate toward the equator. Because it is generally believed that warm-water animal communities are more diverse than cold-water ones one may attempt to test this idea by measuring changes in diversity in both areas during the Ordovician. Among several proposed, the measure of diversity based on Shannon's formula of information content (Berry *et al.* 1979) seems to be one of the simplest and has relatively easily understood ways of inference. The formula may be written in the following way:

$$D = - \sum_{s=1}^{n} C_s \log C_s.$$

Where D=diversity of the assemblage; C_s=relative contribution (decimal fraction) of particular species s to total sample. This is a measure of improbability of a particular composition of the assemblage. The largest value of D characterizes samples with large numbers of equally contributing species, the smallest one samples in which a single species dominates the assemblage.

Index of diversity is calculated here for Oklahoma, Utah (data from Ethington & Clark 1981), Jämtland (data from Löfgren 1978), the Holy Cross Mountains, Västergötland and Öland sections. Data are plotted against time scaled by appearances of zonally (or potentially zonally) diagnostic species (Fig. 12). Results are quite opposite to those expected (Dzik 1984). During the Early Ordovician the diversities of Baltic and Midcontinent assemblages do not differ significantly; both are relatively high. During the Middle Ordovician, plots for studied sections from both areas diverge somewhat, with higher (sic!) diversity in supposedly cold-water Baltic assemblages. In all cases, a significant decrease in diversity is usually connected with shallowing of basins. The difference between Baltic and Midcontinent assemblages may be partially caused by differences in bathymetric characters of the studied sections. Still remaining to be explained is the lack of any significant difference between other, bathymetrically more comparable parts of the sections, which may suggest that conodonts preserved in the sediment represent only a small fraction of the trophic group to which they belonged.

Acknowledgements. – Several Swedish sections of the Ordovician were sampled by me and some samples were processed during a two-month stay at the Department of Palaeobiology, University of

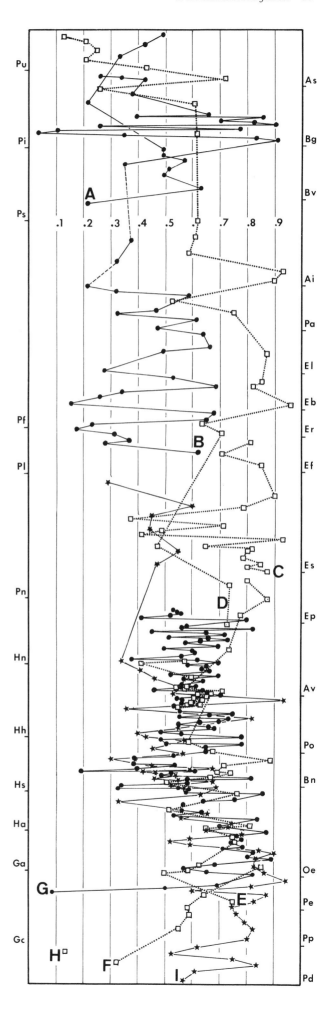

Fig. 12. Index of diversity plotted against time for sections representative of the Baltic and Midcontinent provinces. Appearances of zonal (or potentially) zonal species (same as on Fig. 11) indicated by initials left and right of the diagram. □A. Section of the 'Bromide' Formation near Fittstown, Oklahoma. □B. McLish to Mountain Lake Formations north of Ardmore, Oklahoma. □C. Gammalbodberget section, Jämtland (data from Löfgren 1978). □D. Mójcza Limestone, Holy Cross Mountains, Poland (Dzik 1978). □E. Kalkberget section, Jämtland (Löfgren 1978). □F. Ottenby cliff section, Öland. □G. Top of West Spring Creek to Oil Creek Formations north of Ardmore, Oklahoma. □H. Wysoczki chalcedonite, Holy Cross Mountains, Poland (Szaniawski 1980). □I. Top of House to Crystal Peak Formations from selected outcrops of the Ibex area, Utah (Ethington & Clark 1981).

Uppsala, in 1980. I greatly appreciate the efforts of Professor Anders Martinsson and Dr. Stefan Bengtson in organizing my visit to the Department and the assistance of Dr. Yngve Grahn (Swedish Geological Survey, Uppsala) during my field excursions.

I am especially grateful to Professors Walter C. Sweet and Stig M. Bergström of the Department of Geology and Mineralogy of The Ohio State University who made possible my postdoctoral fellowship at their institution, allowed me to examine their huge collections of Ordovician conodonts, provided many very helpful discussions and finally corrected the English language of this paper. The SEM photographs mounted on Figs. 2B, C and 7 were taken at the Department. Technical assistance of Mrs. Maureen L. Lorenz is gratefully acknowledged.

Baltic and Polish samples were processed at the Institute of Paleobiology of the Polish Academy of Sciences (Zakład Paleobiologii PAN, abbreviated ZPAL) and photographed specimens are housed there. Micrographs other than those enumerated above were taken at the Nencki's Institute of Experimental Biology in Warsaw. Drawings of conodont elements presented in the text are reconstructions based on camera lucida sketches of specimens housed either at the Department of Geology and Mineralogy of The Ohio State University in Columbus, Ohio, USA, or at the Institute of Paleobiology of the Polish Academy of Sciences in Warsaw, Poland.

References

Aldridge, R. J. 1982: A fused cluster of coniform conodont elements from the late Ordovician of Washington Land, western north Greenland. *Palaeontology 25*, 425–430.

An Tai-xiang 1981: Recent progress in Cambrian and Ordovician conodont biostratigraphy of China. *Geological Society of America Special Paper 187*, 209–224.

Baranowski, W. & Urbanek, Z. 1972: Ordovician conodonts from the epimetamorphic complex from Rzeszówek in the Kaczawa Mts. (Western Sudeten). *Bulletin de l'Academie Polonaise des Sciences, Series des Sciences de la Terre 20*, 211–216.

Barnes, C. R. 1977: Ordovician conodonts from the Ship Point and Bad Cache Rapids Formations, Melville Peninsula, southeastern District of Franklin. *Geological Survey of Canada, Bulletin 269*, 99–119.

Barnes, C. R. & Poplawski, M. L. S. 1973: Lower and Middle Ordovician conodonts from the Mystic Formation, Quebec, Canada. *Journal of Paleontology 47*, 780–790.

Barnes, C. R., Rexroad, C. B. & Miller, J. F. 1973: Lower Paleozoic conodont provincialism. *Geological Society of America, Special Paper 141 (1972)*, 157–190.

Barnes, C. R., Kennedy, D. J., McCracken, A. D., Nowlan, G. S. & Tarrant, G. A. 1979: The structure and evolution of conodont apparatuses. *Lethaia 12*, 125–151.

Bergström, S. M. 1964: Remarks on some Ordovician conodont faunas from Wales. *Acta Universitatis Lundensis, Sectio II, 3*, 1–66.

Bergström, S. M. 1971: Conodont biostratigraphy of the Middle and Upper Ordovician of Europe and eastern North America. *Geological Society of America, Memoir 127, (1970)*, 83–157.

Bergström, S. M. 1978: Middle and Upper Ordovician conodont and graptolite biostratigraphy of the Marathon, Texas graptolite zone reference standard. *Palaeontology 21*, 723–758.

Bergström, S. M. 1979: Whiterockian (Ordovician) conodonts from the Hølonda Limestone of the Trondheim Region, Norwegian Caledonides. *Norsk Geologisk Tidsskrift 59*, 295–307.

Bergström, S. M. 1981: Families Drepanoistodontidae and Periodontidae. *In* R. A. Robison (ed.): *Treatise on Invertebrate Paleontology, Part W, Miscellanea (Supplement 2) Conodonta*, W 128–129, W 143–145. Geological Society of America, Boulder, Colorado.

Bergström, S. M. & Carnes, J. B. 1976: Conodont biostratigraphy and paleoecology of the Holston Formation (Middle Ordovician) and associated strata in eastern Tennessee. *Geological Association of Canada, Special Paper 15*, 27–58.

Bergström, S. M. & Cooper, R. A. 1973: *Didymograptus bifidus* and the trans-Atlantic correlation of the Lower Ordovician. *Lethaia 6*, 313–340.

Bergström, S. M., Epstein, A. G. & Epstein, J. B. 1972: Early Ordovician North Atlantic province conodonts in eastern Pennsylvania. *U.S. Geological Survey Professional Paper 800-D*, 37–44.

Bergström, S. M. & Sweet, W. C. 1966: Conodonts from the Lexington Limestone (Middle Ordovician) of Kentucky and its lateral equivalents in Ohio and Indiana. *Bulletins of American Paleontology 50*, 271–441.

Berry, W. B. N., Lawson, D. A & Yancey, E. S. 1979: Species-diversity patterns in some Middle Ordovician communities from California–Nevada. *Palaeogeography, Palaeoclimatology, Palaeoecology 26*, 99–116.

Bohlin, B. 1949: The Asaphus Limestone in northernmost Öland. *Bulletin of the Geological Institution of the University of Uppsala 37*, 79–166.

Bradshaw, L. E. 1969: Conodonts from the Fort Peña Formation (Middle Ordovician) Marathon Basin, Texas. *Journal of Paleontology 43*, 1137–1168.

Bruton, D. L. & Bockelie, J. F. 1980: Geology and paleontology of the Hølonda area, Western Norway – a fragment of North America. *Virginia Polytechnic Institute and State University, Department of Geological Sciences, Memoir 2*, 41–55.

[Carnes, J. B. 1975: Conodont biostratigraphy in the lower Middle Ordovician of the western Appalachian thrust-belts in northeastern Tennessee. Unpublished PhD dissertation, The Ohio State University. 170 pp.]

Cooper, B. J. 1981: Early Ordovician conodonts from the Horn Valley Siltstone, Central Australia. *Palaeontology 24*, 147–183.

Drygant, D. M. (Дрыгант, Д. М.) 1974: Новые среднеордовикские конодонты Северо-Западной Волыни. (New Middle Ordovician conodonts of NW Volhynia.) *Палеонтологический сборник 11*, 54–57.

Dzik, J. 1976: Remarks on the evolution of Ordovician conodonts. *Acta Palaeontologica Polonica 21*, 395–455.

Dzik, J. 1978: Conodont biostratigraphy and paleogeographical relations of the Ordovician Mójcza Limestone (Holy Cross Mts., Poland). *Acta Palaeontologica Polonica 23*, 51–72.

Dzik, J. 1984: Early Ordovician conodonts from the Barrandian and Bohemian–Baltic faunal relationships. *Acta Palaeontologica Polonica 28* (in press).

Dzik, J. & Trammer, J. 1980: Gradual evolution of conodontophorids in the Polish Triassic. *Acta Palaeontologica Polonica 25*, 55–89.

Ethington, R. L. & Clark, D. L. 1981: Lower and Middle Ordovician conodonts from the Ibex area, western Millard County, Utah. *Brigham Young University Geology Studies 28:2*, 1–155.

Ethington, R. L. & Schumacher, D. 1969: Conodonts of the Copenhagen Formation (Middle Ordovician) in central Nevada. *Journal of Paleontology 43*, 440–484.

Fåhraeus, L. E. 1966: Lower Viruan (Middle Ordovician) conodonts from the Gullhögen quarry, southern central Sweden. *Sveriges Geologiska Undersökning, Årsbok 60:5*, 1–40.

Fåhraeus, L. E. & Hunter, D. R. 1981: Paleoecology of selected conodontophorid species from the Cobbs Arm Formation (middle Ordovician), New World Island, north central Newfoundland. *Canadian Journal of Earth Sciences 18*, 1653–1665.

Fåhraeus, L. E. & Nowlan, G. S. 1978: Franconian (Late Cambrian) to Early Champlainian (Middle Ordovician) conodonts from the Cow Head Group, Western Newfoundland. *Journal of Paleontology 52*, 444–471.

Harris, A. G., Bergström, S. M., Ethington, R. L. & Ross, R. J., Jr. 1979: Aspects of Middle and Upper Ordovician conodont biostratigraphy of carbonate facies in Nevada and southeast California and comparison with some Appalachian successions. *Brigham Young University Geology Studies 26:3*, 7–33.

Higgins, A. C. 1967: The age of the Durine Member of the Durness Limestone Formation at Durness. *Scottish Journal of Geology 3*, 382–388.

Hoffman, A. 1979: Community paleoecology as an epiphenomenal science. *Paleobiology 5*, 357–379.

Jaanusson, V. 1957: Unterordovizische Illaeniden aus Skandinavien. Mit Bemerkungen über die Korrelation des Unterordoviziums. *Bulletin of the Geological Institution of the University of Uppsala 37*, 79–166.

Jaanusson, V. 1979: Ordovician. *In* R. A. Robison & C. Teichert (eds.): *Treatise on Invertebrate Paleontology, Part A, Introduction,* A 136–A 166. Geological Society of America, Boulder, Colorado.

Jaanusson, V. & Bergström, S. M. 1980: Middle Ordovician faunal spatial differentiation in Baltoscandia and the Appalachians. *Alcheringa 4,* 89–110.

Jeppsson, L. 1971: Element arrangement in conodont apparatuses of *Hindeodella* type and in similar forms. *Lethaia 4,* 101–123.

Jeppsson, L. 1979: Conodonts. *In* V. Jaanusson, S. Laufeld & R. Skoglund (eds.): Lower Wenlock faunal and floral dynamics. Vattenfallet section, Gotland. *Sveriges Geologiska Undersökning, Ser. C:762,* 225–248.

Kennedy, D. J. 1980: A restudy of conodonts described by Branson & Mehl, 1933, from the Jefferson City Formation, Lower Ordovician, Missouri. *Geologica et Palaeontologica 14,* 45–76.

Kennedy, D. J., Barnes, C. R. & Uyeno, T. T. 1979: A Middle Ordovician conodont faunule from the Tetagouche Group, Camel Back Mountains, New Brunswick. *Canadian Journal of Earth Sciences 16,* 540–551.

Kuwano, Y. 1982: Element composition of some Silurian ozarkodinids. *In* Jeppsson, L. & Löfgren, A. (eds.): Third European Conodont Symposium (ECOS III) Abstracts. *Publications from the Institutes of Mineralogy, Paleontology and Quaternary Geology, University of Lund, Sweden No. 238,* 15–16.

Landing, E. 1976: Early Ordovician (Arenigian) conodont and graptolite biostratigraphy of the Taconic allochton, eastern New York. *Journal of Paleontology 50,* 614–646.

Lindström, M. 1971: Lower Ordovician conodonts of Europe. *Geological Society of Amcerica, Memoir 127 (1970),* 21–61.

Lindström, M. 1976 a: Conodont provincialism and paleoecology – a few concepts. *Geological Association of Canada, Special Paper 15,* 27–58.

Lindström, M. 1976 b: Conodont paleogeography of the Ordovician. *In* M. G. Bassett (ed.): *The Ordovician System,* 501–522. University of Wales Press and National Museum of Wales, Cardiff.

Lindström, M. 1977: Genera *Acodus* Pander, 1856 and *Paltodus* Pander, 1856. *In* W. Ziegler (ed.): *Catalogue of Conodonts III,* 1–20, 415–434. E. Schweizerbart'sche Verlagsbuchhandlung, Stuttgart.

Lindström, M., Racheboeuf, P. R. & Henry, J.-L. 1974: Ordovician conodonts from the Postolonnec Formation (Crozon Peninsula, Massif Armoricain) and their stratigraphic significance. *Geologica et Palaeontologica 8,* 15–28.

Löfgren, A. 1978: Arenigian and Llanvirnian conodonts from Jämtland, northern Sweden. *Fossils and Strata 13.* 129 pp.

MacArthur, R. & Wilson, E. O. 1967: *The Theory of Island Biogeography.* 203 pp. Princeton University Press, Princeton.

[McHargue, T. R. 1975: Conodonts of the Joins Formation (Ordovician), Arbuckle Mountains, Oklahoma. Unpublished Master's Thesis, University of Missouri. 151 pp.]

McHargue, T. R. 1982: Ontogeny, phylogeny, and apparatus reconstruction of the conodont genus *Histiodella,* Joins Fm., Arbuckle Mountains, Oklahoma. *Journal of Paleontology 56,* 1410–1433.

McTavish, R. A. 1973: Prioniodontacean conodonts from the Emanuel Formation (Lower Ordovician) of Western Australia. *Geologica et Palaeontologica 7,* 27–58.

Miller, J. F. 1980: Taxonomic revision of some Upper Cambrian and Lower Ordovician conodonts, with comments on their evolution. *University of Kansas Paleontological Contributions, Paper 99,* 1–44.

Mound, J. F. 1965: A conodont fauna from the Joins Formation (Ordovician), Oklahoma. *Tulane Studies in Geology 4:1,* 1–46.

Mound, J. F. 1968: Conodonts and biostratigraphy of the lower Arbuckle Group (Ordovician), Arbuckle Mountains, Oklahoma. *Micropaleontology 14,* 393–434.

Nowlan, G. S. 1979: Fused clusters of the conodont genus *Belodina* from the Thumb Mountain Formation (Ordovician), Ellesmere Island, District of Franklin. *Geological Association of Canada, Paper 79-1 A,* 213–218.

Orchard, M. J. 1980: Upper Ordovician conodonts from England and Wales. *Geologica et Palaeontologica 14,* 9–44.

Serpagli, E. 1967: I conodonti dell'Ordoviciano superiore (Ashgiliano) delle Alpi Carniche. *Bolletino della Società Paleontologica Italiana 6,* 30–111.

Serpagli, E. 1974: Lower Ordovician conodonts from Precordilleran

Argentina (Province of San Juan). *Bolletino della Società Paleontologica Italiana 13,* 17–98.

Seddon, G. & Sweet, W. C. 1971: An ecologic model for conodonts. *Journal of Paleontology 45,* 969–980.

Stouge, S. 1982: Preliminary conodont biostratigraphy and correlation of Lower to Middle Ordovician carbonates from the St. George Group, Great Northern Peninsula, Newfoundland. *Mineral Development Division, Department of Mines and Energy, Government of Newfoundland and Labrador, Current Research Report 82-3,* 1–59.

Stouge, S. (in press): Conodonts of the Middle Ordovician Table Head Formation in western Newfoundland. *Fossils and Strata 16.*

Sweet, W. C. 1979 a: Conodonts and conodont biostratigraphy of post-Tyrone Ordovician rocks of the Cincinnati Region. *U.S. Geological Survey Professional Paper 1066–G,* 1–26.

Sweet, W. C. 1979 b: Late Ordovician conodonts and biostratigraphy of the western Midcontinent Province. *Brigham Young University Geology Studies 26:3,* 45–74.

Sweet, W. C. 1981 a: Genera *Phragmodus* Branson & Mehl, 1933, *Plectodina* Stauffer, 1935, *Aphelognathus* Branson, Mehl & Branson, 1951, and *Belodina* Ethington, 1959. *In* W. Ziegler (ed.): *Catalogue of Conodonts,* 27–84, 245–290. E. Schweizerbart'sche Verlagsbuchhandlung, Stuttgart.

Sweet, W. C. 1981 b: Macromorphology of elements and apparatuses. *In* R. A. Robison (ed.): *Treatise on Invertebrate Paleontology, Part W, Miscellanea (Supplement 2), Conodonta,* W 5–W 20. Geological Society of America, Boulder.

Sweet, W. C. 1982: Conodonts from the Winnipeg Formation (Middle Ordovician) of the northern Black Hills, South Dakota. *Journal of Paleontology 56,* 1029–1049.

Sweet, W. C. & Bergström, S. M. 1962: Conodonts from the Pratt Ferry Formation (Middle Ordovician) of Alabama. *Journal of Paleontology 36,* 1214–1252.

Sweet, W. C. & Bergström, S. M. 1972: Multielement taxonomy and Ordovician conodonts. *Geologica et Palaeontologica SB 1,* 29–42.

Sweet, W. C. & Bergström, S. M. 1973: Biostratigraphic potential of the Arbuckle Mountains sequence as a reference standard for the Midcontinent Middle and Upper Ordovician. *Geological Society of America, Abstracts with Programs 5, 4,* 355.

Sweet, W. C. & Bergström, S. M. 1974: Provincialism exhibited by Ordovician conodont faunas. *Society of Economic Paleontologists and Mineralogists, Special Publication 21,* 189–202.

Sweet, W. C., Thompson, T. L. & Satterfield, I. R. 1975: Conodont stratigraphy of the Cape Limestone (Maysvillian) of eastern Missouri. *Missouri Geological Survey, Studies in Stratigraphy, Report of Investigations 57,* 1–59.

Sweet, W. C., Turco, C. A., Warner, E. & Wilkie, L. C. 1959: The American Upper Ordovician standard. I. Eden conodonts from the Cincinnati Region of Ohio and Kentucky. *Journal of Paleontology 33,* 1029–1068.

Szaniawski, H. 1980: Conodonts from the Tremadocian chalcedony beds, Holy Cross Mountains (Poland). *Acta Palaeontologica Polonica 25,* 101–121.

Tipnis, R. S., Chatterton, B. D. E. & Ludvigsen, R. 1979: Ordovician conodont biostratigraphy of the southern District of Mackenzie, Canada. *Geological Association of Canda, Special Paper 18 (1978),* 39–91.

van Wamel, W. A. 1974: Conodont biostratigraphy of the Upper Cambrian and Lower Ordovician of northwestern Öland, southeastern Sweden. *Utrecht Micropaleontological Bulletins 10,* 1–126.

Viira, V. (Вийра, В.) 1974: Конодонты ордовика Прибалтики. [Ordovician conodonts of the east Baltic.] 141 pp. Валгус, Таллин.

[Votaw, R. B. 1971: Conodont biostratigraphy of the Black River Group (Middle Ordovician) and equivalent rocks in the eastern Midcontinent, North America. Unpublished PhD dissertation, The Ohio State University. 170 pp.]

Webers, G. F. 1966: The Middle and Upper Ordovician conodont faunas of Minnesota. *Minnesota Geological Survey, Special Publication Series SP-4,* 1–123.

Williams, A. 1976: Plate tectonics and biofacies evolution as factors in Ordovician correlation. *In* M. G. Bassett (ed.): *The Ordovician System,* 29–66. University of Wales Press and National Museum of Wales, Cardiff.

FOSSILS AND STRATA
No. 15, p. 86. Oslo 1983 12 15.

A contribution to the Third European ECOS III
Conodont Symposium, Lund, 1982

Simple-cone studies: some provocative thoughts

LENNART JEPPSSON

When organizing a workshop for ECOS III, Godfrey Nowlan asked me to say something about my work with the genus *Panderodus*. This is a written account of my provocative presentation, partly modified in the light of comments I received at the meeting. I have spent several years studying the genus *Panderodus* and my work is far from finished. The following experiences may have some general relevance to the study of simple cones.

Only some of the eight to ten distinct groups of elements in each apparatus of *Panderodus* can readily be identified as homologous in all the species. Also, two of them are easy to identify as homologous with elements in the apparatuses of other genera. One is the symmetrical element. Symmetrical elements have been described in many simple-cone apparatuses and they have often been identified as Sa elements, implying some kind of homology with such elements in other apparatuses. I doubt many of these identifications. In my opinion the following two requirements are absolutely necessary to identify an element convincingly as a tr or Sa element.

Because the tr element is unpaired, it is the rarest of all types of elements. In some *Panderodus* species its frequency is only about 1% even in a collection recovered from a fine screen. The element is very small in all species of *Panderodus* and thus easily lost through a too coarse screen.

It is symmetrical, not nearly or almost. Earlier accounts of so called symmetrical elements in *Panderodus* referred to specimens with a furrow on one lateral side only; a genuine such element has one on each lateral side.

The only other element whose homologies I am confident about both within the genus and outside it, is the single pair of ne or M elements. These elements are asymmetrical and strongly twisted like the homologous oistodontiform elements. In other respects, such as the small size, they are closely similar to the tr or Sa element. This similarity may or may not be true for related genera.

There are probably only about ten species of *Panderodus* known but about fifty specific names, most of which are based on a few of the species. Four or five of the seven or eight lineages I recognize in the Silurian have now been identified far back in the Ordovician. The differences between an Ordovician and a Silurian population of the same lineage are small compared to the differences between separate lineages. In order to recognize this, each lineage should be treated as one species, subdivided into chronological subspecies. Thus the search for the valid name for a species cannot be limited to one system. For example, the name *Panderodus equicostatus* was based on a Caradoc specimen, but the taxon is unrepresented in the other Ordovician collections of *Panderodus* I have seen. It is common, however, in the Silurian ones. On the other hand, *Panderodus gracilis* probably occurs in many Ordovician collections, but in the Silurian ones I have so far only seen it in some of my own collections from the Leintwardinian Hemse Marl on Gotland. Except for three localities close to each other where good collections may be obtained I have only stray specimens.

The naming of taxa and selecting of types for them are irrevocable, but the consequences of this do not always seem to be realized. Every other piece of information we publish can be and often is overlooked.

Having studied the types for about hundred names, I have a strong feeling that names often are given only as a means of talking conveniently about the fragmentary objects we have found, and that the only recognized longlasting effect of the process is that it perpetuates the name of the author. The particular topic of simple-cone taxonomy is now so difficult that it is necessary to work for several years on each genus and all the types of nominal taxa referred to it and on types resembling any of the elements of its species, to exclude any possibility that there are names already available for use. The situation is now such that a correct nomenclature often requires more work than a good taxonomy. I suppose that the situation is or will soon be similar with other groups of fossils. A major problem is that some types are unavailable without large travel grants. They cannot be borrowed, because the mail is not trusted. However, in my experience, the risk of losing specimens in the mail is small.

Another problem regards the documentation and the deposition of types. Perhaps 90% of all published pictures of *Panderodus* are unidentifiable to species, and sometimes also to genus as the pictures show only one side of a coated element. Further, the types, extracted from a collection whose composition cannot now be checked, are often unidentifiable, if they cannot be compared with a collection of the population. Thus, Branson & Mehl's (1933) Ordovician *Panderodus* species caused me many problems. Some of the types agree reasonably well with those of *Panderodus unicostatus*, and so do some of those described by Stauffer (1935b, 1940). However, in 1981 I got from David Kennedy a number of acetic-acid residues prepared from samples he had collected in connection with his restudy of Branson & Mehl's and Stauffer's localities. In this material I found two species of *Panderodus*, neither of which is *P. unicostatus*. One of them is the same species as I have in the upper Hemse Marl. My notes on the types of *P. gracilis* fit well with elements of this species. In conclusion, in addition to the illustrated type specimens and otherwise selected specimens it is necessary to make available at least one good slide with hundreds of elements representing the whole conodont fauna, so that also those elements that may have been erroneously excluded by the author are represented. Preferably this slide should be that from which the holotype is selected. Regarding all old types the kind of job done by David Kennedy is vital for a stable nomenclature.

References

Branson, E. B. & Mehl, M. G. 1933–1934: Conodont studies. *The University of Missouri Studies 8*, 349 pp.

Stauffer, C. R. 1935: The conodont fauna of the Decorah Shale (Ordovician). *Journal of Paleontology 9*, 596–620.

Stauffer, C. R. 1940: Conodonts from the Devonian and associated clays of Minnesota. *Journal of Paleontology 14*, 417–435.

Lennart Jeppsson, Department of Historical Geology and Palaeontology, Sölvegatan 13, S-223 62 Lund, Sweden; 14th February, 1983.

Conodonts and biostratigraphy in the Ordovician of the Siberian Platform

TAMARA A. MOSKALENKO

FOSSILS AND STRATA

ECOS III

A contribution to the Third European
Conodont Symposium, Lund, 1982

Moskalenko, Tamara A. 1983 12 15: Conodonts and biostratigraphy in the Ordovician of the Siberian Platform. *Fossils and Strata*, No. 15, pp. 87–94. Oslo. ISSN 0300-9491. ISBN 82-0006737-8.

Key Ordovician sections have been studied and sampled along the Kulumbe, Moyero, Podkamennaya Tunguska, Angara and Lena Rivers. The following conodont Assemblage Zones ranging from the upper Lower Ordovician to Upper Ordovician are proposed for the region: *Scolopodus quadraplicatus – Histiodella angulata, Coleodus mirabilis, Cardiodella–Polyplacognathus, Phragmodus flexuosus, Ptiloconus anomalis – Bryantodina lenaica, Polyplacognathus sweeti – Phragmodus inflexus, Belodina compressa – Culumbodina mangazeica, Acanthocordylodus festus, 'Spathognathodus'? dolboricus, Acanthodina nobilis, Aphelognathus pyramidalis*. The conodont faunas under investigation are compared with those of equivalent age in other regions. The marked distinctiveness of the Siberian conodont fauna and its close resemblance to corresponding faunas of the Midcontinent Province are discussed. □*Conodonta, biostratigraphy, biogeography, Ordovician, Siberian Platform.*

Tamara A. Moskalenko, Institute of Geology and Geophysics, Siberian Branch of the USSR Academy of Sciences, Novosibirsk 630090, USSR; 10th July, 1982.

The study of Ordovician conodonts during many years has resulted in much information obtained from various localities on the Siberian Platform (Fig. 1) and particularly from the key sections along the Kulumbe, Moyero, Podkamennaya Tunguska, Angara and Lena Rivers. Separate parts of these sections include the stratotypes for regional units. The conodonts proved to be significant in solving problems of regional stratigraphy and intra-regional and inter-regional correlation as well. These detailed studies have ascertained the succession of distinct conodont assemblages, the taxonomic composition of these assemblages and the assignment of the latter to certain stratigraphic intervals.

In the modern regional scheme for Ordovician stratigraphy on the Siberian Platform the main regional unit corresponds to the regional stage. Fig. 2 shows the regional sequences and the distribution of stratigraphically important conodont species; Figs. 3 and 4 illustrate some specimens of more typical conodont species. All illustrated specimens are housed in the collections of the Geological Museum, Institute of Geology and Geophysics, Siberian Branch of the USSR Academy of Sciences. Collections including them have been assigned catalogue numbers IGG 397, 537, 614, 629.

Conodont assemblages

Conodonts from the Ordovician sections in the region under discussion are distributed irregularly. Thus, the zonation of Lower Ordovician rocks, tentatively assigned to the Lower Tremadocian, is limited because conodonts are rare. In the Nyaian Stage, assigned to the Upper Tremadocian, a representative conodont assemblage containing *Acanthodus lineatus* (Furnish), *Loxodus bransoni* Furnish, *Cordylodus angulatus* Pander, and *C. rotundatus* Pander was recognized in sections along the Lena and Angara Rivers mainly. The first two species are important elements in the Late Tremadocian ACL-fauna, which is characteristic of the Midcontinent Province (Lindström 1976). The latter two species are present all over the world (Lindström 1976:504), a statement that is supported by finds from Europe, Asia, North America, Australia, and New Zealand.

The deposits of the following Ugorian Stage contain an impoverished conodont association, that is the result of wide development of variegated dolomites and sandstones in this interval of the Ordovician sequence on the Siberian Platform. The following species are the most significant: *Scolopodus cornutiformis* Branson & Mehl, *Drepanodus (Drepanoistodus?) costatus* Abaimova.

The Kimaian Stage has yielded a rich conodont assemblage, present in all key sections, including the stratotype of this subdivision on the Kulumbe River (Moskalenko 1982) and in many other localities. *Scolopodus quadraplicatus* Branson & Mehl and *Histiodella angulata* Moskalenko are typical species. *Drepanodus costatus* extends into this interval and *Acodus deltatus* Lindström and some other conodonts make their first appearance; the upper part of the stage contains *Leptochirognathus* spp., solitary *Oistodus multicorrugatus* Harris, *Microzarkodina* ex gr. *flabellum* (Lindström), and elements resembling those of *Oepikodus quadratus* (Graves & Ellison). The assemblage composition is undoubtedly similar to that of Fauna D from North America (Ethington & Clark 1971; Barnes *et al.* 1973), named the *S. quadraplicatus* fauna (Lindström 1976). The upper part of this subdivision contains elements typical of Whiterockian age (Faunas E, 1–4) in North America. It should be emphasized, however, that the Kimaian conodont assemblage remains generally the same. The author agrees with S. Bergström (1977) that rocks containing conodonts of Faunas 1–4 correlate in part or completely with the Arenigian.

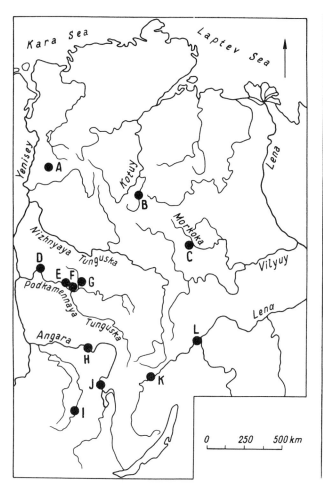

Fig. 1. Sketch map of the Siberian Platform, showing the geographic locations of main sections (marked with circles): □A. The Kulumbe River. □B. The Moyero River. □C. The Morkoka River Basin. □D, E, F, G. The Podkamennaya Tunguska River Basin in the districts of the Stolbovaya River (D), of the Bolshaya Nirunda River (E), near the village Baykit (F), of the Nizhnyaya Chunku River (G). □H, I, J. The Angara River Basin in the districts of the latitudinal flow of this river (H), of the Uda and Biryusa Rivers (I), of the meridional flow of this river (J). □K, L, The Lena River in the district of its upper reaches (K), near the mouth of the Vitim River (L).

The base of the Vihorevian Stage represents a very distinct boundary in the conodont succession. No species crosses this boundary; above it occurs an absolutely distinctive assemblage. It includes various species of *Coleodus*, *Neocoleodus*, *Erismodus*, *Microcoelodus*, *Polycaulodus* and other genera. This assemblage is readily traced in continuous sections on the Moyero River and in the basin of the Angara River (Moskalenko 1970; Kanygin *et al.* 1980). It is similar, even at the species level, to the conodont association from the Dutchtown Formation in Missouri (Cullison 1938), considered to be typical of Fauna 5 in North America (Sweet *et al.* 1971), i.e. Chazyan (Llanvirnian) in age.

The Vihorevian conodont fauna is closely related to that from the Mukteian Stage. The components of the former assemblage continue into the latter, but their importance diminishes. The dominant taxa include various species of *Cardiodella*, *Polyplacognathus* and of some other genera, appearing in this stratigraphic interval. *Cardiodella* n. sp. 1 (the description of the species is in press) and *C. tumida* (Branson & Mehl) are especially common in this interval. Although

numerous to abundant, their vertical range is limited. The deposits corresponding to this stratigraphic level can be identified in various parts of the Siberian Platform, because of this distinctive fauna, particularly in continuous sections on the Moyero River and in the basin of the Angara River. Similar conodont assemblages with *Cardiodella*, *Erismodus* and other genera have been recognized in North America in the Middle Ordovician Joachim Formation in Missouri (Branson & Mehl 1933; Andres 1967), in the McLish and Bromide Formations in Oklahoma (Branson & Mehl 1943), i.e. Chazyan in age.

The base of the Volginian Stage represents a major boundary in the succession of faunas of all kinds on the Siberian Platform. The Volginian conodont assemblage also signifies a new stage in the development of conodontophorids. It is widespread on the Platform being recognized from a number of localities and is well represented in the stratotype section of the stage on the Lena River. *Phragmodus* appears first in this part of the section and immediately becomes an important element in the Volginian and some subsequent faunas. *Phragmodus flexuosus* Moskalenko is a dominant species in the Volginian assemblage. Representatives of this association are known from North America where they characterize Faunas 5 and 6, which are assigned to the Chazyan (Sweet *et al.* 1971).

The base of the Kirenskian–Kudrinian Stage shows an increase in conodont diversity. There appear such typical species as *Bryantodina lenaica* Moskalenko, *Ptiloconus anomalis* (Moskalenko), *Microcoelodus tunguskaensis* Moskalenko, *Oulodus restrictus* (Moskalenko); in the upper part of the stage *Evencodus* and *Stereoconus* begin to play an appreciable role. Great correlative importance will apparently be played by plectodinans eventually, but currently their effective application is hampered due to inadequate taxonomic definition of the multielement genus and its species. This conodont assemblage is well represented in the stratotype section of the stage on the Lena River and persists in all localities with deposits of this interval.

The Chertovskian Stage, established on the Lena River, can be traced in many places on the Siberian Platform, including continuous sections on the Kulumbe, Moyero, and Podkamennaya Tunguska Rivers. *Phragmodus inflexus* Stauffer, *Polyplacognathus sweeti* Bergström, *Oistodus petaloideus* Moskalenko and other species are characteristic of the Chertovskyian assemblage. *Evencodus* and *Stereoconus* continue to occur. *Polyplacognathus sweeti* is rare on the Siberian Platform. However, this species is very important in having wide geographic and limited vertical range. It is known from North America, Europe and Asia. It is confined to a stratigraphic interval that corresponds to the *Nemagraptus gracilis* graptolite zone (Bergström 1971). *Phragmodus inflexus* is an important element in North American Fauna 7, i.e. Black River – Porterfield age (Sweet *et al.* 1971).

The conodont assemblage from the Baksian Stage is significantly different in comparison with the content of the previous one. For the first time *Belodina*, *Culumbodina*, *Acanthocordylodus* and other similar acanthodontid forms appear in the Siberian Ordovician sections. In the lower part of this subdivision *Phragmodus inflexus* continues to occur, but in succeeding strata it is replaced by *Phragmodus undatus* Branson & Mehl. The latter, however, is extremely rare on the Siberian

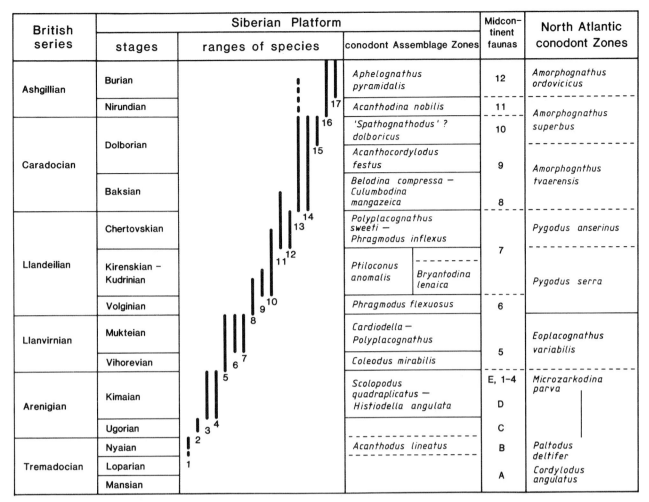

British series	Siberian Platform			Midcontinent faunas	North Atlantic conodont Zones
	stages	ranges of species	conodont Assemblage Zones		
Ashgillian	Burian		Aphelognathus pyramidalis	12	Amorphognathus ordovicicus
	Nirundian		Acanthodina nobilis	11	Amorphognathus superbus
Caradocian	Dolborian		'Spathognathodus'? dolboricus	10	
			Acanthocordylodus festus	9	Amorphognthus tvaerensis
	Baksian		Belodina compressa — Culumbodina mangazeica	8	
Llandeilian	Chertovskian		Polyplacognathus sweeti — Phragmodus inflexus	7	Pygodus anserinus
	Kirenskian – Kudrinian		Ptiloconus anomalis | Bryantodina lenaica		Pygodus serra
	Volginian		Phragmodus flexuosus	6	
Llanvirnian	Mukteian		Cardiodella — Polyplacognathus	5	Eoplacognathus variabilis
	Vihorevian		Coleodus mirabilis		
Arenigian	Kimaian		Scolopodus quadraplicatus — Histiodella angulata	E, 1–4	Microzarkodina parva
				D	
	Ugorian			C	
	Nyaian		Acanthodus lineatus	B	Paltodus deltifer
Tremadocian	Loparian				Cordylodus angulatus
	Mansian			A	

Fig. 2. Stratigraphical division of the Ordovician, vertical distribution of stratigraphically important species and conodont zonation on the Siberian Platform; approximate correlation of the Ordovician subdivisions between the Siberian Platform, North America and Europe. Black vertical lines show the distribution of the following species: (1) *Acanthodus lineatus* (Furnish), (2) *Scolopodus cornutiformis* Branson & Mehl, (3) *Scolopodus quadraplicatus* Branson & Mehl, (4) *Histiodella angulata* Moskalenko, (5) *Coleodus mirabilis* Moskalenko, (6) *Cardiodella*

n. sp. 1, *C. tumida* (Branson & Mehl), (7) *Polyplacognathus* n.sp., (8) *Phragmodus flexuosus* Moskalenko, (9) *Bryantodina lenaica* Moskalenko, (10) *Ptiloconus anomalis* Moskalenko, (11) *Phragmodus inflexus* Stauffer, (12) *Polyplacognathus sweeti* Bergström, (13) *Belodina compressa* (Branson & Mehl), (14) *Acanthocordylodus festus* Moskalenko, (15) *'Spathognathodus'*? *dolboricus* Moskalenko, (16) *Acanthodina nobilis* Moskalenko, (17) *Aphelognathus pyramidalis* (Branson, Mehl & Branson).

Platform. The Baksian conodont fauna has been intensively studied from western and northern sections of the region (Moskalenko 1973). The *Belodina–Panderodus* association makes it possible to correlate approximately the Baksian conodont assemblage with North American Fauna 8, i.e. Trenton (Caradocian) age (Sweet *et al.* 1971; Sweet & Bergström 1976; Bergström 1977).

The conodont assemblage from the Dolborian Stage is closely related to the previous one with many genera and species in common. The main difference lies in the relative abundance of the taxa. Thus, *Belodina* and *Culumbodina* are typical genera of the Baksian assemblage, and *Acanthocordylodus* and other acanthodontids predominate in the Dolborian. New taxa appear only in the upper part of the Dolborian Stage such as *'Spathognathodus'*? *dolboricus* Moskalenko and a peculiar group of conodonts showing considerable morphologic variation, but having the same type of sculpture (Fig. 4Af, Ag). The Dolborian assemblage can be traced in various districts along the Podkamennaya Tunguska River (Moskalenko 1973) and along the Moyero River. It is rather endemic and can only

tentatively be compared to North American Faunas 9 and 10, i.e. upper Trentonian and Edenian age.

The Nirundian Stage has weak paleontologic support; the conodont fauna is also greatly impoverished. It contains rare *Acanthodina nobilis* Moskalenko and some other species. All of them pass into overlying strata of the Burian Stage. Reliable representative localities of the Nirundian Stage are restricted to the basin of the Podkamennaya Tunguska River.

The Burian Stage, on the contrary, is easily distinguished by conodonts. *Aphelognathus pyramidalis* (Branson, Mehl & Branson), a multielement species (Nowlan & Barnes 1981), is the principal component in the Burian assemblage. Its elements – aphelognathiform, trichonodelliform and zygognathiform – make their first appearance at the base of the subdivision and are then constantly present in the limestones, which comprise this subdivision. However, the upper limit of the assemblage has not been determined because of widespread pre-Silurian erosion. This produced the restricted extent of the Burian beds on the Platform. These beds have been well studied only from the basin of the Podkamennaya Tunguska River and some isolated localities in the river basin of the Morkoka. The presence of *Aphelognathus pyramidalis* enables the comparison of the Burian assemblage with Fauna 12 from North America, the stratigraphic range of which is restricted to Maysvillian and Richmondian (Nowlan & Barnes 1981).

Conodont zonation

The data reviewed above establishes the succession and boundaries of conodont assemblage changes in the Ordovician of the Siberian Platform. This allows a conodont zonal scale to be proposed for the upper Lower Ordovician, Middle and Upper Ordovician of this region. The principle of the first appearance of the zonal denominating taxon was commonly used in distinguishing the zones. A zone normally corresponds to the biozone of the zonal index species within the Siberian Platform, but it may be less than the actual stratigraphic range of this species. In the latter case the upper boundary of the zone is controlled by the appearance in the section of the zonal index species of the next zone and by proper change of the conodont association. The proposed zonation by conodonts reflects considerable changes in the composition of the successive assemblages. It has recently become possible to propose the following conodont assemblage zones (Moskalenko 1983):

The *Scolopodus quadraplicatus – Histiodella angulata* Zone is defined by the vertical range of the zonal index species; this range corresponds with that of the Kimaian Stage.

The *Coleodus mirabilis* Zone contains the rocks of the Vihorevian Stage with a conodont fauna sharply differing from the previous one at the generic level. The lower boundary is defined by the appearance of *Coleodus, Neocoleodus, Erismodus, Microcoelodus* and the upper one by the appearance of *Cardiodella* and its associated assemblage.

The *Cardiodella–Polyplacognathus* Zone ranges through the strata of the Mukteian Stage. Taxa such as *Cardiodella, Polyplacognathus* and many other genera are unknown in the underlying as well as in the overlying strata. Thus, the lower and the upper boundaries are clearly determined by the debut and the disappearance of the zonal assemblage.

The *Phragmodus flexuosus* Zone is of wide distribution on the Siberian Platform; its lower boundary is determined by the appearance of the zonal index species and the upper one by the appearance of *Bryantodina lenaica* and associated conodonts. The zone includes the deposits of the Volginian Stage.

The *Ptiloconus anomalis – Bryantodina lenaica* Zone includes the deposits of the Kirenskian–Kudrinian Stage. The lower boundary is drawn at the appearance of *Bryantodina lenaica* and *Ptiloconus anomalis*; the former is typical of the lower half of the zone and therefore it is probably possible to distinguish a *Bryantodina lenaica* Subzone within the zone under discussion. The upper boundary is determined by the change of the conodont assemblage.

The *Polyplacognathus sweeti – Phragmodus inflexus* Zone embraces the deposits of the Chertovskian Stage. The debut of the species, from which the zone was named, marks its lower boundary; the first appearance of *Belodina* and associated assemblage coincides with the upper boundary.

The *Belodina compressa – Culumbodina mangazeica* Zone has a precise lower boundary that coincides with the base of the

Baksian Stage. It embraces the time of flourishing of the zonal index species and apparently is an acme-zone. The zone scope is restricted to the Baksian Stage.

The *Acanthocordylodus festus* Zone corresponds to the time of flourishing of the various acanthodontids. The zone embraces the major part of the Dolborian Stage. The lower boundary needs to be refined; tentatively it is placed at the base of the stage mentioned above; the upper boundary is defined by the appearance of the species typical of the next zone.

The '*Spathognathodus*'? *dolboricus* Zone embraces the beds which have been formed during the existence of the zonal species. It corresponds to the upper part of the Dolborian Stage.

The *Acanthodina nobilis* Zone is poor in conodonts. The lower boundary is drawn at the beginning of the development of *Acanthodina nobilis*, and the upper one is determined by the appearance of *Aphelognathus pyramidalis*. This zone coincides with the Nirundian Stage.

The *Aphelognathus pyramidalis* Zone has precise boundaries and is defined by the vertical range of the zonal index species. It includes the deposits of the Burian Stage.

Provincial features

Biogeographical differentiation of the conodontophorid fauna has occurred throughout the Ordovician. This is reflected in the development of conodont provinces in Europe (North Atlantic Province) and in cratonic North America (Midcontinent Province).

The analysis of the material under investigation proves the Ordovician conodont faunas from Central Siberia to be unique. Their marked distinctiveness is expressed in the presence of a considerable number of endemic taxa. At the same time they show an evident resemblance with contemporaneous conodont faunas from Canada and central parts of North America. This fact gives support for the existence

---

Fig. 3. □A, B. *Acanthodus lineatus* (Furnish). Nyayian Stage, section J; IGG 614/3, 5; ×40. □C. *Loxodus bransoni* Furnish. Nyayian Stage, section J; IGG 614/115; ×45. □D. *Histiodella angulata* Moskalenko. Kimaian Stage, section B; IGG 629/165; ×45. □E, F. *Scolopodus quadraplicatus* Branson & Mehl. Kimaian Stage, section B; IGG 629/368, 369; ×40. □G. *Erismodus typus* Branson & Mehl. Mukteian Stage, section B; IGG 629/147; ×45. □H. *Erismodus asymmetricus* (Branson & Mehl). Vihorevian Stage, section B; IGG 629/136; ×45. □I–K. *Coleodus mirabilis* Moskalenko. □I. Vihorevian Stage, section J; IGG 614/47; ×30. □J. Vihorevian Stage, section B; IGG 629/61; ×35. □K. Mukteian Stage, section H; IGG 614/51; ×45. □L–N. *Neocoleodus dutchtownensis* Youngquist & Cullison. Vihorevian stage. □L. Section B; IGG 629/198; ×45. □M. Section J; IGG 614/125; ×25. □N. Section J; IGG 614/126; ×45. □O–R. *Cardiodella* n. sp. l. Mukteian Stage. □O. Section H; IGG 614/34; ×30. □P. Section B; IGG 629/52; ×45. □Q. Section B; IGG 629/53; ×45. □R. Section B; IGG 629/54; ×45. □S. *Cardiodella tumida* (Branson & Mehl). Mukteian Stage, section H; IGG 614/44; ×45. □T, U. *Ptiloconus? costulatus* Moskalenko. Mukteian Stage, section H. □T. IGG 614/201; ×30. □U. IGG 614/203; ×20. □V–X. *Polyplacognathus* n. sp. Mukteian Stage, section H. □V. IGG 614/162; ×45. □W. IGG 614/166; ×30. □X. IGG 614/165; ×45. □Y–Ab. *Polyplacognathus?* n. sp. Mukteian Stage. □Y, Z. Section J; IGG 614/174, 177; ×45. □Aa, Ab. Section B; IGG 629/286, 298; ×45.

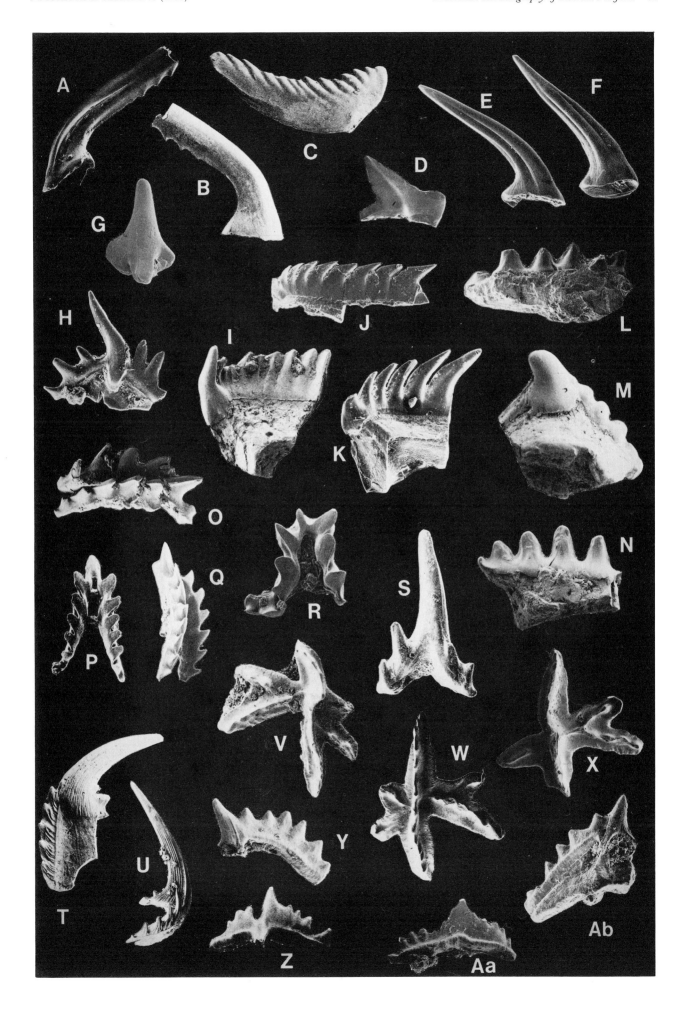

throughout the Ordovician of contacts between the Central Siberian and North American seas. The nature of these contacts, however, did not remain the same; at one time they strengthened, at another they weakened or ceased to exist, probably reflecting transgressive or regressive cycles. During regressive stages the geographic faunal differentiation was more distinct; during transgressions the faunal differences were reduced.

In times of significant transgression the comparatively isolated Central Siberian water area was inhabited by conodontophorid genera and species typical of the Midcontinent Province. Thus, during Kimaian time (late Arenigian) the *S. quadraplicatus* fauna, which is typical of the Midcontinent Province, is of wide occurrence. At Vihorevian–Mukteian time (Llanvirnian?) a new transgression is accompanied by the appearance of the *Coleodus* fauna and following it the *Cardiodella* fauna, which is related to the very shallow water facies of the advancing sea. In the course of the transgression, the maximum of which is in Volginian age (early Llandeilian?) the *Phragmodus* fauna, characteristic of the Midcontinent Province, appears (for the first time in the region) and quickly dominates the faunas. A new outburst in the development of phragmodontids occurred in Chertovskian time (late Llandeilian–early Caradocian). Later on, however, phragmodontids are not of wide distribution in Central Siberia. In North America *Phragmodus undatus* often predominates in Faunas 8–12, whereas on the Siberian Platform this species occurs extremely rarely. The Baksian assemblage shows the predominance of *Belodina–Culumbodina–Panderodus* association, which is a typical component in the contemporaneous fauna from North America, but in a lateral facies series it perhaps takes a position displaced to shallower depths in comparison to the phragmodont community (Barnes & Fåhraeus 1975). The final stage, Burian time (Ashgillian), again demonstrates free biogeographic connections between the Siberian and North American seas, resulting in the presence in the Burian assemblage of typical elements of the Richmond Fauna 12.

Regression stages, on the contrary, involve specialization of the compared faunas and an increase in endemic forms, which hampers their comparison. Most vividly this peculiarity is manifested in the second half of the Middle Ordovician and during most of the Late Ordovician. The representatives of the acanthodontid group begin to play an important role at this time.

A review of laws controlling the distribution of conodonts clearly reveals the existence of Ordovician biogeographical connections between the Central Siberian and North American seas, resulting in the presence of faunas common to both regions. Thus, both biogeographical areas may be considered as parts of a single large biogeographical unit, that is of the Midcontinent Province.

Acknowledgements. – I am indebted to Galina V. Lugovtsova, Institute of Geology and Geophysics, Siberian Branch of the USSR Academy of Sciences, for the English translation of the manuscript. I am most grateful to Dr. Anita Löfgren from the Lund University for suggesting the preparation of this paper and for constructive criticism. I would like to thank Dr. Christopher R. Barnes from the Department of Earth Sciences, Memorial University of Newfoundland, for critically reading the manuscript and offering very valuable comments.

References

Andrews, H. E. 1967: Middle Ordovician conodonts from the Joachim Dolomite of Eastern Missouri. *Journal of Paleontology 41*, 881–901.

Barnes, C. R. & Fåhraeus, L. E. 1975: Provinces, communities and the proposed nektobenthic habit of Ordovician conodontophorids. *Lethaia 8*, 133–149.

Barnes, C. R., Rexroad, C. B. & Miller, J. F. 1973: Lower Paleozoic conodont provincialism. *In* Rhodes, F. H. T. (ed.): Conodont paleozoology. *Geological Society of America Special Paper 141*, 147–190.

Bergström, S. M. 1971: Conodont biostratigraphy of the Middle and Upper Ordovician of Europe and Eastern North America. *In* Sweet, W. C. & Bergström, S. M. (eds.): Symposium on conodont biostratigraphy. *Geological Society of America Memoir 127*, 83–167.

Bergström, S. M. 1977: Early Paleozoic conodont biostratigraphy in the Atlantic Borderlands. *In* Swain, F. M. (ed.): *Stratigraphic Micropaleontology of Atlantic Basin and Borderlands*, 85–110. Elsevier, Amsterdam.

Branson, E. B. & Mehl, M. G. 1933–1934: Conodont studies. *University of Missouri Studies 8*, 1–349.

Branson, E. B. & Mehl, M. G. 1943: Ordovician conodont faunas from Oklahoma. *Journal of Paleontology 17*, 374–387.

Cullison, J. S. 1938: Dutchtown fauna of southeastern Missouri. *Journal of Paleontology 12*, 219–228.

Ethington, R. L. & Clark, D. L. 1971: Lower Ordovician conodonts in North America. *In* Sweet, W. C. & Bergström, S. M. (eds.): Symposium on conodont biostratigraphy. *Geological Society of America Memoir 127*, 63–82.

Kanygin, A. V., Moskalenko, T. A. & Yadrenkina, A. G. (Каныгин, А. В., Москаленко, Т. А. и Ядренкина, А. Г.) 1980: О пограничных отложениях нижнего и среднего ордовика на Сибирской платформе. [On the boundary deposits of the Lower and Middle Ordovician of the Siberian Platform.] *Геология и геофизика 1980:6*, 13–19.

Lindström, M. 1976: Conodont palaeogeography of the Ordovician. *In* Bassett, M. G. (ed.): *The Ordovician System. Proceedings of a Palaeontological Association Symposium, Birmingham, Sept. 1974*, 501–

Fig. 4. □A–D. *Phragmodus flexuosus* Moskalenko. Volginian Stage. □A–C. Section J; IGG 614/143 a, 145, 143; ×30. □D. Section B; IGG 629/245; ×45. □E–G. *Ptiloconus anomalis* (Moskalenko). Kirenskian–Kudrinian Stage, section B; IGG 629/323, 324, 325; ×45. □H–J. *Bryantodina lenaica* Moskalenko. Kirenskian–Kudrinian Stage, section B; IGG 629/50, 49, 51; ×45. □K. *Stereoconus bicostatus* Moskalenko. Chertovskian Stage, section J; IGG 614/236; ×45. □L–N. *Phragmodus inflexus* Stauffer. Chertovskian Stage, section K; IGG 537/55, 56, 57; ×45. □O. *Oistodus petaloideus* Moskalenko. Chertovskian Stage, section D; IGG 397/201; ×45. □P. *Polyplacognathus sweeti* Bergström. Chertovskian Stage, section K; IGG 537/58; ×45. □Q–S. *Belodina compressa* (Branson & Mehl). □Q. Baksian Stage, section D; IGG 397/53; ×30. □R. Baksian Stage, section E; IGG 397/54; ×30. □S. Dolborian Stage, section B; IGG 629/39; ×50. □T. *Belodina diminutiva* (Branson & Mehl). Baksian Stage, section B; IGG 629/41; ×45. □U, V. *Culumbodina mangazeica* Moskalenko. □U. Dolborian Stage, section B; IGG 629/81; ×45. □V. Baksian Stage, section D; IGG 397/62; ×30. □W, X. *Phragmodus undatus* Branson & Mehl. Baksian Stage, section D; IGG 397/202, 203; ×45. □Y. *Pseudooneotodus mitratus* (Moskalenko). Baksian Stage, section D; IGG 397/166; ×30. □Z. *Phragmodus?* (*Spinodus?*) *tunguskaensis* Moskalenko. Dolborian Stage, section B; IGG 629/262; ×30. □Aa. *Acanthodina nobilis* Moskalenko. Burian Stage, section E; IGG 397/92; ×30. □Ab. *Acanthocordylodus festus* Moskalenko. Dolborian Stage, section B; IGG 629/1; ×30. □Ac. *Acanthocordylodus prodigialis* Moskalenko. Dolborian Stage, section B; IGG 629/2; ×30. □Ad, Ae. '*Spathognathodus*'? *dolboricus* Moskalenko. Dolborian Stage, section B; IGG 629/378, 379; ×45. □Af, Ag. N. gen. et n. sp. Dolborian Stage, section B; IGG 629/91, 92. □Af. ×30. Ag₁. ×30. Ag₂. ×180. □Ah–Am. *Aphelognathus pyramidalis* (Branson, Mehl & Branson). Burian stage, section E; IGG 397/195, 196, 197, 198, 199, 200; ×45.

552. University of Wales Press and National Museum of Wales, Cardiff.

Moskalenko, T. A. (Москаленко, Т. А.) 1970: Конодонты криволуцкого яруса (средний ордовик) Сибирской платформы. [Conodonts of the Krivoluk stage (Middle Ordovician) on the Siberian Platform.] 1–116. 'Nauka', Moscow. Moscow.

Moskalenko, T. A. (Москаленко, Т. А.) 1973: Конодонты среднего и верхего ордовика Сибирской платформы. [Conodonts of the Middle and Upper Ordovician on the Siberian Platform.] 1–144. 'Nauka', Novosibirisk.

Moskalenko, T. A. (Москаленко, Т. А.) 1982: Конодонты. *В кн.* Соколов, Б. С. (ред.): Ордовик Сибирской платформы. Опорный разрез на р. Кулюмбе. [Conodonts. *In* Sokolov, B. S. (ed.): The Ordovician of the Siberian Platform. Key section on the Kulumbe River.] 100–144, 182–190. 'Nauka', Moscow.

Moskalenko, T. A. (Москаленко, Т. А.) 1983: Закономерности развития и биогеографические связи ордовикских конодонтофорид на Сибирской плат-

форме. | *В кн.* | Бетехтина, О. А. | и Журавлева, И. Т. (ред.): Среда и жизнь в геологическом прошлом. [Course of the development and the biogeographic relationships of the conodontophorids on the Siberian Platform. *In* Betekhtina, O. A. & Zhuravleva, I. T. (eds.): Environment and life in the geological past.] 76–97. 'Nauka', Novosibirsk.

Nowlan, G. S. & Barnes, C. R. 1981: Late Ordovician conodonts from the Vauréal Formation, Anticosti Island, Quebec. *Geological Survey of Canada Bulletin 329,* 1–49.

Sweet, W. C., Ethington, R. L. & Barnes, C. R. 1971: North American Middle and Upper Ordovician conodont faunas. *In* Sweet, W. C. & Bergström, S. M. (eds.): Symposium on conodont biostratigraphy. *Geological Society of America Memoir 127,* 163–193.

Sweet, W. C. & Bergström, S. M. 1976: Conodont biostratigraphy of the Middle and Upper Ordovician of the United States Midcontinent. *In* Basset, M. G. (ed.): *The Ordovician System. Proceedings of a Palaeontological Association Symposium, Birmingham, Sept. 1974,* 121–151. University of Wales Press and National Museum of Wales, Cardiff.

Early Silurian conodonts of eastern Canada

GODFREY S. NOWLAN

FOSSILS AND STRATA

ECOS III

A contribution to the Third European Conodont Symposium, Lund, 1982

Nowlan, Godfrey S. 1983 12 15: Early Silurian conodonts of eastern Canada. *Fossils and Strata*, No. 15, pp. 95–110. Oslo. ISSN 0300-9491. ISBN 82-0006737-8.

Recently, extensive collections for conodonts have been made in Llandovery strata of eastern Canada. The Ordovician–Silurian boundary is readily recognizable on the basis of conodonts and complete sequences across it occur on Anticosti Island and in Gaspé Peninsula. Early Llandovery shallow water sequences are dominated by *Panderodus, Icriodella* and *Oulodus?,* whereas deep water sequences contain *Panderodus* and small, generalized species of *Ozarkodina.* Late Llandovery sequences are dominated by *Panderodus, Ozarkodina* and new species of *Pterospathodus.* The Llandovery–Wenlock boundary is difficult to define precisely in the region because of the scarcity of diagnostic conodonts which are probably ecologically excluded by a marked shallowing event near the series boundary. Lower Silurian lithologic units across eastern Canada are described briefly and correlated using conodonts and key brachiopod occurrences. The greatest potential for biostratigraphic zonation lies in the strata of Anticosti Island and the Chaleurs Bay region. □*Conodonta, biostratigraphy, palaeoecology, Silurian, eastern Canada.*

Godfrey S. Nowlan, Geological Survey of Canada, 601 Booth Street, Ottawa, Ontario, Canada, K1A 0E8; 1st August, 1982.

There are very few published reports of conodonts from Lower Silurian strata of eastern Canada; those that are available are of very recent origin. This paper is an attempt to review the recently published information, to summarize current research and its preliminary results and to outline the potential for further work in the region. A map showing the main outcrop areas including strata of Early Silurian age in the region (Fig. 1) and a table outlining the current biostratigraphic interpretation of the formations and significant faunas (Fig. 2) are provided. Within this framework the significance and potential of each area for Lower Silurian biostratigraphy is discussed.

The biostratigraphic utility of conodonts in the Silurian has been under study for only two decades, but already numerous zonations have been proposed based on geographically widespread sequences. These zonal schemes are the basis for Silurian conodont biostratigraphy, but no single standard zonation is widely accepted yet. The Lower Silurian (Llandovery) is the least well known part of the system mainly because the interval is poorly represented in several of the classic areas on which Silurian conodont biostratigraphy is based. Eastern Canada has several very extensive sections of Llandovery strata, particularly on Anticosti Island and the Gaspé Peninsula and these have proven potential for conodont zonation.

The first conodont-based zonation of the Silurian was made by Walliser (1962, 1964). His work was based mainly on studies of strata exposed at Cellon in the Carnic Alps, and he was able to subdivide the Silurian and basal Devonian into 11 successive conodont zones. Unfortunately the Cellon section is unsatisfactory for the early part of the Silurian because, as Schönlaub (1971) has demonstrated, much of the Llandovery is missing. Revisions to Walliser's (1964) zonation were first made by Nicoll & Rexroad (1969) and Pollock *et al.* (1970) based on conodont faunas obtained from the midcontinent of

North America and by Aldridge (1972, 1975) based on faunas from the United Kingdom. These authors showed that there were Silurian conodonts older than those recognized at Cellon and zonations were proposed for the North American and British Silurian sequences. Conodonts from the earliest part of the Silurian were unknown until the recent work of McCracken & Barnes (1981a, b) in Anticosti Island and Nowlan (1981) in Gaspé; in both places conodonts are associated with Llandovery A brachiopods. In most areas that have been studied in detail (e.g., Cellon, Arbuckle Mountains of Oklahoma, Nevada) only the upper part of the Llandovery is represented, the oldest strata bearing conodonts of the *P. celloni* Zone.

Two conodont zonal schemes are used in the correlation chart (Fig. 2): one based on North American faunas described by McCracken & Barnes (1981a), Nicoll & Rexroad (1969) and Pollock *et al.* (1970) and the other based on British Silurian faunas described by Aldridge (1972). All are described as assemblage zones, although those of Aldridge (1972) are based on first appearance. In North America the oldest Silurian conodont zone is the *Oulodus? nathani* Zone erected by McCracken & Barnes (1981a) based on material from the upper part of the Ellis Bay Formation at Anticosti Island, which has been revised as part of the overlying Becscie Formation by Petryk (1981). The zonal species was first described from Anticosti Island and its only other occurrence is in the Clemville Formation of Gaspé Peninsula (Figs. 1 and 2). The forms assigned to *Oulodus?* are almost certainly representatives of a new genus and their biostratigraphic potential has yet to be determined fully. Younger representatives of the genus including *O.? kentuckyensis* (Branson & Branson) and other as yet unnamed species occur through the Becscie and Gun River formations on Anticosti Island (McCracken & Barnes 1981a; Fåhraeus & Barnes 1981). Study of faunas from the Clemville Formation of Gaspé

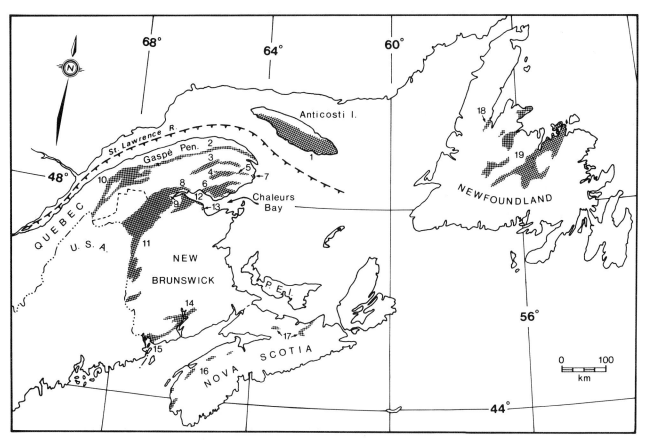

Fig. 1. Map of eastern Canada showing main outcrop areas with Lower Silurian strata. In some areas most of the outcrop is Lower Silurian whereas in others Lower Silurian strata may form only a small portion of the outcrop area. Numbered areas correspond to numbered columns on Fig. 2 and each area is discussed in the text.

Peninsula has shown that the key taxa of the *I. discreta* – *I. deflecta* Assemblage Zone (Aldridge 1972) extend to the base of the Silurian and overlap the *O.? nathani* Zone.

Strata at the base of the Silurian on the northern margin of the Michigan Basin were assigned to the *Panderodus simplex* Zone by Pollock *et al.* (1970). The top of this zone was defined as being immediately below the first occurrence of the genus *Icriodina* or of *Icriodella discreta* (Pollock *et al.* 1970:746). A zone based on species of *Panderodus* has little application and the interval can only be regarded as unzoned due to lack of key taxa. The *O.? nathani* Zone appears to be a suitable replacement and in any case *Icriodella* species have now been shown to extend downward to the base of the Silurian.

In North America the *O.? nathani* Zone is followed by the *Distomodus kentuckyensis* Zone (Cooper 1975) a 'synonym' of the *Icriodina irregularis* Zone of Nicoll & Rexroad (1969) because *I. irregularis* s.f. is the platform element of multielement *D. kentuckyensis*. The representative taxa of this zone are virtually the same as those of the *I. discreta* – *I. deflecta* Assemblage Zone. The *D. kentuckyensis* Zone is succeeded by the *P. celloni* Zone of Nicoll & Rexroad (1969). It is unfortunate that this zone covers a larger stratigraphic interval than the zone of the same name erected by Walliser (1964, 1972) in the Carnic Alps.

In Britain the *I. discreta* – *I. deflecta* Assemblage Zone is succeeded by the *D. staurognathoides* Zone (Aldridge 1972), the base of which coincides with the *D. staurognathoides* Datum Plane of Cooper (1980). The datum plane concept used by Cooper (1980) has limited application to the Llandovery

because only two such planes are represented. *D. staurognathoides* is present in several areas of eastern Canada (see below).

The *D. staurognathoides* Assemblage Zone is followed by the *Icriodella inconstans* Assemblage Zone in Britain (Aldridge 1972) and its top is marked by the first appearance of *Pterospathodus amorphognathoides*. In contrast, *P. amorphognathoides* appears in the top of the *Pterospathodus celloni* in North America, hence the dashed line joining the tops of the *P. celloni* and *I. inconstans* zones on Fig. 2. These two zones are followed in most areas by the *P. amorphognathoides* Zone, the base of which coincides with the *P. amorphognathoides* Datum Plane of Cooper (1980). The *P. amorphognathoides* Zone spans the Llandovery–Wenlock boundary (Walliser 1964; Aldridge 1972; Barrick & Klapper 1976). Specimens of *P. amorphognathoides* are rare in eastern Canada and the taxon seems to be replaced by new species of *Pterospathodus* (Fig. 4).

Studies of Lower Silurian strata and conodonts in eastern Canada have only just begun but it is clear that considerable potential exists for refinement of Early Silurian biostratigraphy. The strong potential of the area lies in the fact that there is a remarkably complete Late Ordovician – Early Silurian sequence of carbonates on Anticosti Island (Nowlan & Barnes 1981; McCracken & Barnes 1981a, b; Fåhraeus & Barnes 1981; Uyeno & Barnes 1981), as well as correlative, thick sequences in both shallow and deep water carbonate-clastic facies to the south on the Gaspé Peninsula (Nowlan 1981). The combination of these two areas provides an insight into environmental preference and geographic distribution of conodont species that should permit a more comprehensive and realistic biozonation.

Geologic setting

Areas of outcrop that include Lower Silurian strata are illustrated on Fig. 1; in some of the areas (e.g. Anticosti Island) the entire shaded portion is Lower Silurian whereas in others the proportion of Lower Silurian strata may be small or uncertain. The numbered areas on the map (Fig. 1) correspond to the numbered columns on the correlation chart (Fig. 2) which are representative but not necessarily comprehensive stratigraphic sections for each area. It is not possible to show all stratigraphic relationships because of the small size of the diagram and the complexity of the relationships between basal Silurian and underlying rocks. In view of the large size of the region under review (a land area of approximately 330,000 km²) only brief comments are possible for each area and the reader is referred to appropriate publications for more complete descriptions both in this review and in the section on biostratigraphy.

A number of distinct lithofacies belts can be identified in the Lower Silurian of eastern Canada. These belts parallel the general northeasterly trend of the Appalachian Orogen. To the west and north of the main Appalachian belt are the virtually flat-lying, thick carbonate sediments of Anticosti Island (1). The strata of this region have been described by Bolton (1972) and Petryk (1979, 1981). The Lower Silurian is represented by about 540 m of shallow marine limestone and shale of the Becscie, Gun River, Jupiter and Chicotte formations. Anticosti Island is part of the eastern segment of the St. Lawrence Platform, bounded to the north and west by the Canadian Shield and to the south and east by the Appalachian Orogen; its unique setting as a cratonic margin basin has resulted in a thick, fossiliferous sequence of shallow-water carbonate rocks.

The rest of the areas shown on Figs. 1 and 2 lie to the south and east of a major structural lineament known as Logan's Line which separates platformal areas from the Appalachian Orogen. The paleogeographic and structural relationships between areas on either side of the lineament during the Early Silurian are uncertain, but a dramatic change in facies can be observed across the lineament.

During the late Early and early Middle Ordovician a number of collisional events took place in the Appalachian Orogen (e.g. Williams & Hatcher 1982). These include the collision of an island arc with the North American craton, related obduction of ophiolite complexes in Québec and Newfoundland and the deformation of a volcanic arc on the eastern side of the orogen. The net result of these Ordovician tectonic events was an apparent fragmentation into regional depressions and uplifted blocks. Deposition in the troughs carried on throughout the late Middle Ordovician and Early Silurian, whereas the uplifted areas were only gradually covered by later Silurian sediments. Hence, the nature of the contact between basal Silurian and older strata varies widely across the region.

Lower Silurian strata of Gaspé Peninsula, the Matapedia Valley and western and northern New Brunswick (2–13) can

be considered as a broad package of lithofacies belts. In the basinal areas (3, 4, 5, 7, 8, 9, 11) conformable contacts exist at the base of the Silurian, whereas in uplifted areas (2, 6, 10, 12, 13) the lowest Silurian strata may be of earliest Llandovery age or younger. Strata in the basinal areas generally show evidence of upward shallowing as the basins were filled with sediment, whereas there is usually a transition from non-marine or shallow marine clastics to more open marine limestones in the uplifted areas.

Lower Silurian strata of a completely different type are exposed in southwestern New Brunswick (14, 15). These outcrop areas are preserved as fault-bound blocks within older strata and they are locally intruded by granite. Thick volcanic flows, tuff and breccia with minor rhyolite and interbedded slate, mudstone and argillaceous limestone characterize the Silurian of this region which is on the eastern side of the Appalachian Orogen.

Two distinct belts of strata of Silurian age crop out in Nova Scotia (16, 17). The Arisaig belt of central Nova Scotia (17) is a continuous sequence ranging from earliest Silurian to early Devonian age (Boucot *et al.* 1974). The Arisaig Group comprises mainly alternating beds of mudstone and siltstone with subsidiary fine-grained sandstone and brachiopod-rich limestone. The unit disconformably overlies volcanic rocks of Ordovician age.

The White Rock Formation of the Annapolis Valley area (16) is characterized by thick quartz sandstone units and mafic to felsic volcanics (Crosby 1962). The unit is underlain by slate of Ordovician age and overlain by the Kentville Formation of Wenlock–Ludlow age. It was probably deposited under shallow marine conditions (Schenk *et al.* 1980) and is part of the Meguma Zone (Williams 1978), a tectonolithologic zone that cannot be traced outside of Nova Scotia in the Appalachian Orogen. The Meguma Zone is interpreted as a segment of a continental embankment of a Precambrian continent, possibly Africa (Schenk 1971; Schenk *et al.* 1980).

The Lower Silurian strata of Newfoundland cannot be related easily to lithofacies belts on the mainland. Fossils are rare and much of the sequence may be non-marine. In the White Bay area (18) the Sops Arm Group comprises ash flow tuffs and rhyolite interbedded into slate, conglomerate, sandstone and local limestone. The unit unconformably overlies part of the Taconic allochthon but, in most places, it is in thrust contact with the underlying rocks (e.g. Smyth & Schillereff 1982).

The Botwood Group (19) of central Newfoundland (Williams 1962) comprises fine- to coarse-grained sandstone interbedded with siltstone and shale, minor limestone and calcareous sandstone. The unit may be as much as 4 500 m thick but only the lower portion is of Early Silurian age. Also included under 19 on Figs. 1 and 2 is an area around Notre Dame Bay, where the Springdale Group, a presumed correlative of the Botwood Group crops out. It is made up mainly of volcanic flows and pyroclastic rocks with some interbedded sandstone and conglomerate that may be non-marine.

Biostratigraphy

Many studies are currently in progress on the Early Silurian of eastern Canada, and this paper is intended as a summary of

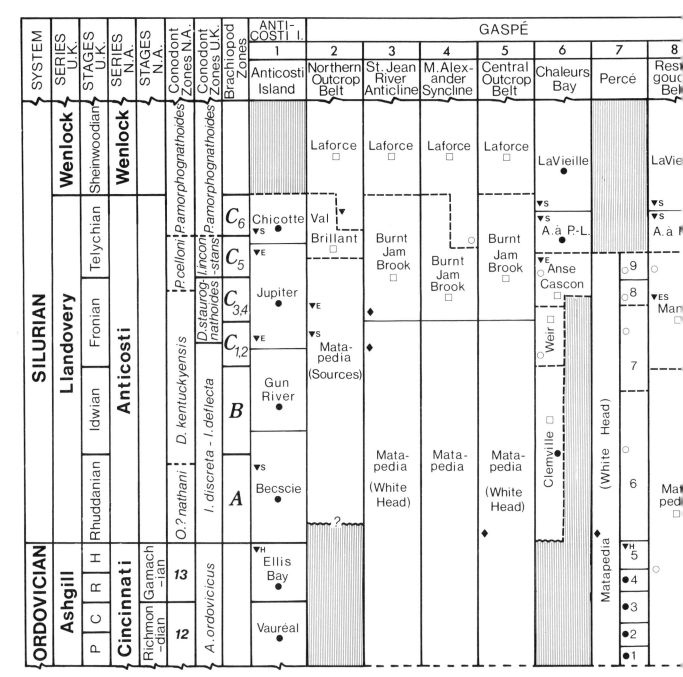

Fig. 2. Correlation chart for Lower Silurian strata of eastern Canada. Derivation of conodont zones is discussed in text. Solid circles indicate presence of conodonts throughout a unit; open circles indicate isolated occurrences of conodonts at the level indicated; triangles indicate levels from which diagnostic brachiopods are known (E is eocoeliid, S is stricklandid and H is Hirnantian); squares indicate that the unit is dominantly composed of clastic rocks and a V within the square indicates that the unit is composed of volcanic and clastic rock. N. B. is New Brunswick; N. S. is Nova Scotia.

the current state of knowledge. The correlation chart (Fig. 2) provides information on the main occurrences of biostratigraphically useful brachiopods and graptolites known from the Lower Silurian of eastern Canada, but its main purpose is to show the occurrences of conodonts recovered to date. Some of the diagnostic conodonts are illustrated on Figs. 3 and 4. Illustrated specimens are deposited in the National Type Fossil Collection at the Geological Survey of Canada (GSC) in Ottawa. Each column of the correlation chart will be discussed in terms of conodont biostratigraphy and the occurrences of diagnostic brachiopods and graptolites. The sources for the

conodont zones on the left side of Fig. 2 have been discussed in the introduction.

Anticosti Island (1)

Anticosti Island has been the subject of intensive study in recent years mainly because it exposes a superb section of Lower Silurian strata. McCracken & Barnes (1981a) have defined the position of the Ordovician–Silurian boundary and they erected a new basal Silurian Zone, the *Oulodus? nathani* Zone. More recently, Nowlan (1982) has pinpointed the Ordovician–Silurian boundary at the eastern end of the island

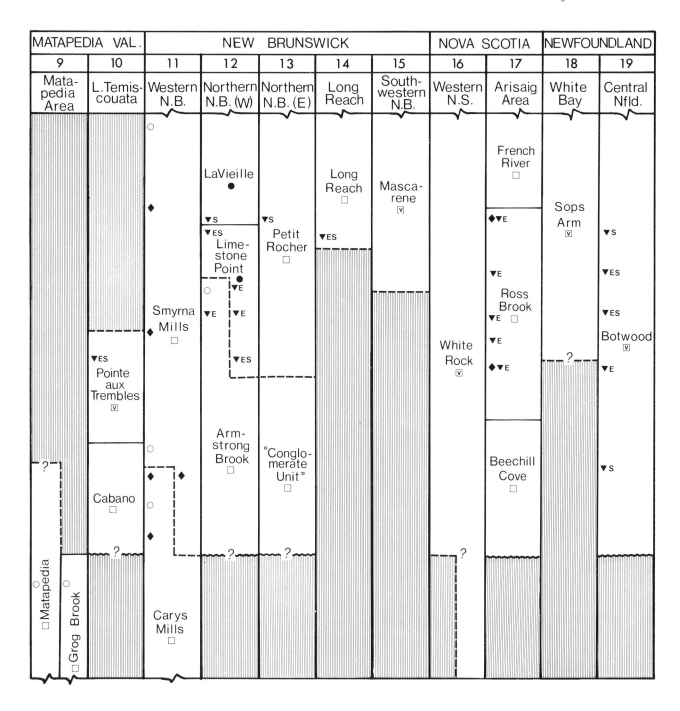

using conodonts in a section originally described by Cocks & Copper (1981); the boundary recognized by the latter authors on the basis of brachiopods is virtually identical to that recognized on the basis of conodonts.

A review of the paleontology of Anticosti Island was compiled in 1981 for a volume to accompany the field guide to a meeting of the IUGS Subcommission on Silurian Stratigraphy and the Ordovician–Silurian Boundary Working Group (Lespérance 1981). In that volume, McCracken & Barnes (1981b) reviewed their earlier work on the Ordovician–Silurian boundary and Fåhraeus & Barnes (1981) and Uyeno & Barnes (1981) provided summaries on conodont biostratigraphy of the middle and upper part of the Lower Silurian, respectively. The *O.? nathani* Zone is succeeded by the *D. kentuckyensis* Zone which also yields *O.? kentuckyensis*, a possible descendant of *O.? nathani*. Faunas from the upper Becscie and

Gun River formations are numerically dominated by elements of *Panderodus*; other common taxa are: *O.? kentuckyensis*, *Ozarkodina* sp. cf. *O. gulletensis* and *Ozarkodina* n. sp. A (Fåhraeus & Barnes 1981). The overall fauna is little changed from that of the lower part of the Becscie Formation, but potential for further zonation lies in specimens of the genus *Oulodus?*.

The fauna from the Jupiter and Chicotte formations has been reported by Uyeno & Barnes (1983). *D. staurognathoides* appears about 18 m above the base of the Jupiter Formation; it occurs throughout and ranges into the lower part of the Chicotte Formation. The lower 14 m of the Jupiter are thus assigned to the *I. discreta–I. deflecta* Assemblage Zone, and almost all the rest of it to the *D. staurognathoides* Zone. Uyeno & Barnes (1983) further subdivided the *D. staurognathoides* Zone into an informal lower *D. staurognathoides* fauna and upper

Ozarkodina aldridgei fauna. About 8 m below the top of the Jupiter Formation *Aulacognathus bullatus* appears together with *Oulodus petilus*. A little higher *Icriodella inconstans* and *O. gulletensis* make their first appearance. Uyeno & Barnes (1983) assigned the upper 2 m of the Jupiter Formation and the lower 24 m of the Chicotte Formation to the *I. inconstans* Assemblage Zone. *Pterospathodus celloni* occurs within this interval. The upper part of the Chicotte Formation is assigned to the *P. amorphognathoides* Zone because of the occurrence of the nominal taxon 24 m above the base of the Chicotte Formation. The Llandovery–Wenlock boundary as determined by conodonts may lie at the top of, or in the upper part of the Chicotte Formation.

Anticosti Island is the region in eastern Canada with the highest potential for refinement of Early Silurian biostratigraphy. An Ordovician–Silurian boundary stratotype has been proposed for the area and the re-establishment of the Anticosti Series has also been advocated (Barnes & McCracken 1981). Field work is in progress in 1982 to try to refine and expand paleontological collections in the systemic boundary interval and in the Lower Silurian strata.

Northern Outcrop Belt (2)

The Sources Formation of the Matapedia Group has not been collected for conodonts, but stricklandid and eocoeliid brachiopods (*S. lens progressa* and *E. intermedia*) have been reported from different levels in the formation (Bourque 1977). The quartzite of the Val Brillant Formation is unfossiliferous. The Laforce Formation has not been collected for conodonts but the lower part has yielded *Pentamerus oblongus* (Bourque 1977) and the ostracode *Zygobolba* (M. J. Copeland, personal communication, 1982). The stratigraphy of this belt has been described in detail by Bourque (1977).

St. Jean River Anticline (3)

No samples for conodonts have been collected from this area. Graptolites from the upper part of the White Head Formation of the Matapedia Group (*Monograptus* cf. *M. sedgwicki*) indicate a Llandovery C_{1-2} age. The occurrence of monograptids assignable to the *M. turriculatus* Zone in the Burnt Jam Brook Formation indicates a Llandovery C_{2-3} age (Bourque 1977). The stratigraphy of this region has been described in detail by Bourque (1977).

Mount Alexander Syncline (4)

Two samples from the lower part of the Laforce Formation have yielded conodonts; these were collected by P.-A. Bourque and the specimens were identified by C. B. Rexroad (Bourque & Lachambre 1981:122). A sample from 4 m above the base of the formation yielded *Pterospathodus pennatus* and *Distomodus staurognathoides* and another from 7 m above the base of the formation yielded *Pterospathodus celloni* and *Ozarkodina* cf. *O. sagitta*. These collections indicate a Llandovery C_{5-6} age for the lower part of the Laforce Formation. The stratigraphy has been described in detail by Bourque & Lachambre (1981).

Central Outcrop Belt (5)

No samples for conodonts have been collected from the Central Outcrop Belt but the Percé area just to the east has been extensively sampled (Nowlan 1981). Graptolites

assigned to *Diplograptus* cf. *D. modestus* and *Climacograptus* cf. *C. scalaris normalis* have been recovered from the middle part of the White Head Formation of the Matapedia Group (Bourque & Lachambre 1981). These indicate an Early Llandovery age.

Chaleurs Bay (Gaspé Peninsula) (6)

Extensive collections for conodonts have been made by the author from the Lower Silurian of the northern shore of Chaleurs Bay in 1979 and 1981. Results from the 1979 collections have been published (Nowlan 1981) but only about half of the 1981 collections have been processed so far. The stratigraphy of this region has been described in detail by Bourque (1975) and Bourque & Lachambre (1981).

Strata at the base of the Silurian are referred to the Clemville Formation (110 m) and brachiopods of the *Stricklandia lens*

Fig. 3. Early Silurian conodonts of eastern Canada; index of sections provided in appendix. □A, D. *Oulodus? kentuckyensis* (Branson & Branson), lateral views. □A. GSC 72367; Section 1A, 17.5 m above base of Clemville Formation; ×50. □D. GSC 72368, Section 1A, 95 m above base of Clemville Formation; ×60. □B *Oulodus? nathani* McCracken & Barnes, lateral view, GSC 66501; Section 1A, 94 m above base of Clemville Formation; ×50. □C, G, K. *Ozarkodina oldhamensis* (Rexroad). □C. Lateral view, *g* element, GSC 72369; Section 1A, 17 m above base of Clemville Formation; ×85. □G. Lateral view, *f* element, GSC 72370; Section 1B, 7.8 m above base of Clemville Formation; ×70. □K. Posterior view, *e* element, GSC 72371; locality as C; ×105. □E, F. *Distomodus* n. sp. A. □E. Oral view, *g* element, GSC 72372; Section 1B, 1 m above base of Clemville Formation; ×60. □F. Posterior view, *b* element, GSC 72373; locality as E; ×50. □H. *Ambalodus galerus* Walliser s. f., lateral view, GSC 72374; Section 9, 25 m above base of Limestone Point Formation; ×50. □I, L, M, O, P. *Ozarkodina* aff. *O. hadra* (Nicoll & Rexroad), lateral views. Note the considerable stratigraphic range of these specimens: M occurs about 17 m below beds containing *O. sagitta rhenana*. All differ from typical *O. hadra* in the more anterior situation of the basal cavity apex. □I. *g* element; GSC 72375; Section 8, 47 m above base of Limestone Point Formation; ×50. □L. *g* element, GSC 72376; Section 7, 35m above base of Limestone Point Formation; ×50. □M. *g* element, GSC 72377; Section 8, 83 m above base of La Vieille Formation; ×40.□O. *g* element, GSC 72378; Section 8, 54.5 m above base of Limestone Point Formation; ×70. □P. *f* element, GSC 72379; locality as O; ×105. □J. *Ozarkodina* cf. *O. gulletensis* (Aldridge), lateral view, *g* element, GSC 72380; Section 6, Limestone Point Formation, 51 m above base of section; ×35. This specimen differs from typical *O. gulletensis* in that it is not arched and the posterior denticles are as high as those on the anterior. □N, Q, U. *Ozarkodina excavata excavata* (Branson & Mehl), lateral views. □N. *f* element, GSC 72381; locality as M; ×100. □Q. *g* element of the type that Rhodes (1953) described as *Prioniodella inclinata*, GSC 66558; Section 2A, 25 m above base of Anse à Pierre-Loiselle Formation; ×70. □U. *g* element of the type that Branson & Mehl (1933) described as *Ozarkodina simplex*, GSC 72382; locality as M; ×55. □R, S, T. *Ozarkodina gulletensis* (Aldridge). Note that these forms are very late Llandovery because of the occurrence of Wenlock conodonts only about 17 m above. □R. Lateral view, *g* element, GSC 72383; locality as M; ×45. □S. Lateral view, *f* element, GSC 72384; locality as M; ×55. □T. Posterior view, *e* element, GSC 72385; locality as M; ×65. □V, W. *Kockelella ranuliformis* (Walliser), lateral and oral views, *g* element, GSC 72386; Section 10, Limestone Point Formation, 70 m above base of section; ×80. □X. *Ozarkodina* n. sp. A Nowlan 1981, lateral view, GSC 66525; Section 3, top of unit 7, White Head Formation; ×105. □Y, Z. *Icriodella inconstans* Aldridge, oral views, *g* elements. □Y. GSC 72387; Section 3, base of unit 8, White Head Formation; ×60. □Z. GSC 72388; locality as J; ×60. □Aa. *Icriodella deflecta* Aldridge, oral view, *g* element, GSC 66488; Section 1B, 108 m above base of Clemville Formation; ×80. □Bb. *Icriodella discreta* Pollock, Rexroad & Nicoll, oral view, *g* element, GSC 72389; locality as A; ×85.

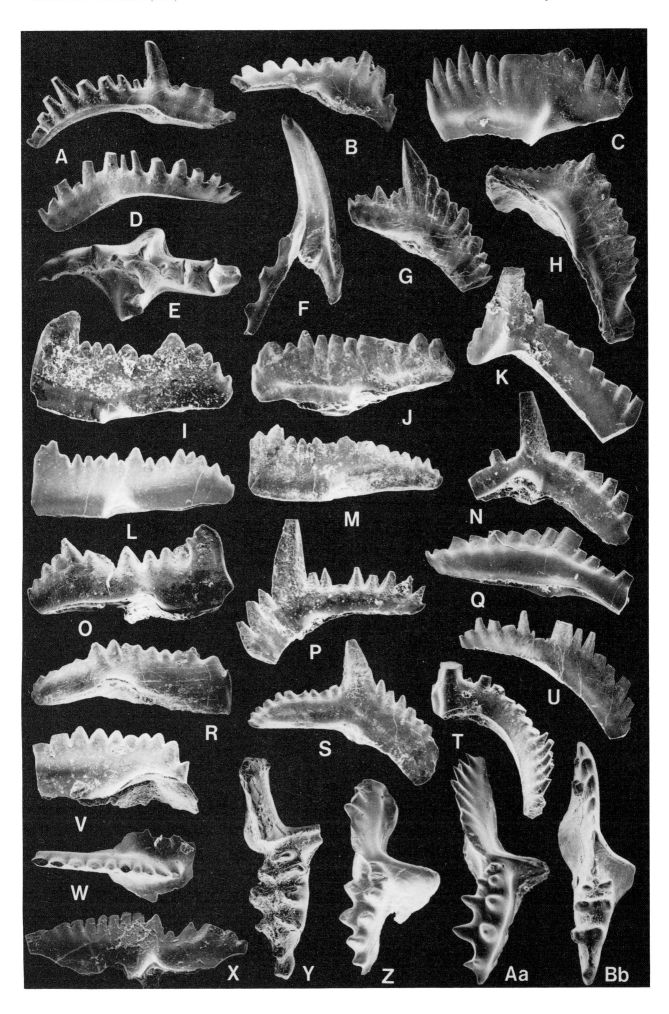

typica Phylozone (Llandovery A$_{3-4}$) have been reported from the upper two thirds of the unit (Boucot & Bourque 1981). The age of the lowest third of the formation is uncertain based on brachiopods.

Conodonts are present throughout the formation but samples from the lowest 3 m yield only simple cones and in one case a probable new species of *Distomodus* (Fig. 3 E, F). The following species are common in the lower 30 m of the formation: *Icriodella discreta* (Fig. 3 Bb; first appearance 4 m above base), *I. deflecta* (Fig. 3 Aa), *Ozarkodina hassi, O. oldhamensis* (Fig. 3 C, G, K) and simple cones of the genus *Panderodus*. Rare components of the fauna are *Oulodus? nathani* (Fig. 3 B), *Oulodus? kentuckyensis* (Fig. 3 A, D) and elements of the simple cone genera *Decoriconus* and *Walliserodus*. Nowlan (1981, Fig. 5) illustrated the stratigraphic ranges of conodont taxa in the Clemville Formation recovered from the 1979 collections, and examination of the 1981 collections yielded few exceptions besides the presence of *Distomodus* n. sp. A at 1 m above the base of the formation and the extension of the range of *O.? kentuckyensis* into the lower part of the formation. Over 1200 conodont specimens have now been recovered from the Clemville Formation and the faunas are numerically dominated by specimens of *Icriodella, Panderodus* and *Ozarkodina*.

The presence of a few specimens of *O.? nathani* indicates an earliest Llandovery age for the base of the sequence and the overall faunal content is very similar to that of the lowest Silurian strata on Anticosti Island (see McCracken & Barnes 1981a, b). The fauna does not change much through the Clemville Formation, except that *Distomodus kentuckyensis* appears 2 m from the top. The appearance of this taxon is consistent with Llandovery A$_4$ age indicated by the brachiopods (Boucot & Bourque 1981).

The overlying Weir Formation (45 m) is composed mainly of arkose but three units of calcarenite have been collected from near the base. These yield essentially the same fauna as that of the upper part of the Clemville Formation. Specimens of *Icriodella* and *Panderodus* dominate the fauna. Nowlan (1981, Pl. 4:6) indicated the presence of specimens of *I.* aff. *I. malvernesis* in the Weir Formation which may suggest an early Llandovery C age. The Weir Formation does not contain diagnostic brachiopods although those of the immediately underlying Clemville Formation are of Llandovery A$_4$ age. Strata of Llandovery B age cannot be identified on the basis of brachiopods in the Chaleurs Bay region of Gaspé Peninsula (Boucot & Bourque 1981).

The succeeding Anse Cascon Formation is composed mainly of sandstone, and only a thin interval of carbonate rock occurs within the formation. Samples from this carbonate interval yield mainly specimens of *Panderodus*. Specimens possibly assignable to *Ozarkodina gulletensis* were illustrated by Nowlan (1981). In addition, specimens of *I. inconstans* are known from the same level, indicating a probable early Telychian age for the strata. Boucot & Bourque (1981) recognize the *Eocoelia curtisi* Phylozone in the middle part of the Anse Cascon which indicates a Llandovery C$_{4-5}$ age (late Fronian–early Telychian).

The Anse à Pierre-Loiselle Formation is abundantly fossiliferous. *Costistricklandia lirata* (Llandovery C$_6$ – early Wenlock) is present in the upper part of this formation and ranges into the lower part of the overlying La Vieille Formation.

Specimens of *Panderodus* are numerically dominant in samples from the Anse à Pierre-Loiselle Formation but several other taxa are represented. Nowlan (1981) has illustrated several of the species recovered. The diverse fauna includes *Apsidognathus tuberculatus, Carniodus carnulus, Distomodus staurognathoides, Pterospathodus pennatus, P.* n. sp. A, *P.* n. sp. B., *Ozarkodina excavata excavata* and *O.* cf. *O. gulletensis*. The forms assigned to *Pterospathodus* n. sp. A are those referred to *P.* aff. *P. amorphognathoides* by Nowlan (1981). Hundreds of specimens have been recovered since the initial report, and it is clear that this species with a tridenticulate lateral process is a new species. It is closely related to *P.* n. sp. B which will be discussed under northern New Brunswick (column 12).

Regionally the Anse à Pierre-Loiselle Formation is correlative with part of the Limestone Point Formation on the south shore of Chaleurs Bay (Fig. 2, column 12) and also with the Chicotte Formation of Anticosti Island (Uyeno & Barnes 1981).

The only specimen of *P. amorphognathoides* recovered from eastern Canada outside of Anticosti Island comes from the overlying La Vieille Formation. *P. amorphognathoides* is remarkable by its general absence and it seems to be replaced by *Pterospathodus* n. sp. B which ranges into strata of probable early Wenlock age. The lower part of the La Vieille Formation is probably of late Llandovery age but the upper part is of definite early Wenlock age (Nowlan 1981:273). The definition of the Llandovery–Wenlock boundary is not possible based on brachiopods or conodonts because the key taxa cross the boundary. A brief discussion of the Llandovery–Wenlock boundary in the region is provided below.

Percé (7)

Conodonts from the Percé region have been reported in some detail by Nowlan (1981). Additional collections have been made since the preparation of that article and some new results are available.

The base of the Silurian at Percé is taken to lie just above unit 5 which contains a Hirnantian fauna (see Lespérance & Sheehan 1981). Extensive and closely spaced sampling through the lower part of unit 6 undertaken in 1981 has proved fruitless. Not a single conodont element has been recovered from 10 large (5 kg) samples from the lower 10 m of unit 6. The upper half of unit 6 (240 m) yields *Distomodus kentuckyensis* and an assortment of simple cones (Nowlan 1981; Fig. 4). Unit 7 (35 m) has yielded *D. kentuckyensis*, some diminutive representatives of generalized *Ozarkodina* species and a diversity of simple cone elements (Nowlan 1981:267). The succeeding unit 8 bears an impoverished fauna of diminutive *Ozarkodina* sp. and simple cones.

The age of the uppermost unit was in doubt until Nowlan (1981) recovered a rather poor specimen of *Aulacognathus bullatus* from the unit. Additional collecting has produced several well preserved specimens of *A. bullatus* (Fig. 4 E, O, P) confirming an upper age limit of Llandovery C$_5$ (early Telychian) for the White Head Formation. *A. bullatus* is also known from the upper part of the Jupiter Formation on Anticosti Island (Uyeno & Barnes 1981).

Skidmore & Lespérance (1981) have pointed out the marked lithologic resemblance of unit 7 to the Burnt Jam Brook Formation of the interior and the resemblance of the base of unit 8 to the base of the Laforce Formation. It can be pointed out that conodonts from unit 7 could correlate with graptolites recovered from the Burnt Jam Brook Formation (Fig. 2, columns 3, 4) and that conodonts from unit 9 (*A. bullatus*) correlate well with those from the base of the Laforce Formation in the Mount Alexander Syncline (*P. celloni*, see above). If the age of the base of the Laforce is regionally lowered as it must be in the Mount Alexander Syncline, then it is possible that the coarse conglomerate and limestone at the base of unit 9, rather than those at the base of unit 8, correlate with strata of similar lithology at the base of the Laforce Formation. Further information is required on conodonts from the base of the Laforce Formation.

Restigouche Belt (8)

The Matapedia Group has been collected in the area of the Restigouche Belt but all samples are of Late Ordovician age (Nowlan, unpublished data). The only Silurian conodonts known from the Restigouche Belt are those identified by C. B. Rexroad from the Mann Formation and reported in Bourque & Lachambre (1981, Fig. 58). The fauna comprises: *Carniodus* sp., *Diadelognathus* sp., *Icriodella* cf. *I. inconstans* and *Ligonodina variabilis*. This may suggest a Llandovery C_5 (Telychian) age for part of the Mann Formation. *Costistricklandia lirata*(= *C. gaspeensis*) has been recorded from the upper part of the Anse à Pierre-Loiselle Formation and lower part of the La Vieille Formation by Bourque & Lachambre (1981). These two formations have not been collected for conodonts in this region, except for a single sample of Wenlock age reported by Bourque & Lachambre (1981: 118) from the upper part of La Vieille Formation, which yielded specimens assigned by C. B. Rexroad to *Spathognathodus primus*, *Trichonodella excavata* and *Ozarkodina? simplex*.

Matapedia Area (9)

Conodonts frm the Matapedia and Grog Brook groups of the Matapedia area in Quebec and New Brunswick are of Late Ordovician age (Nowlan 1981; in press). No Silurian conodonts have been recovered from the Matapedia Group in this area despite extensive sampling. Silurian megafossils and ostracodes have been reported (but not illustrated) from the area.

Lake Temiscouata (10)

No collections have been made for conodonts from the Lake Temiscouata region. The Cabano Formation is of early Llandovery age based on the occurrence of the brachiopods *Protatrypa* and *Fardenia* (see Lajoie *et al.* 1968: 624) in the upper part of the unit. The succeeding Pointe aux Trembles Formation is of about Llandovery C_{1-2} age based on the presence of the brachiopods *Stricklandia lens* cf. *progressa* and *Eocoelia hemisphaerica* (Berry & Boucot 1970; Lajoie *et al.* 1968).

Western New Brunswick (11)

The Carys Mills Formation is a southwesterly extension of the Matapedia Group and it has been dated on the basis of graptolites (Pavlides 1968; Pavlides & Berry 1966). Berry (*in* Pavlides 1968, Table 3) assigned some graptolites from the Carys Mills Formation to zone 13, the *O. truncatus intermedius* Zone of Caradoc age, and others he considered to be of Early

Silurian age. Recently, Rickards & Riva (1981) have demonstrated that the graptolites assigned to zone 13 are in fact of basal Silurian age and consist of only one species, *Glyptograptus? persculptus*. This determination places all of the Carys Mills faunas in the Lower Silurian, but it is possible that the formation extends down into the Upper Ordovician.

The only conodonts recovered from an outcrop believed to be of the Carys Mills Formation were obtained by the author from exposures on the Trans Canada Highway, 1 km south of the town of Grand Falls, New Brunswick. This region has not been mapped recently but lies on strike with the Carys Mills Formation of northeastern Maine as delineated by Pavlides (1968). The possibility remains that these exposures will be reassigned to the Smyrna Mills Formation when the area is mapped in detail. The faunules recovered from the area are poorly preserved and fragmentary but include representatives of *Icriodella deflecta*, *Walliserodus curvatus*, *Panderodus* spp. and a species of *Ozarkodina* represented by ozarkodiniform elements quite unlike those of *O. hassi* or *O. oldhamensis* and more like those assigned to a younger species such as *O. gulletensis*. This sparse evidence suggests an early Llandovery age (A–C_1) for these outcrops. This is consistent with graptolite evidence from the Carys Mills Formation.

The Smyrna Mills Formation has been dated by means of graptolites (Pavlides & Berry 1966, see Table 2). The graptolites indicate a range in age from early Llandovery to early Ludlow. Two samples from the Smyrna Mills Formation in western Brunswick have yielded conodonts. No systematic attempt has been made yet to collect the formation in detail. A sample (GSC Loc. 98599) collected from strata just west of Payton Lake, 13 km NNW of Woodstock, New Brunswick yielded fragmentary specimens of *Icriodina* and '*Drepanodus suberectus*' *sensu* Rexroad (1967). This assemblage is suggestive of an early Llandovery age. A second sample (GSC Loc. 97517) from an outcrop along a tributary of the North Becaguimec River, 4 km NNW of Carlisle, New Brunswick, yielded conodonts of Wenlock age including specimens of *Ozarkodina sagitta* that most closely resemble *O. s. rhenana* (see St. Peter 1982).

Northern New Brunswick (west part) (12)

More than 160 samples have been processed for conodonts from Lower Silurian strata of northern New Brunswick. The following sections have been collected (the references in parentheses direct the reader to an illustration of each of the sections): Flanagans, Quinn Point, Hendry Brook and Limestone Point (Noble 1976, Fig. 2, Sections 2, 3, 6 and 8); Point La Roche and Dickie Cove–Black Point (Lee & Noble 1977, Fig. 6). Over 10,000 conodont specimens have been recovered from the Armstrong Brook, Limestone Point and La Vieille formations.

The Armstrong Brook Formation is mostly composed of non-marine clastic rocks but a few thin limestone units are present near the top of the formation in some places. Samples are numerically dominated by specimens of *Panderodus*, but *Pterospathodus* n. sp. B and *Ozarkodina* aff. *O. hadra* have also been recovered. The nature of the latter two species is discussed below. Noble (1976) has identified *Eocoelia intermedia* from the upper part of the Armstrong Brook Formation, suggesting a Llandovery C_{3-4} age. The species recovered from the unit range up into the overlying Limestone Point Formation.

The Limestone Point Formation has yielded the brachiopod *E. intermedia* near the base (at Hendry Brook) and bears *Eocoelia sulcata* and *Costistricklandia gaspensis* (= *C. lirata*) at the top in most sections suggesting an upper limit of Llandovery C₆–Wenlock age. Conodonts are abundant throughout the unit.

Conodonts occurring at the base of the Limestone Point Formation include *Apsidognathus tuberculatus* (Fig. 4 A, B), *Distomodus staurognathoides*, *Ozarkodina* cf. *O. gulletensis* (Fig. 3 J), *O. hadra*, *O.* aff. *O. hadra* (Fig. 3 I, L, M, O, P), *Pterospathodus* n. sp. A (Fig. 4 T, W, Z) and *P.* n. sp. B. (Fig. 4 J, L, Q, R, U). Specimens of *Panderodus* numerically dominate the samples, but specimens of *Pterospathodus* n. sp. B and *Ozarkodina* spp. are also very common. *A. tuberculatus* and *D. staurognathoides* are locally abundant. This assemblage suggests a Telychian (Llandovery C₅₋₆) age for the base of the Limestone Point Formation. The base of the formation may be older than this at the Hendry Brook section because of the recovery of *E. intermedia*. The conodont fauna typical of the base of the Limestone Point Formation in most sections, does not appear until 40 m above the base of the Hendry Brook section also suggesting an older base to the formation at this locality. Samples from the lower 40 m of the Limestone Point Formation at Hendry Brook are overwhelmingly dominated by diminutive specimens of *Panderodus* with only sparse representatives of *Oulodus?* sp. and *Pterospathodus* n. sp. B. The base of the Limestone Point Formation is therefore probably diachronous.

The specimens assigned to *Pterospathodus* n. sp. A are equivalent to specimens illustrated as *P.* aff *P. amorphognathoides* from the Anse à Pierre-Loiselle Formation (Fig. 2, column 6) by Nowlan (1981). This species bears three denticles on the inner lateral process and has now been recovered in sufficient quantity to indicate that it bears little relationship to *P. amorphognathoides* but rather it is a new species.

The specimens assigned to *Pterospathodus* n. sp. B are similar to those of *P.* n. sp. A but there is a very high degree of variability. Typical specimens have a trilobate inner lateral process with prominent denticle rows on each lobe; usually one of the denticle rows is shorter than the other two. One to three low nodes are present on an expansion of the base on the outer side of the element. The ozarkodiniform element that belongs in this apparatus is very similar in lateral profile to the pterospathodiform element, and also very similar to the ozarkodiniform element of *Pterospathodus* sp. A (see Fig. 4 R, T).

A wide diversity of spathognathodiform elements belonging to species of *Ozarkodina* are present (together with other elements of the apparatus) in the Limestone Point Formation. Some are readily assignable to *O. gulletensis* (Fig. 3 R, S, T) and *O. hadra* whereas others differ in some respects from both of these species. Those referred to *O.* aff. *O. hadra* are similar to *O. hadra* except that the basal cavity is centrally to anteriorly located like that in *O. confluens*. The diversity of elements is sufficient to draw analogies with each of the morphotypes of *O. confluens* described by Klapper & Murphy (1975). This broad range of forms is present from the base to the top of the

formation and cast some doubt on the potential utility of species of *Ozarkodina* for biostratigraphy. *Ozarkodina excavata excavata* is a distinctive species present in many samples of the Limestone Point Formation.

Rare components of the fauna from the Limestone Point Formation include *Ambalodus galerus* s.f. (Fig. 3 H), *Carniodus* sp., *Kockelella ranuliformis* (Fig. 3 V, W), *Pterospathodus* cf. *P. celloni* (Fig. 4 V, X, Y) and *P. pennatus* (Fig. 4 S). *K. ranuliformis*, most common in the *P. amorphognathoides* Zone, is present near the top of the formation at Hendry Brook, suggesting that the upper part of the unit is Llandovery C₆–Wenlock in age. Specimens assigned to *P.* cf. *P. celloni* occur rarely in samples from the middle part of the formation. They are similar to *P. celloni* but lack basal expansion.

Aulacognathus bullatus is very abundant in the Dickie Cove section and it occurs there together with *A. kuehni* (Fig. 4 I) and forms transitional between the two species. The Dickie Cove section also provides the only occurrence of *Icriodella inconstans* where it is present in considerable numbers. The same section has yielded a few specimens of *Astropentagnathus irregularis* (Fig. 4 D).

The simple cone fauna of the Limestone Point Formation is dominated by specimens of *Panderodus*, with rare occurrences

---

Fig. 4. Early Silurian conodonts of eastern Canada; index of sections provided in appendix. □A, B. *Apsidognathus tuberculatus* Walliser. □A. Oral view, *g* element, GSC 72390; Section 9, 110 m above base of Limestone Point Formation; ×35. □B. Aboral view, *f* element, GSC 72391; Section 10, Limestone Point Formation, about 130 m above base of section. □C. *Apsidognathus* aff. *A. walmsleyi* Aldridge, oral view, *g* element, GSC 72392; Section 6, Limestone Point Formation, 51 m above base of section; ×75. □D. *Astropentagnathus irregularis* Mostler, oral view, *g* element, GSC 72393; locality as C; ×50. □E, O, P. *Aulacognathus bullatus* (Nicoll & Rexroad). □E. Oral view, *g* element, GSC 72394; Section 4, 40 m above base of unit 9, White Head Formation; ×60. □O, P. Aboral and oral views, *g* element, GSC 72395; locality as E; ×60. □F-H. *Distomodus staurognathoides* (Walliser), oral views, *g* elements. □F. GSC 72396; Section 3, 0.8 m above base, unit 8, White Head Formation; ×60. □G. GSC 72397; Section 2B, 78 m above base of Anse à Pierre-Loiselle Formation; ×50. □H. GSC 72398; locality as C; ×40. □I. *Aulacognathus kuehni* Mostler, oral view, *g* element, GSC 72399; locality as C; ×65. □J, L, Q, R, U. *Pterospathodus* n. sp. B. □J. Oral view, *g* element, GSC 72400; Section 8, 15 m above base of Limestone Point Formation; ×70. □L. Oral view, *g* element, GSC 72401; locality as J; ×90. □Q. Lateral view, *g* element, GSC 72402; Section 8, 70 m above base of La Vieille Formation; ×65. □R. Lateral view, *f* element, GSC 72403; Section 8, 46.7 m above base of Limestone Point Formation; ×35. □U. Oral view, *g* element, GSC 72404; locality as R; ×80. □K. *Pterospathodus amorphognathoides* Walliser, oral view, *g* element, GSC 66570; Section 5, 70 m above base of La Vieille Formation; ×50. □M, N. *Distomodus kentuckyensis* Branson & Branson. □M. Oral view, *g* element, GSC 66512; Section 1B, 12.7 m above base of Clemville Formation; ×70. □N. Inner lateral view, *e* element, GSC 66508; locality as M; ×45. □S. *Pterospathodus pennatus* (Walliser), oral view, *g* element, GSC 66566; Section 2A, 25 m above base of Anse à Pierre-Loiselle Formation; ×100. □T, W, Z. *Pterospathodus* n. sp. A (= *P.* aff. *P. amorphognathoides* of Nowlan 1981). □T. Lateral view, *f* element, GSC 66568; locality as S; ×70. □W. Lateral view, *g* element, GSC 66567; locality as S; ×100. □Z. Oral view, *g* element, GSC 66569; locality as S; ×90. □V, X, Y. *Pterospathodus* cf. *P. celloni* (Walliser). Elements of this species are rare and none shows the asymmetrical flaring of the base typical of *P. celloni* s.s. □V. Lateral view, *f* element, GSC 72405; Section 10, Limestone Point Formation, 96 m above base of section; ×85. □X. Oral view, *g* element, GSC 72406; locality as V; ×85. □Y. Lateral view, *g* element, GSC 72407; locality as V; ×60.

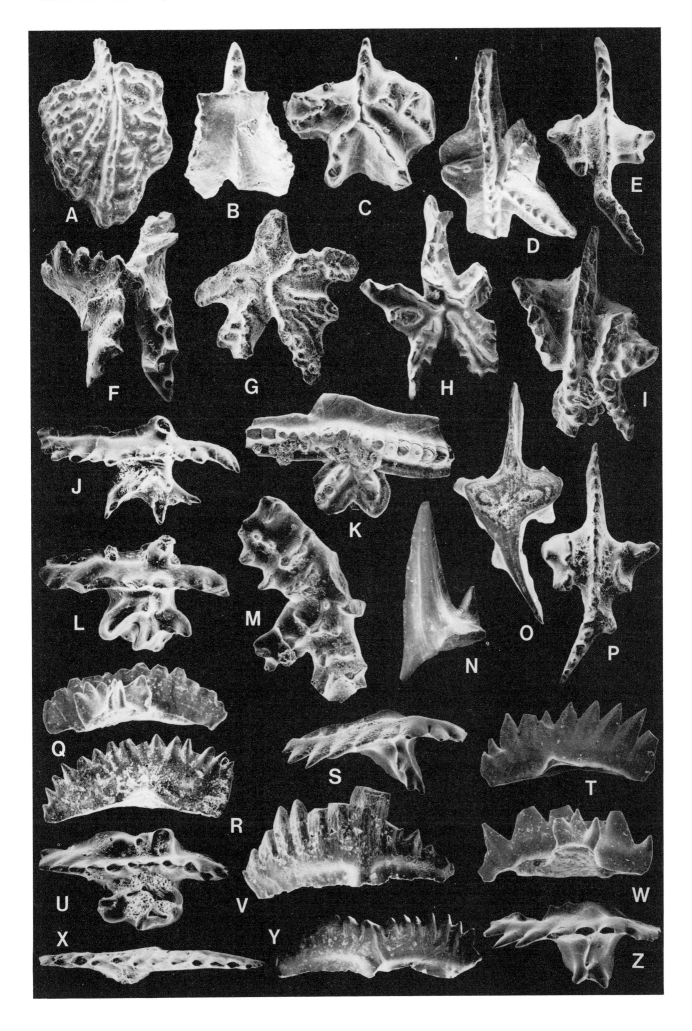

of *Pseudooneotodus beckmanni*. No specimens of *P. bicornis* have been recovered so far.

The fauna of the lower part of the overlying La Vieille Formation is similar in some respects to that of the upper part of the Limestone Point Formation, but it is much less diverse and most samples yield only sparsely. Specimens of *Panderodus* and *Ozarkodina* dominate the fauna. *D. staurognathoides* and *Pterospathodus* n. sp. B are present in a few samples near the base of the formation and occur sporadically through the lower 70 m of the formation. The absence of biostratigraphically diagnostic forms in this part of the sequence makes placement of the Llandovery–Wenlock boundary difficult.

Conodonts of diagnostic Wenlock age occur at a level estimated to be about 100 m above the base of the La Vieille Formation (total 300 m) in a railroad cutting inland from Quinn Point. *Ozarkodina sagitta rhenana* occurs in reasonable abundance at this level and it indicates an early Wenlock, Sheinwoodian age. It co-occurs with specimens referable to *O. confluens*.

It is concluded that the La Vieille Formation in northern New Brunswick is correlative with strata assigned to the same formation on the north side of Chaleurs Bay and with part of the Smyrna Mills Formation in western New Brunswick. The problem of the position of the Llandovery–Wenlock boundary is discussed below.

Northern New Brunswick (east) (13)

Noble (1976) has pointed out that easterly exposures of Lower Silurian rocks in northern New Brunswick differ markedly in lithology from those in the centre and west, and he interpreted this as deepening eastward into a basin. The Petit Rocher Formation lies in the same stratigraphic position as the Limestone Point and La Vieille Formations to the west. Noble & Howells (1979) interpreted the Petit Rocher Formation as being of shelf basin to slope origin. *Costistricklandia lirata* has been recovered about 300 m above the base of the formation and two samples for conodonts from this level failed to yield any diagnostic forms. Only a few fragmentary specimens of *Panderodus* were recovered. The samples were taken from the Pointe Rochette Section (Noble 1976, Fig. 2, Section 10). A considerable thickness of strata remain to be examined in detail.

In this region other relatively unfossiliferous sequences have been described as slope to basin facies equivalents of the Limestone Point and La Vieille Formation. One of these has been briefly described by Noble (1980:6) as the LaPlante Section, and it is located on Highway 11, 4 km inland (west) from the Pointe Rochette section just mentioned. It is interesting to note that this section yielded conodonts of Late Silurian–Early Devonian age including *Ozarkodina remscheidensis* and a single specimen assignable to *Icriodus woschmidti?*. Clearly this section is part of another basin of much younger age and cannot be interpreted as representing basinal equivalents of the La Vieille Formation.

Southwestern New Brunswick (14, 15)

No samples for conodonts have been collected from this region. The Long Reach Formation comprises mostly volcanic flows and breccias with minor sediments. Berry & Boucot (1970) report the occurrence of *Costistricklandia* and

Eocoelia sulcata from the lower part of the formation, suggesting a late Llandovery–early Wenlock age. The upper part of the formation may be as young as Ludlow. About 80 km to the southwest, the Mascarene Group (15) also comprises thick volcanic flows, tuff and breccia (Smith 1966). The lower part of the unit is unfossiliferous in Canada, but the probably correlative Quoddy Formation in adjacent Maine has yielded fossils of late Llandovery–Wenlock age (Bastin & Williams 1914; Berry & Boucot 1970).

Western Nova Soctia (16)

The White Rock Formation of the Annapolis Valley is of uncertain age. The only fossil reported from it is a brachiopod that is Caradoc or younger (Lane 1975:47). Several samples of carbonate sediments associated with pillow basalts in the Fales River Section (Schenk *et al.* 1980: 49, 50) of the White Rock Formation have been processed for conodonts but failed to yield any specimens.

Arisaig Area (17)

The only conodonts described from the Arisaig area are from the Stonehouse Formation of Late Silurian age (Legault 1968). The substantial thickness of Lower Silurian sediments remains uncollected. The age of the sediments in the region is well established on the basis of brachiopods (Boucot *et al.* 1974; Harper 1973). The Ross Brook Formation contains *Eocoelia hemisphaerica, E. intermedia, E. curtisi* and *E. sulcata* indicating a Landovery C_1–C_6 age. It also contains graptolites indicative of a range from zones 19 to about 25 (Berry & Boucot 1970). The succeeding French River Formation is probably of Wenlock age but it does not contain diagnostic fossils.

White Bay (18)

The lower part of the Sops Arm Group comprises volcanics, but carbonate sediments occur in the middle and upper parts of the formation (Lock 1972; Smyth & Schillereff 1982). Lock (1972: 322) makes passing mention of poorly preserved fossils indicating middle to late Silurian ages; neither the nature of the fossils nor the source of identification is provided. Smyth & Schillereff (1982: 90) indicate that gastropods, crinoids, corals and rare brachiopods occur in the Simms Ridge Formation in the middle part of the Sops Arm Group. They also report Silurian conodonts from the Natlins Cove Formation at the top of the Sops Arm Group (Smyth & Schillereff 1982: 91), but no precise age is indicated and no faunal list is provided. It is clear that potential for conodont recovery in the Sops Arm Group is good but they may all be younger than Early Silurian.

Central Newfoundland (19)

The age of the Botwood Group in central Newfoundland is reasonably well established on the basis of brachiopods and graptolites (Berry & Boucot 1970: 125, 126) and it ranges in age from early Llandovery to Ludlow. Little collecting has been done for conodonts. The age of the Springdale Group in the western part of central Newfoundland is less well known because it is made up of non-marine clastic sediments. A single sample from the Botwood Group near Glendale, Newfoundland (GSC Locality 99226) contains only fragments of *Panderodus* and spathognathodiform elements referable to a species of *Ozarkodina*. The unit is largely made up of clastic sediments unsuitable for acid digestion.

Ordovician–Silurian Boundary

Continuous stratigraphic sequences across the Ordovician–Silurian boundary are present in two different facies in eastern Canada. On Anticosti Island (1) McCracken & Barnes (1981a, b) have described and discussed the conodont faunas from a shallow carbonate facies. They have shown that Ordovician and Silurian conodont faunas are distinguishable at both the generic and specific level with many Ordovician genera being extinguished across the boundary. Only simple cone taxa including, *Decoriconus costulatus, Panderodus gibber, P. serratus, Pseudooneotodus beckmanni* and *Walliserodus* cross the boundary (McCracken & Barnes 1981a, Fig. 12).

Conodonts and megafauna occur in continuous sequences across the boundary in the deeper water facies of the Matapedia Group in Gaspé Peninsula (3, 4, 5, 7, 8). These sections provide useful correlation between the predominantly North American megafaunas of Anticosti Island and the generally European megafaunas of the Matapedia Group, but unfortunately the presumed earliest Silurian units are virtually unfossiliferous. The details of the Ordovician–Silurian boundary in this region have been reviewed by Nowlan (1981).

Elsewhere in eastern Canada, the position of the boundary is either in doubt or else strata were not deposited in the late Ordovician and early Silurian because of the irregular post-Taconic (Middle Ordovician) topography.

Llandovery–Wenlock Boundary

It is difficult to define faunally the Llandovery–Wenlock boundary in eastern Canada. The diagnostic brachiopod *Costistricklandia lirata* and conodont *Pterospathodus amorphognathoides* both cross the series boundary. Conodonts of definite Wenlock age such as *Ozarkodina sagitta rhenana* occur in the sequence but these do not serve to define the *base* of the Wenlock.

On Anticosti Island, Uyeno & Barnes (1983) reported *P. amorphognathoides* from the upper part of the Chicotte Formation and concluded that most and probably all of the Chicotte is of Llandovery age. The general scarcity of *P. amorphognathoides* in the region is probably a result of ecologic control (see below) so that the range of the species may be considerably restricted in this region. It is not possible, however, to identify the Llandovery–Wenlock boundary on the basis of conodonts on Anticosti Island. Barnes & McCracken (1981) have proposed that the Anticosti Series be reconsidered as the basal Silurian series and therefore it is important to try to identify the base of the Wenlock on Anticosti Island. If the Anticosti series is adopted as the lower Silurian series, we cannot be certain how its top relates to the base of the Wenlock. Detailed study of faunas from the Chicotte Formation should be undertaken soon.

In the Chaleurs Bay region, a general depletion in diversity and abundance takes place across the contacts of the Limestone Point Formation to the south, and the Anse à Pierre-Loiselle Formation to the north, with the overlying La Vieille Formation. This is coincident with a general coarsening of the limestone units across the contacts. *P. amorphognathoides* is known from only a few specimens at a single locality in the region despite the fact that almost two hundred samples have been collected from strata considered to be late Llandovery and/or early Wenlock in age. *C. lirata* appears just below the formational contact and becomes abundant in the basal La Vieille. Definitive Wenlock conodonts, such as *O. sagitta* do not occur until a level about 100 m above the base of the La Vieille Formation. Conodonts in the basal 100 m of the formation are sparse but comprise representatives of *Panderodus, Ozarkodina* and *Pterospathodus* including the only occurrence of *P. amorphognathoides*.

The lithologic change from fine grained carbonates into coarse calcarenite that corresponds to the base of the La Vieille Formation can also be noted in other areas. On Anticosti Island the contact between the Jupiter (fine grained) and the Chicotte (crinoidal calcarenite) formations occurs at about the same level. The base of the Laforce Formation (Fig. 2, columns 2–5) is well dated in some places, giving a late Llandovery age, and comprises coarse calcarenites.

In the deep-water basins of Gaspé Peninsula such as that represented at Percé (Fig. 2, column 7) there is a shallowing upwards to produce coarse crinoidal limestone and conglomerate at the base of unit 9. This level is also approximately the same age (late Llandovery).

There is therefore a marked coarsening of sediment in the late Llandovery of eastern Canada represented in the shallow carbonate (Anticosti Island), deep-water carbonate (e.g. Percé) and shallow carbonate–clastic (Chaleurs Bay region) facies, probably representing shallowing at this time. It is perhaps this event that makes correlation using conodonts difficult, because the shallow-water conditions appear to exclude biostratigraphically useful conodont genera (e.g. *Kockelella*, with the possible exception of *K. ranuliformis*, see below). It may be necessary to use *Ozarkodina* for biostratigraphy in this interval. *O. confluens* is known only from strata of Wenlock age (e.g. Aldridge 1975) but its usefulness is in doubt, paradoxically because of the immense diversity of morphology noted in blade elements of *Ozarkodina* in the late Llandovery that makes it difficult to separate *O. confluens* from ancestral species. *O. sagitta* is biostratigraphically the most promising form because of its distinctiveness and reasonably widespread distribution. The regional base of the Wenlock may have to be defined using this taxon, although it does not occur in basal Wenlock strata (Aldridge 1975). The Chaleurs Bay region has the best potential in eastern Canada for studies of early Wenlock conodonts, because the La Vieille Formation is comprised of limestone and its upper half is of Wenlock age.

Paleoecology

Previous studies of the paleoecology of Llandovery conodonts have been made by Aldridge (1976) and Aldridge & Mabillard (1981). A study conducted by Le Fèvre *et al.*(1976) on Ordovician and Silurian conodonts of Hudson Bay also discussed briefly the paleoecology of middle and upper Llandovery conodonts.

Unlike studies of the British Early Silurian (Aldridge 1976; Aldridge & Mabillard 1981) no rigorous approach to the study of diversity and distribution has been undertaken on eastern Canadian samples as yet. A purely intuitive approach is unsatisfactory, but in the case of this review, some comments can be made on distribution of conodonts in the markedly different facies belts, namely shallow carbonate, deep carbonate and shallow carbonate–clastic facies.

Conodonts of Llandovery A age are rare in most places but in the shallow-water carbonate strata of Anticosti Island and the shallow carbonate–clastic strata of the Chaleurs Bay region they are abundant. In both areas, specimens of *Panderodus* numerically dominte the fauna, but specimens of *Icriodella* and *Oulodus?* are also abundant. In younger strata *Icriodella* has been shown to have a preference for nearshore environments (see, e.g., Aldridge & Mabillard 1981). Diminutive specimens of species of *Ozarkodina*, such as *O. hassi* and *O. oldhamensis*, are also abundant in both shallow carbonate and shallow clastic–carbonate facies. The interpretation of the Clemville Formation as of very shallow water origin conflicts with the offshore origin ascribed to it by Bourque (1981) based on sedimentary structures.

Earliest Llandovery strata of deep-water origin, such as unit 6 of the White Head formation (Fig. 2, column 7), are virtually devoid of conodonts. Species that are present include: *Decoriconus costulatus, Distomodus kentuckyensis, Panderodus gracilis, P. serratus* and *Walliserodus curvatus*. In addition to these species, elements of diminutive, generalized species of *Ozarkodina,* such as *O. hassi* and *O. spp. A* and *B* of Nowlan (1981), are also present. Aldridge (1976) and Aldridge & Mabillard (1981) have pointed out the preference of *Decoriconus* and *Walliserodus* for more offshore environments in younger strata.

A similar contrast exists in younger Llandovery (B–C$_{1-2}$) strata with species of *Icriodella* and *Oulodus?* present together with abundant *Panderodus* in the shallow-water sediments of the Gun River Formation on Anticosti Island and the upper part of the Clemville and Weir formations in the Chaleurs Bay region. The composition of deep water faunas of the White Head Formation remains essentially unchanged from lower strata.

Member 3 of the Jupiter Formation is shaly and probably of deeper-water origin than other strata on Anticosti Island. It bears representatives of *D. staurognathoides* and small specimens of generalized species of *Ozarkodina,* including *Ozarkodina aldridgei* and *Ozarkodina pirata* of Uyeno & Barnes (1983). The latter species is similar to *O. hassi* and to *Ozarkodina* sp. B of Nowlan (1981). A similar fauna has been recovered from deep water strata of units 7 and 8 of the White Head Formation (Nowlan 1981). Aldridge (1976) considered an association of *O. gulletensis* and *I. inconstans* to be indicative of shallow-water conditions in Telychian strata of Britain. A similar association has been recovered from the demonstrably shallow-water Anse Cascon Formation (Nowlan 1981; Bourque 1981) and from the uppermost part of the Jupiter Formation of Anticosti Island (Uyeno & Barnes 1981). No Telychian strata of deep-water origin were deposited in eastern Canada; in the Percé region depositional environments shallow towards the top of the White Head Formation. The general scarcity of elements of *Pterospathodus celloni* in eastern Canada may be the result of environmental conditions that were too shallow for its widespread development. Aldridge (1976) indicated that *P. celloni* is more common in offshore environments in the British Lower Silurian, and LeFèvre et al. (1976) came to a similar conclusion for the Lower Silurian of the Hudson Bay region.

In a discussion of the distribution of conodonts in the time interval representative of the *P. amorphognathoides* Zone, Aldridge & Mabillard (1981) concluded that *Apsidognathus, Icriodella, Kockelella ranuliformis* and species of *Ozarkodina* including *O. excavata* were prominent components of nearshore faunas. In eastern Canada, the Limestone Point Formation and Anse à Pierre-Loiselle Formation on the south and north sides of Chaleurs Bay, respectively, contain a fauna dominated numerically by *Panderodus* but characterized by species of *Apsidognathus, Ozarkodina* and *Pterospathodus*. This assemblage may suggest a nearshore but not shallowest nearshore environment. *Icriodella*, although locally abundant in northern New Brunswick, is not common. This interpretation fits well with the intermediate environment suggested for the Anse à Pierre-Loiselle Formation by Bourque (1981). The absence or extreme scarcity of *P. amorphognathoides, Carniodus carnulus, Dapsilodus* and *Decoriconus* certainly suggests that it was not a markedly offshore environment.

The domination of faunas from the La Vieille Formation by species of *Panderodus* and *Ozarkodina* also suggests a shallow-water environment according to the distribution reported by Aldridge & Mabillard (1981). This is supported by the shallowing interpreted from sedimentary structures in the lower part of the La Vieille Formation by Bourque (1981). This shallowing episode culminated in the development of patch reefs and algal flats in the middle part of the La Vieille Formation.

Samples from the lower part of the Chicotte Formation, Anticosti Island, include *Apsidognathus tuberculatus, Pterospathodus pennatus procerus* and *Carniodus carnulus* (Uyeno & Barnes 1983). A nearshore environment is suggested on the basis of *A. tuberculatus*, but a more offshore environment is suggested by *Carniodus*. *P. amorphognathoides* makes its only appearance near the top of the Chicotte Formation, but this appearance may be environmentally controlled by the shallowness of the environment of deposition of much of the Chicotte Formation.

As a result of this preliminary and admittedly intuitive survey of the distribution of conodonts in the Lower Silurian of eastern Canada several tentative conclusions can be drawn:

(1) *Icriodella* and several species of *Ozarkodina* inhabit regions of shallow water, a fact that has long been known from the work of Aldridge (1976).

(2) Small, generalized species of *Ozarkodina,* such as *O. hassi* and *O. spp. A* and *B* of Nowlan (1981), inhabited both nearshore and offshore environments in the earliest Silurian but shifted their preference to offshore environments in the middle Llandovery.

(3) An assemblage dominated by *Panderodus* but characterized by species of *Apsidognathus, Pterospathodus* and *Ozarkodina* represents an intermediate to nearshore environment in the Llandovery C$_{5-6}$ interval in the Chaleurs Bay region.

(4) The scarcity of zonal indicators such as *P. celloni, P. amorphognathoides* and *C. carnulus* is probably a result of environmental conditions that were too shallow.

(5) *Panderodus* is abundant in all environments in the Lower Silurian of eastern Canada and is therefore probably a widespread pelagic form.

(6) *Dapsilodus,* an apparent indicator of deep offshore environments (Aldridge & Mabillard 1981; Amsden et al.

1980) is rare in the Lower Silurian of eastern Canada despite deep water deposition in the early and middle Llandovery, although it is present in the middle part of the Jupiter Formation on anticosti Island (Uyeno & Barnes 1983).

References

Aldridge, R. J. 1972: Llandovery conodonts from the Welsh Borderland. *Bulletin of the British Museum (Natural History) Geology 22*, 127–231.

Aldridge, R. J. 1975: The stratigraphic distribution of conodonts in the British Silurian. *Journal of the Geological Society 131*, 607–618.

Aldridge, R. J. 1976: Comparison of macrofossil communities and conodont distribution in the British Silurian. *In* Barnes, C. R. (ed.): Conodont Paleoecology, 91–104. *Geological Association of Canada, Special Paper 15*.

Aldridge, R. J. & Mabillard, J. E. 1981: Local variations in the distribution of Silurian conodonts: an example from the *amorphognathoides* interval of the Welsh Basin. *In* Neale, J. W. and Brasier, M. D. (eds.): *Microfossils from Recent and Fossil Shelf Seas*, 10–17. Ellis Horwood Ltd., Chichester, England.

Amsden, T. W., Toomey, D. F., & Barrick, J. E. 1980: Paleoenvironment of Fitzhugh Member of Clarita Formation (Silurian, Wenlockian) southern Oklahoma. *Oklahoma Geological Survey, Circular 83*. 54 pp.

Barnes, C. R. & McCracken, A. D. 1981: Early Silurian chronostratigraphy and a proposed Ordovician–Silurian boundary stratotype, Anticosti Island, Québec. *In* Lespérance, P. J. (ed.): *Subcommission on Silurian Stratigraphy, Ordovician–Silurian Boundary Working Group. Field Meeting, Anticosti–Gaspé, Québec, 1981, Vol. II: Stratigraphy and Paleontology*, 71–79.

Barrick, J. E. & Klapper, G. 1976: Multielement Silurian (late Llandoverian–Wenlockian) conodonts of the Clarita Formation, Arbuckle Mountains, Oklahoma, and the phylogeny of *Kockelella*. *Geologica et Palaeontologica 10*, 59–100.

Bastin, E. S. & Williams, H. S. 1914: Easport Folio. *United States Geological Survey, Folio 192*.

Berry, W. B. N. & Boucot, A. J. 1970: Correlation of the North American Silurian rocks. *Geological Society of America, Special Paper 102*. 289 pp.

Bolton, T. E. 1972: Geological map and notes on the Ordovician and Silurian litho- and biostratigraphy, Anticosti Island, Quebec. *Geological Survey of Canada, Paper 71–19*. 45 pp.

Boucot, A. J. & Bourque, P.-A. 1981: Brachiopod bistratigraphy of the Llandoverian rocks of the Gaspé Peninsula. *In* Lespérance, P. J. (ed.): *Subcommission on Silurian Stratigraphy, Ordovician–Silurian Boundary Working Group. Field Meeting, Anticosti–Gaspé, Québec, 1981, Vol. II: Stratigraphy and Paleontology*, 315–321.

Boucot, A. J., Dewey, J. F., Dineley, D. L., Fletcher, R., Fyson, W. K., Griffin, J. G., Hickox, C. F., McKerrow, W. S., & Ziegler, A. M. 1974: Geology of the Arisaig area, Antigonish County, Nova Scotia. *Geological Society of America, Special Paper 139*. 191 pp.

Bourque, P.-A. 1975: Lithostratigraphic framework and unified nomenclature for Silurian and basal Devonian rocks in eastern Gaspé Peninsula, Québec. *Canadian Journal of Earth Sciences 12*, 858–872.

Bourque, P.-A. 1977: Silurian and basal Devonian of northeastern Gaspé Peninsula. *Ministère des Richesses Naturelles, Québec, Étude Scientifique 29*. 232 pp.

Bourque, P.-A. 1981: Baie des Chaleurs Area. *In* Lespérance, P. J. (ed.): *Subcommission on Silurian Stratigraphy, Ordovician–Silurian Boundary Working Group. Field Meeting, Anticosti–Gaspé, Québec, 1981, Vol. I: Guidebook*, 42–56.

Bourque, P.-A. & Lachambre, G. 1981: Stratigraphie du Silurien et du Dévonien basal du sud de la Gaspésie. *Ministère de l'Energie et des Resources, Québec, Etude Scientifique 30*. 123 pp. [Imprint 1980.]

Branson, E. B. & Mehl, M. G. 1933: Conodont Studies, Number 1. *The University of Missouri Studies 8*, 1–72.

Cocks, L. R. M. & Copper, P. 1981: The Ordovician–Silurian boundary at the eastern end of Anticosti Island. *Canadian Journal of Earth Sciences 18*, 1029–1034.

Cooper, B. J. 1975: Multielement conodonts from the Brassfield Limestone (Silurian) of southern Ohio. *Journal of Paleontology 49*, 984–1008.

Cooper, B. J. 1980: Toward an improved Silurian conodont biostratigraphy. *Lethaia 13*, 209–227.

Crosby, D. G. 1962: Wolfville map-area, Nova Scotia (21H/1). *Geological Survey of Canada, Memoir 325*. 67 pp.

Fåhraeus, L. E. & Barnes, C. R. 1981: Conodonts from the Becscie and Gun River Formations (Lower Silurian) of Anticosti Island, Québec. *In* Lespérance, P. J. (ed.): *Subcommission on Silurian Stratigraphy, Ordovican–Silurian Boundary Working Group. Field Meeting, Anticosti–Gaspé, Québec, 1981, Vol. II: Stratigraphy and Paleontology*, 165–172.

Harper, C. W., Jr. 1973: Brachiopods of the Arisaig Group (Silurian–Lower Devonian) of Nova Scotia. *Geological Survey of Canada, Bulletin 215*. 163 pp.

Klapper, G. & Murphy, M. A. 1975: Silurian–Lower Devonian conodont sequence in the Roberts Mountain Formation of central Nevada. *University of California Publications in Geological Sciences 111*, 1–62. [Imprint 1974.]

Lajoie, J., Lespérance, P. J., & Béland, J. 1968: Silurian stratigraphy and paleogeography of Matapedia–Temiscouata region, Quebec. *American Association of Petroleum Geologists Bulletin 52*, 615–640.

Lane, T. E. 1975: Stratigraphy of the White Rock Formation. *In* Harris, I. M. (ed.): Ancient sediments of Nova Scotia. *Field Trip Guide, Eastern Section, Society of Economic Paleontologists and Mineralogists, 1975*, 43–62.

Lee, H. J. & Noble, J. P. A. 1977: Silurian stratigraphy and depositional environments: Charlo–Upsalquitch Forks area, New Brunswick. *Canadian Journal of Earth Sciences 14*, 2533–2542.

LeFèvre, J., Barnes, C. R., & Tixier, M. 1976: Paleoecology of Late Ordovician and Early Silurian conodontophorids, Hudson Bay Basin. *In* Barnes, C. R. (ed.): Conodont Paleoecology, 70–89. *Geological Association of Canada, Special Paper 15*.

Legault, J. A. 1968: Conodonts and fish remains from the Stonehouse Formation, Arisaig, Nova Scotia. *Geological Survey of Canada, Bulletin 165*, 1–45.

Lespérance, P. J. 1981 (ed.): *Subcommission on Silurian Stratigraphy, Ordovician–Silurian Boundary Working Group. Field Meeting, Anticosti–Gaspé, Québec, 1981, Vol. II: Stratigraphy and Paleontology*. 321 pp. Université de Montréal, Département de Géologie, 1981.

Lespérance, P. J. & Sheehan, P. M. 1981: Hirnantian fauna in and around Percé, Québec. *In* Lespérance, P. J. (ed.): *Subcommission on Silurian Stratigraphy, Ordovician–Silurian Boundary Working Group. Field Meeting, Anticosti–Gaspé, Québec, 1981, Vol. II: Stratigraphy and Paleontology*, 247–256.

Lock, B. E. 1972: Lower Paleozoic history of a critical area; eastern margin of the St. Lawrence Platform in White Bay, Newfoundland, Canada. *International Geological Congress, Montreal, 1972, Proceedings, Section 6*, 310–324.

McCracken, A. D. & Barnes, C. R. 1981 a: Conodont biostratigraphy and paleoecology of the Ellis Bay Formation, Anticosti Island, Québec, with special reference to Late Ordovician and Early Silurian chronostratigraphy and the systemic boundary. *Geological Survey of Canada Bulletin 329*, 51–134.

McCracken, A. D. & Barnes, C. R. 1981 b: Conodont biostratigraphy across the Ordovician–Silurian boundary, Ellis Bay Formation, Anticosti Island, Québec. *In* Lespérance, P. J. (ed.): *Subcommission on Silurian Stratigraphy, Ordovician–Silurian Boundary Working Group. Field Meeting, Anticosti–Gaspé, Québec, 1981, Vol. II: Stratigraphy and Paleontology*, 61–69.

Nicoll, R. S. & Rexroad, C. B. 1969: Stratigraphy and conodont paleontology of the Salamonie Dolomite and Lee Creek Member of the Brassfield Limestone (Silurian) in southeastern Indiana and adjacent Kentucky. *Indiana Geological Survey Bulletin 40*. 73 pp. [Imprint 1968.]

Noble, J. P. A. 1976: Silurian stratigraphy and paleogeography, Pointe Verte area, New Brunswick, Canada. *Canadian Journal of Earth Sciences 13*, 537–546.

Noble, J. P. A. 1980: Llandovery–Wenlock facies in northern New Brunswick. *In* Pickerill, R. K. (ed.): Ordovician, Silurian and Devonian strata of northern New Brunswick and southern Gaspé. *Canadian Paleontology and Biostratigraphy Seminar, Fredericton, 1980*, 1–4.

Noble, J. P. A. & Howells, K. D. M. 1979: Early Silurian biofacies and lithofacies in relation to Appalachian basins in north New Brunswick. *Bulletin of Canadian Petroleum Geology 27*, 242–265.

Nowlan, G. S. 1981: Late Ordovician–Early Silurian conodont biostratigraphy of the Gaspé Peninsula – a preliminary report. *In* Lespérance, P. J. (ed.): *Subcommission on Silurian Stratigraphy, Ordovician–Silurian Boundary Working Group. Field Meeting, Anticosti–Gaspé, Québec 1981, Vol. II: Stratigraphy and Paleontology*, 257–291.

Nowlan, G. S. 1982: Conodonts and the position of the Ordovician–Silurian boundary at the eastern end of Anticosti Island, Québec, Canada. *Canadian Journal of Earth Sciences 19*, 1332–1335.

Nowlan, G. S. 1983: Biostratigraphic, paleogeographic and tectonic implications of late Ordovician conodonts from the Grog Brook Group, northwestern New Brunswick. *Canadian Journal of Earth Sciences 20*, 651–671.

Nowlan, G. S. & Barnes, C. R. 1981: Late Ordovician conodonts from the Vauréal Formation, Anticosti Island, Québec. *Geological Survey of Canada Bulletin 329*, 1–49.

Pavlides, L. 1968: Stratigraphic and facies relationships of the Carys Mills Formation of Ordovician and Silurian age, northeast Maine. *United States Geological Survey, Bulletin 1264*. 44 pp.

Pavlides, L. & Berry, W. B. N. 1966: Graptolite-bearing Silurian rocks of the Houlton–Smyrna Mills area, Aroostook County, Maine. *United States Geological Survey, Professional Paper 550-B*, B 51–B 61.

Petryk, A. A. 1979: Stratigraphie revisée de l'Ile d'Anticosti. *Ministère de l'Energie et des Ressources, Québec, DPV 711*.

Petryk, A. A. 1981: Stratigraphy, sedimentology and paleogeography of the Upper Ordovician–Lower Silurian of Anticosti Island, Québec. *In* Lespérance, P. J. (ed.): *Subcommission on Silurian Stratigraphy, Ordovician–Silurian Boundary Working Group. Field Meeting, Anticosti–Gaspé, Québec, 1981, Vol. II: Stratigraphy and Paleontology*, 11–39.

Pollock, C. A., Rexroad, C. B., & Nicoll, R. S. 1970: Lower Silurian conodonts from northern Michigan and Ontario. *Journal of Paleontology 44*, 743–764.

Rhodes, F. H. T. 1953: Some British Lower Palaeozoic conodont faunas. *Philosophical Transactions of the Royal Society, Series B, 237*, 261–334.

Rickards, R. B. & Riva, J. 1981: *Glyptograptus? persculptus* (Salter), its tectonic deformation, and its stratigraphic significance for the Carys Mills Formation of N. E. Maine, U.S.A. *Geological Journal 16*, 219–235.

Schenk, P. E. 1971: Southeastern Atlantic Canada, northwestern Africa, and continental drift. *Canadian Journal of Earth Sciences 8*, 1218–1251.

Schenk, P. E., Lane, T. E., & Jensen, L. R. 1980: Trip 20: Paleozoic history of Nova Scotia—a time trip to Africa (or South America?). *Field Trip Guidebook, Geological Association of Canada—Mineralogical Association of Canada, Halifax, 1980*. 82 pp.

Schönlaub, H. P. 1971: Zur Problematik der Conodonten–Chronologie an der Wende Ordoviz/Silur mit besonderer Berücksichtigung der Verhältnisse im Llandovery. *Geologica et Palaeontologica 5*, 35–77.

Skidmore, W. B. & Lespérance, P. J. 1981: Percé area. *In* Lespérance P. J. (ed.): *Subcommission on Silurian Stratigraphy, Ordovician–Silurian Boundary Working Group. Field Meeting, Anticosti–Gaspé, Québec, 1981, Vol. I: Guidebook*, 31–40.

Smith, J. C. 1966: Geology of southwestern New Brunswick. *In* Poole, W. H. (ed.): Geology of parts of Atlantic Provinces, Guidebook. *Geological Association of Canada—Mineralogical Association of Canada, Field Trip Guides, Halifax*, 1–12.

Smyth, W. R. & Schillereff, H. S. 1982: The pre-Carboniferous geology of southwest White Bay. *Department of Mines and Energy, Newfoundland and Labrador, Mineral Development Division, Report 82–1*, 78–98.

St. Peter, C. 1982: Geology of Juniper–Knowlesville–Carlisle area, New Brunswick, Map-areas I-16, I-17, and I-18 (parts of 21J/11 and 21J/6). *New Brunswick Department of Natural Resources, Geological Survey Branch, Map Report 82–1*. 82 pp.

Uyeno, T. T. & Barnes, C. R. 1981: A summary of Lower Silurian conodont biostratigraphy of the Jupiter and Chicotte Formations, Anticosti Island, Québec. *In* Lespérance, P. J. (ed.): *Subcommission on Silurian Stratigraphy, Ordovician–Silurian Boundary Working Group.*

Field Meeting, Anticosti–Gaspé, 1981, Vol. II: Stratigraphy and Paleontology, 173–184.

Uyeno, T. T. & Barnes, C. R. 1983: Conodonts of the Jupiter and Chicotte Formations (Lower Silurian), Anticosti Island, Québec. *Geological Survey of Canada Bulletin 355*. 49 pp.

Walliser, O. H. 1962: Conodontenchronologie des Silurs (Gotlandiums) und des tieferen Devons mit besonderer Berücksichtigung der Formationsgrenze. *In* Erben, H. K. (ed.): *Symposiums-Band der 2 internat. Arbeitstagung über die Silur/Devon-Grenze und die Stratigraphie von Silur und Devon, Bonn, Bruxelles, 1960*, 281–287. Schweizerbartsche Verlag, Stuttgart.

Walliser, O. H. 1964: Conodonten des Silurs. *Abhandlungen des Hessischen Landesamtes für Bodenforschung 41*. 106 pp.

Walliser, O. H. 1972: Conodont apparatuses in the Silurian. *Geologica et Palaeontologica, SB 1*, 75–79.

Williams, H. 1962: Botwood (west half) map area, Newfoundland. *Geological Survey of Canada, Paper 62–9*. 16 pp.

Williams, H. 1978: Tectonic lithofacies map of the Appalachian Orogen. *Department of Geology, Memorial University of Newfoundland, Map 1*.

Williams, H. & Hatcher, R. D. jr. 1982: Suspect terranes and accretionary history of the Appalachian orogen. *Geology 10*, 530–536.

Appendix

Location of sections from which illustrated specimens (Figs. 3, 4) were obtained. Sections 1A–2B, 5 are from Chaleurs Bay (area/column 6, Figs. 1, 2); sections 3, 4 are from the Percé area (area/column 7, Figs. 1, 2); sections 6–10 are from northern New Brunswick (area/column 12, Figs. 1, 2).

Section 1A: Type section of Clemville Formation on Petite Port–Daniel River, west of the village of Clemville, southern Gaspé Peninsula, Quebec. Latitude 48°10′34″; longitude 65°01′02″W. See Bourque & Lachambre (1981, Figs. 2, 16).

Section 1B: North flank of the Clemville Anticline on Petite Port–Daniel River, west of the village of Clemville, southern Gaspé Peninsula, Quebec. Latitude 48°10′38″N; longitude 65°01′25″W. See Bourque & Lachambre (1981, Figs. 2, 16).

Section 2A: Roadcut on Highway 132 near Gascons-Est, southern Gaspé Peninsula, Québec. Latitude 48°12′20″N; longitude 64°49′30″W. See Bourque & Lachambre (1981, Figs. 2, 12).

Section 2B: Railroad cut 500 m southeast of section 2 A. Latitude 48°12′13″N; longitude 64°49′05″W. See Bourque & Lachambre (1981, Figs. 2, 12).

Section 3: Roadcuts along Flynn road, Percé, eastern Gaspé Peninsula, Quebec. Latitude 48°30′25″N; longitude 64°14′05″W. See Skidmore & Lespérance (1981, Fig. 25).

Section 4: Bed of 'Cannes-des-Roches' Brook, about 3 km northwest of section 3. Latitude, 48°31′24″N; longitude 64°16′12″W. See Skidmore & Lespérance (1981, Fig. 25) and Nowlan (1981, Fig. 3).

Section 5: Black Cape section, at Howatson Point near New Richmond, southern Gaspé Peninsula, Quebec. Latitude 48°08′30″N; longitude 65°50′19″W. See Bourque & Lachambre (1981, Figs. 2, 24).

Section 6: Mouth of Dickie Cove Brook, west of Jacquet River, northern New Brunswick; south side of Chaleurs Bay. Latitude 47°57′07″N; longitude 66°07′45″W. See Lee & Noble (1977, Figs. 1, 2).

Section 7: Flanagans, northern New Brunswick, just west of Quinn Point (Section 8), south side of Chaleurs Bay. Latitude 47°55′06″N; longitude 65°57′18″W. See Noble (1976, Figs. 1, 2).

Section 8: Quinn Point, northern New Brunswick, south side of Chaleurs Bay. Latitude 47°55′11″N; longitude 65°56′42″W. See Noble (1976, Figs. 1, 2).

Section 9: Limestone Point, northern New Brunswick, type section of Limestone Point Formation; south side of Chaleurs Bay. Latitude 47°48′54″N; longitude 65°43′34″W. See Noble (1976, Figs. 1, 2).

Section 10: Coastal section at mouth of Hendry Brook, northern New Brunswick, south side of Chaleurs Bay. Latitude 47°53′06″N; longitude 65°48′23″W. See Noble (1976, Figs. 1, 2).

Silurian conodonts from Severnaya Zemlya

PEEP MÄNNIK

FOSSILS AND STRATA

ECOS III

A contribution to the Third European Conodont Symposium, Lund, 1982

Männik, Peep 1983 12 15: Silurian conodonts from Severnaya Zemlya. *Fossils and Strata*, No. 15, pp. 111–119. Oslo. ISSN 0300-9491. ISBN 82-0006737-8.

A short description of the Silurian of Severnaya Zemlya consisting of the Vodopad, Golomyannyi, Srednii, Samoilovich, Ustspokoinaya and Krasnaya Bukhta Formations is given. The conodont fauna is discussed. The *Icriodella discreta – I. deflecta* and the *Icriodella inconstans* Assemblage Zones of Britain are recognizable in the lower part of the Silurian on Severnaya Zemlya. The age of the upper part of the Silurian sequence cannot be established by conodonts at the present time. The lack of the widespread species *Pterospathodus celloni* and *P. amorphognathoides* is noteworthy. Five new multielement combinations, *Oulodus?* sp. A, *Oulodus?* sp. B, *Oulodus?* sp. C, *Ozarkodina?* sp. A and *Ozarkodina?* sp. B are proposed. □*Conodonta, biostratigraphy, Silurian, Severnaya Zemlya.*

Peep Männik, Institute of Geology, Estonia puiestee 7, CCCP-200101 Tallinn, Estonia, USSR; 10th August, 1982.

This paper represents a preliminary report on a study of conodonts from Severnaya Zemlya. Previous studies of the geology of this region include those by Urvantzev (1933) and Egiazarov (1959). Detailed research on paleontology and stratigraphy began in 1978, and the first results were reviewed by Menner *et al.* (1979, 1982). Klubov *et al.* (1980) first determined Silurian and Devonian conodonts from the area.

The majority of the material for this study was collected during field work in 1979. The author also had an opportunity to use the samples collected by Dr. E. Kurik in 1974 and by Dr. V. Menner in 1978. Silurian conodonts have been studied from five sections on October Revolution Island and from one on Srednii Island.

The Paleozoic section of October Revolution Island ranges from Cambrian to Upper Devonian. The oldest strata are exposed in the eastern part of the island, which is mostly covered by continental glaciers. The Cambrian–Silurian sediments in the middle and eastern parts of the island are deformed, with the Silurian outcrops situated on the limbs of a northwest–southeast trending anticline, the core of which is formed by Ordovician rocks. Lithologically the Silurian is represented by fossiliferous shallow-water carbonates with thin interbeds of sandstone at some levels. The Silurian marine carbonate rocks are underlain by variegated sandstones and siltstones of Ordovician age. The Ordovician–Silurian boundary is lithologically easily recognizable, but there is no evidence of a hiatus. The system boundary is taken at the top of the uppermost sandstone layer, which is overlain by limestone. The upper part of the Silurian was eroded before Devonian sedimentation in the central and western part of the region, but is mostly preserved in the east. The sequences available for study are exposed in river valleys perpendicular to the outcrop belts on October Revolution Island, and in the cliffs along the coast on Srednii Island (Fig. 1).

In this article a short description of the Silurian deposits is presented, and the conodont faunas recovered by dissolving 230 samples are discussed. The conodont elements are described in terms of the notation proposed by Cooper (1975).

The term s. f. (*sensu formo*) is used to denote species regarded as form taxa. The study of Silurian conodonts from Severnaya Zemlya has resulted in the recognition of faunas and zones known elsewhere; systematic descriptions will be given in future publications.

The conodont elements from Srednii Island are generally amber in color although some variation is noted. On October Revolution Island the conodonts are brown to dark brown. With reference to the conodont color alteration index, the conodonts from Srednii Island have a CAI of about 1–1,5 and those from October Revolution Island 2–3, reflecting burial temperatures of less than 90°C and of 60–200°C, respectively (see Epstein *et al.* 1977).

All the figured specimens are deposited at the Institute of Geology of the Academy of Sciences of the Estonian SSR.

Vodopad Formation

The Vodopad Formation is the lowest unit of the Silurian on Severnaya Zemlya. It was erected by Menner (Menner *et al.* 1979) and is named after the waterfall (Russian водопад) on the Matusevich River on October Revolution Island, where the type section is situated (Fig. 1). The unit, 250–300 m in thickness, is composed of sediments of a major marine transgression, comprising grey and dark grey dolomitic limestones, with siliceous nodules in the upper part. Corals (tabulates, rugosans), stromatoporoids, brachiopods, ostracodes and crinoids are numerous, particularly in the upper part. The lowermost and uppermost parts of the formation are rich in pentamerid brachiopods. The megafauna indicates a Llandovery age (Menner *et al.* 1982).

Silurian conodonts appear in the lowermost beds of the formation. *Ozarkodina oldhamensis* (Rexroad 1967), *O.* aff. *hassi* (Pollock, Rexroad & Nicoll 1970), *Ozarkodina* cf. sp. A Nowlan 1981, *Icriodella* cf. *deflecta* Aldridge 1972, *Oulodus?* cf. *kentuckyensis* (Branson & Branson 1947), *Distomodus* cf. *kentuckyensis* Branson & Branson 1947, *Ambalodus anapetus* Pollock, Rexroad

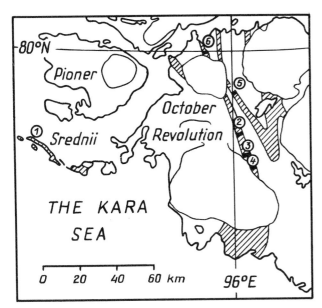

Fig. 1. Sketch map showing the locations of sections collected for conodonts. Diagonal pattern denotes Silurian rocks. 1–Srednii; 2–Matusevich River; 3–Ushakov R.; 4–Stroinaya R.; 5–Spokoinaya R.; 6–Obryvistaya R.

& Nicoll 1970 s. f., *Amorphognathus tenuis* Aldridge 1972 s. f., and *Pseudooneotodus beckmanni* (Bischoff & Sannemann 1958) s. f. are common (Fig. 2). This fauna is very similar to the faunal assemblage that Aldridge (1972) referred to the *Icriodella discreta – I. deflecta* Assemblage Zone in the Welsh Borderland. This suggests an Idwian (B_{1-3}) to early Fronian (C_{1-2}) age. The assemblage on Severnaya Zemlya is also similar to the fauna recovered from the Brassfield Limestone of Indiana and Ohio (Branson & Branson 1974; Rexroad 1967; Cooper 1975). McCracken & Barnes (1981 a, b) reported similar early Silurian conodont faunas including a new species, *Oulodus*? *nathani*, from Member 6 of the Ellis Bay Formation, Anticosti Island. Nowlan (1981) noted that the Clemville Formation, Gaspé Peninsula, yields conodonts identical to those in the Ellis Bay fauna, together with brachiopods of Rhuddanian (Llandovery A) age. According to Nowlan, the beds containing the earliest Silurian faunas on Anticosti as well as in the Gaspé Peninsula are probably equivalent in age to the Rhuddanian Stage. The faunas described by McCracken & Barnes and by Nowlan are assigned to the *O.*? *nathani* Zone, which precedes the *Distomodus kentuckyensis* Zone in North America and the *Icriodella discreta – I. deflecta* Assemblage Zone of Britain (Nowlan 1981). The assemblage from the lower part of the Vodopad Formation does not contain *O.*? *nathani*. This fauna may be younger than that of the *O.*? *nathani* Zone and belong to the *Icriodella discreta – I. deflecta* Assemblage Zone (Fig. 3).

In the upper part of the Vodopad Formation conodonts are rare. Samples from this interval yielded *Ozarkodina* aff. *hassi*, *O.* aff. *broenlundi* Aldridge 1979, *Ambalodus* sp. s. f., *Apsidognathus* sp. A s. f., *Icriodina*? sp. s. f. and gen. nov. A s. f. (Fig. 2). The main changes noted from the earlier fauna are the addition of *O.* aff. *broenlundi* and gen. nov. A s. f., and the absence of *I.* cf. *deflecta* and *O. oldhamensis*. *O.* aff. *broenlundi* (Fig. 5 R) is similar to the species described by Aldridge (1979) from Peary Land, eastern North Greenland. In Peary Land *O. broenlundi* occurs

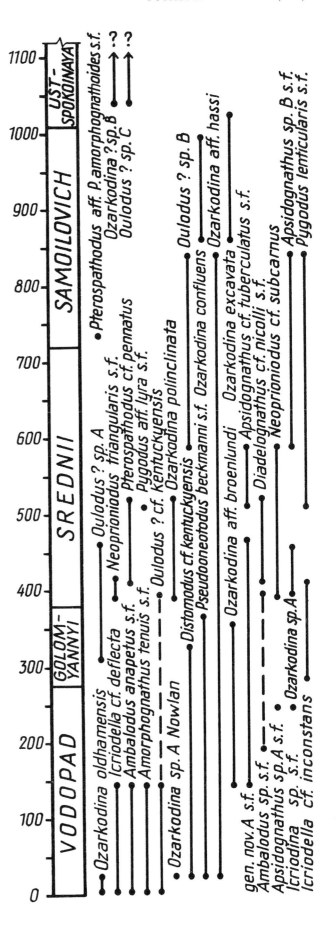

Fig. 2. Stratigraphic ranges of conodonts in the Silurian of Severnaya Zemlya.

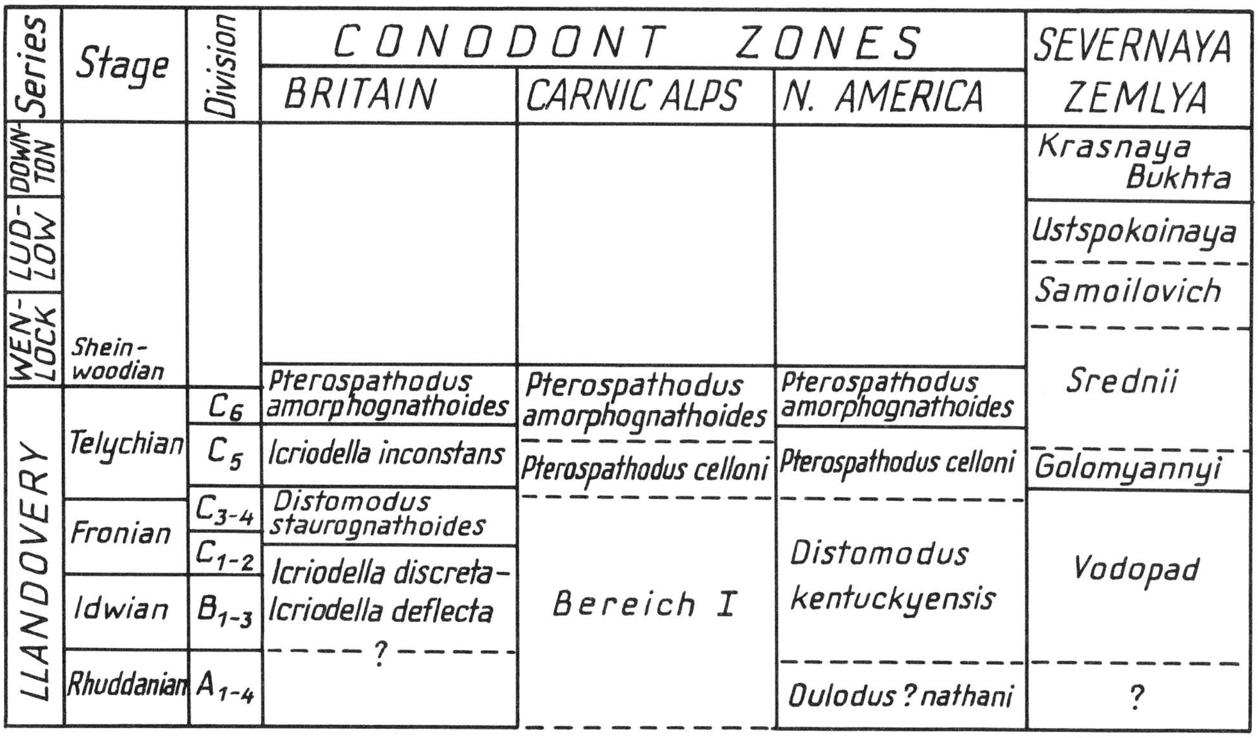

Series	Stage	Division	CONODONT ZONES			SEVERNAYA ZEMLYA
			BRITAIN	CARNIC ALPS	N. AMERICA	
WEN-LOCK / LUD-LOW / DOWN-TON						Krasnaya Bukhta
						Ustspokoinaya
						Samoilovich
	Shein-woodian		Pterospathodus amorphognathoides	Pterospathodus amorphognathoides	Pterospathodus amorphognathoides	Srednii
LLANDOVERY	Telychian	C_6				
		C_5	Icriodella inconstans	Pterospathodus celloni	Pterospathodus celloni	Golomyannyi
	Fronian	C_{3-4}	Distomodus staurognathoides	Bereich I	Distomodus kentuckyensis	Vodopad
	Idwian	B_{1-3}	Icriodella discreta– Icriodella deflecta – – – – ? – – – –			
	Rhuddanian	A_{1-4}			Oulodus ? nathani	?

Fig. 3. Correlation chart of formations on Severnaya Zemlya with the standard conodont zonal sequences of Europe and North America. Data on Britain from Aldridge (1972, 1975); on the Carnic Alps from Walliser (1964); and on North America from Nicoll & Rexroad (1968), Pollock *et al.* (1970), Cooper (1975), McCracken & Barnes (1981 a, b), and Nowlan (1981).

together with *Pterospathodus* (Walliser 1964). The *P. celloni* Zone is approximately equivalent to the *Icriodella inconstans* Assemblage Zone defined in Britain (Aldridge 1972). The material from the upper part of the Vodopad Formation does not allow the age of this interval of the unit to be firmly established. However, the appearance higher in the section of *I.* cf. *inconstans* suggests that it is older in age than the *I. inconstans* Assemblage Zone of Britain, and probably equivalent to the British *Distomodus staurognathoides* Assemblage Zone (Fig. 3). The preservation of the other conodonts from this interval is too poor to give any valuable information.

Golomyannyi Formation

The formation was named by Menner (Menner *et al.* 1979) after one of the islands of the Sedov Archipelago, although the type section of the formation is situated at the Matusevich River on October Revolution Island. The Golomyannyi Formation, 95–115 m thick, represents a phase of regression in the marine basin. The formation is built up of thin-bedded clayey limestones with interbeds of sandstone, stromatolitic limestone, with dolomite at some levels. In the clayey limestones ostracodes, gastropods, small brachiopods and trilobites are abundant, whereas rugosans, tabulates and stromatoporoids are very rare. The brachiopod *Dubaria tenera* Nikiforova & Modzalevskaya indicates that the age of the formation is Llandovery. In some areas of October Revolution and on other islands of Severnaya Zemlya the Vodopad and the Golomyannyi Formations are united in Sneznyi Formation (Menner *et al.* 1982).

The following conodonts have been identified in the

formation: *Icriodella* cf. *inconstans* Aldridge 1972, *O.* aff. *hassi*, *O.* aff. *broenlundi*, *Oulodus*? sp. A, *Ambalodus* sp. s. f. and gen. nov. A s. f. (Fig. 2). The new components in the fauna are *I.* cf. *inconstans* and *Oulodus*? sp. A. *I.* cf. *inconstans* (Fig. 4 T) appears in the lowermost beds of the formation. It is similar to *I. inconstans* Aldridge, although the lanceolate shape of the platform in oral view is less developed. In this formation *Oulodus*? sp. A is represented mainly by its trichonodelliform (Sa) element (Fig. 4 V), which is easily recognizable, with lateral processes that are directed only slightly downwards and are almost straight.

The occurrence of *I.* cf. *inconstans* in the Golomyannyi Formation confirms a Llandovery C_5 age for the unit and enables correlation with the *I. inconstans* Assemblage Zone of Aldridge (1972) in Britain (Fig. 3). *I. inconstans* occurs in Britain in association with *Pterospathodus celloni*, and the *I. inconstans* Assemblage Zone is probably equivalent to the upper part of Walliser's *P. celloni* Zone and may include the lower part of his *P. amorphognathoides* Zone (Aldridge 1972). Aldridge (1972) correlated the *I. inconstans* Assemblage Zone with the upper part of the *Neospathognathodus celloni* Assemblage Zone of North America. At present only *I.* cf. *inconstans* is known from Severnaya Zemlya; *P. celloni* has not been found.

Srednii Formation

The Srednii Formation (300–400 m) was named by Menner (Menner *et al.* 1979) after Srednii Island in the Sedov Archipelago. The lithology is very diverse and represents a new major transgression. In its type section on the Matusevich River the formation is comprised of intercalations of tabulate–

stromatoporoidal limestones, sometimes silicified, with stromatoporoidal bioherms and biostromes, and thick layers of brownish-grey limestones with ostracodes, gastropods, crinoids, and rare tabulates. Interbeds of greenish-grey thin-bedded dolomitic limestones lack fauna. Rare beds with stromatolites are common in the upper part of the formation. The megafauna suggests a Wenlock age, but the basal beds may still belong to the Llandovery (Menner *et al.* 1982).

Conodonts are most abundant in the lower part of the formation. This interval is the richest in conodonts in the whole Silurian sequence on Severnaya Zemlya. Specimens of *Icriodella* cf. *inconstans* and *Oulodus?* sp. A are present in great numbers and in some samples they are the main components of the fauna. *O.* aff. *hassi* and gen. nov. A s. f. also occur. The new elements in the fauna are *O. polinclinata* (Nicoll & Rexroad 1968), *Ozarkodina* sp. A, *Pterospathodus* cf. *pennatus* (Walliser 1964), *Neoprioniodus* cf. *triangularis* Walliser 1964 s. f, *N.* cf. *subcarnus* Walliser 1964 s. f., *Diadelognathus* cf. *nicolli* Aldridge 1972 s. f., *Pygodus* aff. *lyra* Walliser 1964 s. f., *P. lenticularis* Walliser 1964 s. f. and *Apsidognathus* cf. *Tuberculatus* Walliser 1964 s. f. (Fig. 2). *A.* cf. *tuberculatus*, which is represented by broken platform elements (Fig. 4 U), was first described by Walliser 1964 s. f. and *Apsidognathus* cf. *tuberculatus* Walliser the *P. celloni* and *P. amorphognathoides* Zones. It was subsequently reported from Britain (Aldridge 1972), North America (Nicoll & Rexroad 1968; Helfrich 1980) and East Canada (Nowlan 1981). Specimens assigned by Helfrich (1980) and Nowlan (1981) to *A. tuberculatus* differ from those described by Walliser (1964), as they are more rounded in oral view and have one dominant straight high ridge or blade of denticles on one side of which is situated a smooth groove. *A. tuberculatus*, described by Walliser, lacks the groove and is more elongated anteriorly and posteriorly. The free blade is curved and two rows of nodes diverge from it on each side. *A.* cf. *tuberculatus* s. f. on Severnaya Zemlya is similar to forms described by Walliser. In the upper part of the formation a conodont herein called *Apsidognathus* sp. B s. f. (Fig. 5 Y) is found in association with *A.* cf. *tuberculatus*. It is identical to the form assigned by Helfrich (1980, Pl. 1:29) and Nowlan (1981, Pl. 7:7, 12, 13) to *A. tuberculatus*.

P. cf. *pennatus* is represented only by a few delicate specimens, none of them having an expanded blade or nodose denticles (Fig. 4 W); only the Pa element is recognised. It is similar to those described by Nowlan from the Anse á Pierre–Loiselle Formation of the Gaspé Peninsula (Nowlan 1981, Pl. 7:1, 4), although my specimens have only one or two denticles on the lateral process. *P. pennatus* ranges from late Telychian to early Sheinwoodian in Britain (Aldridge 1972, 1975) and is of similar age in North America (Nicoll & Rexroad 1968).

Three species assignable to the multielement genus *Ozarkodina* have been identified from the Srednii Formation. The specimens of *O.* aff. *hassi* are similar to those found from the Vodopad and Golomyannyi Formations (Fig. 5 I, J, K, N, Q). The Pa, Pb, and M elements of *O. polinclinata* are common and easily recognised. In *Ozarkodina* sp. A the spathognathodiform (Pa) element (Fig. 4 M) resembles the equivalent in *O.* aff. *hassi*, with a similar wide and rounded basal cavity, but the denticles are larger and more robust. The element has a large apical denticle, but on some specimens one or two denticles at the anterior end of the blade are of the same size as the cusp.

The Pb element (Fig. 4 N) has a large cusp and the denticles on the anterior blade are much higher than those on the posterior. The basal cavity is wide and rounded beneath the cusp. The possible Sa, Sb, Sc elements of the apparatus (Fig. 4 O, P, Q) are common.

The Sa elements of *Oulodus?* sp. A appear in the Golomyannyi Formation, but the species is most numerous in the lower part of the Srednii Formation, where all the elements have been found. The first P (Pa?) element (Fig. 4 R) is somewhat similar to its equivalent in *Oulodus? nathani* McCracken & Barnes 1981b, but differs in being shorter and in having distinct denticles fused only in their lower parts. The posterior blade is twisted to the outer side. The P (Pb?) elements (Fig. 4 Z') are similar to specimens referred to *Lonchodina walliseri* Ziegler 1960 s. f., but they are very varied. The M element has not been positively identified. Probably, some specimens in my material similar to P (Pb?) elements might occupy the M position in the apparatus. They have a very short anterior process with two or three small discrete denticles (Fig. 4 Z). The Sc element (Fig. 4 S) resembles *Ligonodina kentuckyensis* Branson & Branson 1947 s. f. The elements assumed to be Sb (Fig. 4 X) are somewhat similar to *Lonchodina greilingi* s. f., described by Walliser (1964).

Conodonts assignable to *Oulodus?* cf. *kentuckyensis* are found together with *I.* cf. *inconstans* in the lowermost beds of the Srednii Formation (Figs. 2, 4 F, G, H, J, K). On the Gaspé Peninsula, Canada, Nowlan (1981) described *O?* cf. *kentuckyensis* from the Clemville Formation, assigned to the successive *O.? nathani* and *D. kentuckyensis* Zones (Llandovery A?–C₂) of the North American continent and to the *Icriodella discreta – I.*

Fig. 4. □A, B, C. *Icriodella* cf. *deflecta* Aldridge. □A. Oral view of Pa element, ×75, Cn 5000. Vodopad Formation, Srednii Island. □B. Lateral view of M element, ×80, Cn 5001. □C. Lateral view of S element, ×75, Cn 5002. B and C from the Vodopad formation, October Revolution Island, Stroinaya River. □D, E, I. *Distomodus* cf. *kentuckyensis* Branson & Branson. All specimens from the Golomyannyi Formation, Srednii Island. □D. Lateral view of Pb element, ×40, Cn 5003. □E. Lateral view of Sc element, ×45, Cn 5004. □I. Posterior view of Sa element, ×70, Cn 5005. □F, G, H, J, K. *Oulodus?* cf. *kentuckyensis* (Branson & Branson). All specimens from the Srednii Formation, Srednii Island. □F. Lateral view of first P (Pa?) element, ×75, Cn 5006. □G. Posterior view of Sb element, ×50, Cn 5007. □H. Lateral view of Sc element, ×80, Cn 5008. □J. Lateral view of P (Pb?) element, ×85, Cn 5009. □K. Posterior view of Sa element, ×80, Cn 5010. □L. *Ozarkodina* sp. A Nowlan; lateral view of Pa element, ×75, Cn 5011, Vodopad Formation, Srednii Island. □M, N, O, P, Q. *Ozarkodina* sp. A. All specimens from the Srednii Formation, Srednii Island. □M. Lateral view of Pa element, ×40, Cn 5012. □N. Lateral view of Pb element, ×45, Cn 5013. □O. Posterior view of Sb element, ×55, Cn 5014. □P. Lateral view of Sc element, ×55, Cn 5015. □Q. Posterior view of Sa element, ×40, Cn 5016. □R, S, V, X, Z, Z'. *Oulodus?* sp. A. All specimens from the Srednii Formation, Srednii Island. □R. Lateral view of first P (Pa?) element, ×50, Cn 5017. □S. Lateral view of Sc element, ×45, Cn 5018. □V. Posterior view of Sa element, ×55, Cn 5019. □X. Posterior view of Sb element, ×50, Cn 5020. □Z. Lateral view of M(?) element, ×50, Cn 5021. □Z'. Lateral view of P (Pb?) element, ×55, Cn 5022. □T. *Icriodella* cf. *inconstans* Aldridge; oral view of Pa element, ×45, Cn 5023, Srednii Formation, Srednii Island. □U. *Apsidognathus* cf. *tuberculatus* Walliser s. f.; oral view, ×45, Cn 5024, Srednii formation, October Revolution Island, Ushakov River. □W. *Pterospathodus* cf. *pennatus* (Walliser); lateral view of Pa element, ×85, Cn 5025, Srednii Formation, Srednii Island. □Y. gen. nov. A s. f.; lateral view, ×50, Cn 5026, Golomyannyi Formation, October Revolution Island, Ushakov River.

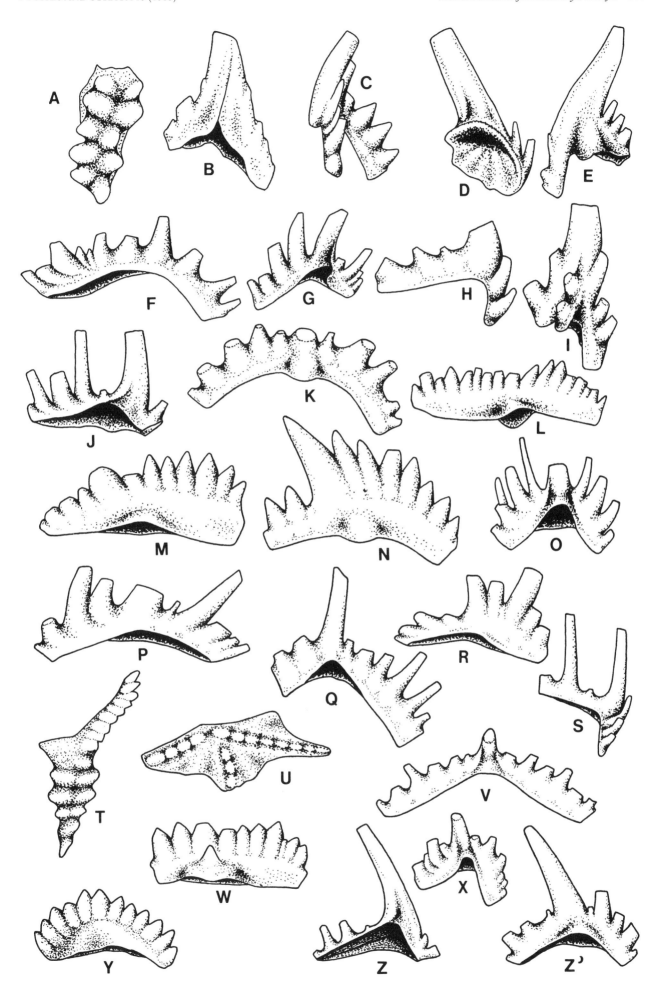

deflecta Assemblage Zone of Britain. On Severnaya Zemlya *O.?* cf. *kentuckyensis* is found together with forms characteristic of the *Icriodella discreta – I. deflecta* Assemblage Zone in the Vodopad Formation, but also occur in the lower part of the Srednii Formation. The age of this horizon may be late Llandovery, as it contains *Apsidognathus* cf. *tuberculatus* s. f. Several other taxa, *P.* cf. *pennatus, O. polinclinata* and *Pygodus* species, indicate a late Llandovery – earliest Wenlock age. According to Menner (Menner *et al.* 1982) the megafauna indicates a late Llandovery or earliest Wenlock age for the lower part of the formation.

The upper part of the Srednii Formation has yielded few conodonts (Fig. 2). *Ozarkodina* aff. *hassi* is still the dominant species. The main new multielement component is *Oulodus?* sp. B (Fig. 5 S, T, W, X) and the fauna also includes some specimens of *Neoprioniodus* cf. *subcarnus* s. f., *Apsidognathus* cf. *tuberculatus* s. f. and *Apsidognathus* sp. B s. f. Only one P (Pb?) element of *Oulodus?* sp. B has been recognised; it is very similar to *Lonchodina detorta* Walliser 1964 s. f. and is characteristic. The M, Sc, Sb and Sa elements of the apparatus are common. Walliser (1964) described *L. detorta* s. f. as an Upper Silurian species, occurring in the Ludlow *Polygnathoides siluricus* Zone and higher levels in the Carnic Alps. The other elements assigned to *Oulodus?* sp. B are known from Wenlock strata in the Carnic Alps (Walliser 1964). Aldridge (1972) has recorded similar conodonts from the Idwian and lower Wenlock. On Severnaya Zemlya *Oulodus?* sp. B is found together with a megafauna regarded as Wenlock in age (Menner *et al.* 1982). *Neoprioniodus* cf. *subcarnus* s. f. and *Apsidognathus* cf. *tuberculatus* s. f. are common in the upper Llandovery and lower Wenlock

(Walliser 1964; Nicoll & Rexroad 1968; Aldridge 1972). *Apsidognathus* sp. B s.f. is known at present only from the upper Llandovery (Helfrich 1980; Nowlan 1981).

Samoilovich Formation

The Samoilovich Formation, erected by Menner (Menner *et al.* 1979), was named after Samoilovich Island. This formation and both the succeeding formations are composed of regressive sediments. The unit is 260–320 m thick and consists of a rhythmic intercalation of stromatolitic, oolitic and oncolitic limestones with greenish-grey thin-bedded clayey limestones. In the latter lithology gastropods, brachiopods, ostracodes and trilobites are numerous. The boundary with the underlying unit is placed at the top of the uppermost thick (2–3 m) layer of limestones with abundant stromatoporoids characteristic of the Srednii Formation. In the lower part of the formation brachiopods *Hyattidina parva* Nikiforova, *Anabaria rara* (Nikiforova) and *Protatrypa lepidota* Nikiforova & Modzalevskaya have been found that indicate a Wenlock age. The number of megafossils decreases towards the top of the sequence, but an interval rich in tabulates of Ludlow age occurs in the upper part of the formation. On the Ushakov River at the same level the vertebrates *Logania* ex. gr. *ludlowensis* Gross, *Thelodus* sp. nov. and *Darthmutia* have been found. The upper limit of the formation is taken at the basal layer of the greenish-grey sandstones of the overlying Ustspokoinaya Formation. The position of the Lower–Upper Silurian boundary is unclear at present (Menner *et al.* 1982).

The lower part of the Samoilovich Formation yields a conodont fauna similar to that of the upper part of the Srednii Formation (Fig. 2). *Ozarkodina* aff. *hassi, Oulodus?* sp. B, *Apsidognathus* sp. B s.f. and *Pygodus* cf. *lenticularis* s. f. are common. A new component is *Pterospathodus* aff. *P. amorphognathoides* (Pa element) Walliser 1964 (Fig. 5 P), which is identical to the forms illustrated by Nowlan (1981, Pl. 7:2, 5) from the Anse à Pierre–Loiselle Formation. Nowlan remarked that *Pterospathodus* aff. *P. amorphognathoides* probably represents a separate species that may be an ancestor of *P. amorphognathoides*, which was found in the overlying La Vieille Formation. *Pterospathodus* aff. *P. amorphognathoides* is of latest Llandovery age on the Gaspé Peninsula. *Apsidognathus* sp. B s. f., described by Nowlan from the Anse à Pierre–Loiselle Formation as platform element of *Apsidognathus tuberculatus*, is also of the same age. *Pygodus* cf. *lenticularis* s. f. is found from the Telychian in Britain (Aldridge 1972, 1975) and is of the same age in the Carnic Alps (Walliser 1964). All these taxa indicate a late Llandovery age. Only *Oulodus?* sp. B has elements, similar to those known from beds younger than Llandovery. It is noteworthy that Llandovery conodonts here range into strata that may be of late Wenlock age and occur with forms of megafauna known from Wenlock and younger beds. A similar phenomenon, where *O. hassi* ranges up to the Lower–Upper Silurian boundary is known from the Timan–Petchora Province (S. V. Melnikov, personal communication).

Samples from the upper part of the formation yielded only *Ozarkodina confluens* (Branson & Mehl 1933) and *O. excavata* (Branson & Mehl 1933). *O. confluens* ranges elsewhere from late Sheinwoodian to Downtonian (Aldridge 1975), but here

---

Fig. 5. □A, B, C, D. *Ozarkodina oldhamensis* (Rexroad). All specimens from the Vodopad Formation, October Revolution Island, Stroinaya River. □A. Lateral view of Pa element, ×75, Cn 5027. □B. Lateral view of Pb element, ×80, Cn 5028. □C. Posterior view of Sb element, ×140, Cn 5029. □D. Lateral view of M element, ×55, Cn 5030. □E. *Amorphognathus tenuis* Aldridge s. f.; oral view, ×40, Cn 5031, Vodopad Formation, October Revolution Island, Spokoinaya River. □F. *Ambalodus anapetus* Pollock, Rexroad & Nicoll s. f.; lateral view, ×75, Cn 5032, Vodopad Formation, October Revolution Island, Spokoinaya River. □G, H, L, M, U. *Oulodus?* sp. C. All specimens from the Ustspokoinaya Formation, October Revolution Island, Matusevich River. □G. Lateral view of Sc element, ×75, Cn 5033. □H. Posterior view of Sb element, ×40, Cn 5034. □L. Posterior view of Sb element, ×40, Cn 5035. □M. Posterior view of Sb element, ×50, Cn 5036. □U. Posterior view of Sa element, ×40, Cn 5038. □O, V. *Ozarkodina?* sp. B. Both specimens from the Ustspokoinaya Formation, October Revolution Island, Matusevich River. □O. Lateral view of Pa element, ×80, Cn 5037. □V. Lateral view of Pb element, ×40, Cn 5038. □I, J, K, N, Q. *Ozarkodina* aff. *hassi* (Pollock, Rexroad & Nicoll). All specimens from the Srednii Formation, October Revolution Island, Ushakov River. □I. Lateral view of M element, ×75, Cn 5040. □J. Lateral view of Sc element, ×75, Cn 5041. □K. Lateral view of Pa element, ×75, Cn 5042. □N. Posterior view of Sa element, ×80, Cn 5043. □Q. Lateral view of Pb(?) element, ×80, Cn 5044. □P. *Pterospathodus* aff. *amorphognathoides* Walliser s. f.; lateral view, ×155, Cn 5045, Samoilovich Formation, October Revolution Island, Matusevich River. □R. *Ozarkodina* aff. *broenlundi* Aldridge; lateral view of Pa element, ×85, Cn 5046, Golomyannyi Formation, October Revolution Island, Ushakov River. □S, T, W, X. *Oulodus?* sp. B. All specimens from the Srednii Formation, October Revolution Island, Ushakov River. □S. Lateral view of M element, ×155, Cn 5047. □T. Lateral view of P (Pb?) element, ×80, Cn 5048. □W. Posterior view of Sb element, ×85, Cn 5049. □X. Lateral view of Sc element, ×70, Cn 5050. □Y. *Apsidognathus* sp. B s.f.; oral view, ×85, Cn 5051, Samoilovich Formation, October Revolution Island, Matusevich River.

the megafauna indicates a Ludlow age (Menner *et al.* 1982). *O. excavata* is known as a widespread species ranging from Fronian to Downtonian (Aldridge 1975).

Ustspokoinaya Formation

Menner erected the Ustspokoinaya (Menner *et al.* 1982), formerly the Izluchina (Menner *et al.* 1979), Formation, for a sequence of violet-grey, green-grey and red-brown marlstones and clayey limestones with numerous ostracodes. The lowermost part of the formation contains interbeds of sandstone. Thin beds of dolomite are common. The red colour of the sediments increases up the sequence. Only in the eastern part of October Revolution Island, near the mouth of the Spokoinaya River, is the whole sequence of the formation (200–250 m) available for study. On the Matusevich and Ushakov Rivers the upper part of the formation was eroded before Devonian sedimentation. The vertebrates *Tremataspis* sp. nov., *Thelodus* cf. *schmidti* (Pander), *Logania martinssoni* Gross, indicating a Ludlow age, have been found in this formation.

The conodonts in the Ustspokoinaya Formation are known only from the sequences on the Matusevich and Spokoinaya Rivers. On the Matusevich River the lowermost beds of the formation contain some elements of *Ozarkodina excavata*. Two undescribed species designated *Ozarkodina?* sp. B and *Oulodus?* sp. C (Fig. 2) appear higher in the sequence. A few elements of these species have also been recovered from the sequences on the Spokoinaya River. *Ozarkodina?* sp. B is in majority represented by Pa and Pb elements. A few Sc and M elements occur on the Matusevich River. The anterior denticles of the Pa element of *Ozarkodina?* sp. B (Fig. 5 O) are fused nearly to their apices and form a gradually deepening row to the cusp, which is situated near the posterior. At the posterior end the height of the denticles fall rapidly. The basal cavity has flared lips at the midlength of the unit under the posteriorly inclined cusp. The arched Pb element (Fig. 5 V) has a strong cusp situated posterior of midlength and inclined posteriorly. The denticles of the posterior blade are much lower than those of the anterior. The posterior blade is turned outwards and strongly twisted inwards. The basal cavity is widest beneath the cusp. The M element has a long downwards directed posterior process. The anterior one is shorter and twisted to the outer side. The Sc element is hindeodelliform.

The first P (Pa?) element of *Oulodus?* sp. C has laterally compressed and nearly to their apices fused denticles on its anterior blade. The posterior process is laterally bowed and aborally arched. Denticles on this process are discrete with U-shaped interspaces and twisted outwards. The basal cavity is widely flared, aboral surfaces of processes are widely excavated or inverted. The Sc element (Fig. 5 G) is ligonodiniform, and somewhat similar to that of *Ligonodina elegans* Walliser 1964, *sensu* Jeppsson 1969. The asymmetrical Sb element is represented by two forms. They are similar, but one of them has a very strongly posteriorly expanded basal cavity beneath the cusp (Fig. 5 L). The cusp itself is also strongly inclined. The other Sb element has a less expanded basal cavity and its cusp is straighter (Fig. 5 M). A gradual transition between these different forms may be observed (Fig. 5 L, H, M). The Sa element (Fig. 5 U) has two bars which form a deep symmetrical arch. The basal cavity is strongly expanded under the posteriorly inclined cusp and continues as a wide groove on the aboral surfaces of the processes.

The uppermost Silurian formation on Severnaya Zemlya, the Krasnaya Bukhta Formation, has a megafauna indicative of Downton age (Menner *et al.* 1982). The few samples from this formation did not yield conodonts.

Summary and conclusions

This study is based on the examination of 230 samples from the Srednii and October Revolution Islands of Severnaya Zemlya that collectively yielded over 11,000 conodont elements. Conodonts have been recovered from each of the formations studied, except the uppermost Krasnaya Bukhta Formation. Some units have been sampled in much more detail than others, so additional collecting is needed. Future publications will detail the taxonomy of the conodont faunas. On the basis of the information recovered so far several conclusions can be made.

(1) The lower part of the Vodopad Formation yields a conodont fauna that may be assigned to the *Icriodella discreta – I. deflecta* Assemblage Zone (Llandovery B_{1-3}, C_{1-2}) of Britain and *Distomodus kentuckyensis* Zone of North America.

(2) Conodonts from the Golomyannyi Formation suggest a Telychian (Llandovery C_5) age, based on the presence of *Icriodella* cf. *inconstans*.

(3) Conodonts from the lower part of the Srednii Formation are abundant and similar to those known from the late Llandovery (Telychian C_{5-6}) and earliest Wenlock (Sheinwoodian) elsewhere. The age of the upper part of the formation, based on the presence of *Oulodus?* sp. B and on the megafauna, is probably Wenlock.

(4) The absence of *Pterospathodus celloni* and *P. amorphognathoides*, widespread in Europe, America and Asia, is noteworthy.

(5) The lower part of the Samoilovich Formation is dated as Wenlock by megafauna, but yields typical Llandovery conodonts together with *Oulodus?* sp. B. The upper part of the formation contains *Ozarkodina confluens* and *O. excavata*, and the megafauna indicates a Ludlow age.

(6) Conodonts from the Ustspokoinaya Formation do not allow determination of its age. The megafauna indicates a Ludlow age.

Acknowledgments. – I would like to thank Drs. Elga Kurik and Vladimir Menner (junior) for making their samples available for examination; Drs. Elga Kurik, Einar Klaamann and Viive Viira for criticism of the text; Mr. Udo Veske and Mr. Jevgeni Klimov for the photographs and Miss Kaie Vallimäe for the figures.

References

Aldridge, R. J. 1972: Llandovery conodonts from the Welsh Border-land. *Bull. Br. Mus. Nat. Hist. (Geol.) 22*, 125–231.

Aldridge, R. J. 1975: The stratigraphic distribution of conodonts in the British Silurian. *J. Geol. Soc. 131*, 607–618.

Aldridge, R. J. 1979: An Upper Llandovery conodont fauna from Peary Land, eastern North Greenland. *Rapp. Grønlands Geol. Unders. 91*, 7–23.

Bischoff, G. & Sannemann, D. 1958: Unterdevonische Conodonten aus dem Frankenwald. *Notizbl. Hess. L.-Amt. Bodenforsch. 86*, 87–110.

Branson, E. B. & Branson, C. C. 1974: Lower Silurian conodonts from Kentucky. *J. Paleont. 21*, 549–556.

Cooper, B. J. 1975: Multielement conodonts from the Brassfield Limestone (Silurian) of southern Ohio. *J. Paleont. 49*, 984–1008.

Egiazarov, V. H. (Егиазаров, Б. Х.) 1959: Геологическое строение архипелага Северная Земля. [The geological structure of Severnaya Zemlya Archipelago.] *НИИГА, труды 94*. 137 pp.

Epstein, A. G., Epstein, J. B. & Harris, L. D. 1977: Conodont color alteration – an index to organic metamorphism. *Geol. Survey Prof. Paper 995*. 27 pp.

Helfrich, C. T. 1980: Late Llandovery – Early Wenlock conodonts from the upper part of the Rose Hill and the basal part of the Mifflintown Formations, Virginia, West Virginia, and Maryland. *J. Paleont. 54*, 557–569.

Jeppsson, L. 1974: Aspects of Late Silurian conodonts. *Fossils and Strata 6*. 78 pp.

Klubov, B. A., Kachanov, E. I., Karatajūte-Talimaa, V. N., (Клубов, Б. А., Качанов, Е. И., Каратаюте-Талимаа, В. Н.) 1980: Стратиграфия силура и девона о. Пионер (Северная Земля). [The Silurian and Devonian stratigraphy of Pioneer Island (Severnaya Zemlya).] *Изв. АН СССР, серия геологическая 1980:II*, 50–56.

McCracken, A. D. & Barnes, C. R. 1981a: Conodont stratigraphy across the Ordovician–Silurian boundary, Ellis Bay Formation, Anticosti Island, Quebec. *In* Lespérance, P. J. (ed.): *Subcommission on Silurian Stratigraphy, Ordovician–Silurian Boundary Working Group. Field Meeting, Anticosti–Gaspé, Quebec 1981. Vol. 2, Stratigraphy and Paleontology*, 61–69.

McCracken, A. D. & Barnes, C. R. 1981b: Conodont biostratigraphy and paleoecology of the Ellis Bay Formation, Anticosti Island, Quebec, with special reference to Late Ordovician – Early Silurian chronostratigraphy and the systemic boundary. *Geol. Survey of Canada Bull. 329*, 51–134.

Menner, V. V., Matuhin, R. G., Kursh, V. M., Talimaa, V. N., Samoilovich, Yu. G. & Hapilin, A. F. (Меннер, В. В., Матухин, Р. Г., Куршс, В. М., Талимаа, В. Н., Самойлович, Ю. Г. & Хапилин, А. Ф.) 1979: Литолого-фациальные особенности силурийско-девонских-отложений Северной Земли и северо-запада Сибирской платформы. [Lithological and facial peculiarities of the Silurian–Devonian deposits of Severnaya Zemlya and the North-West Siberian Platform.] *In* Matuhin, R. G. (Матухин, Р. Г.) (ed.): Литология и палеогеография Сибирской платформы. *СНИИГГиМС, труды 269*, 39–55.

Menner, V. V., Kurik, E. J., Kursh, V. M., Markovski, V. A., Matuhin, R. G., Modzalevskaya, T. L., Patrunov, D. K., Samoilovich, Yu. G., Smirnova, M. A., Talimaa, V. N., Hapilin, A. F., Cherkesova, S. V. & Abushik, A. F. (Меннер, В. В., Курик, Э. Ю., Куршс, В. М., Марковский, В. А., Матухин, Р. Г., Модзалевская, Т. Л., Патрунов, Д. К., Самойлович, Ю. Г., Смирнова, М. А., Талимаа, В. Н., Хапилин, А. Ф., Черкесова, С. В. & Абушик, А. Ф.) 1982: К стратиграфии силура и девона Северной Земли. [Silurian and Devonian stratigraphy of Severnaya Zemlya.] *In* Yuferev, O. V. (Юферев, О. В.) (ed.): Стратиграфия и палеонтология девона и карбона. 65–73. Москва.

Nicoll, R. S. & Rexroad, C. B. 1968: Stratigraphy and conodont paleontology of the Salamonie Dolomite and Lee Creek Member of the Brassfield Limestone (Silurian) in southeastern Indiana and adjacent Kentucky. *Indiana Geol. Survey Bull. 40*. 87 pp.

Nowlan, G. S. 1981: Late Ordovician – Early Silurian conodont biostratigraphy of the Gaspé Peninsula – a preliminary report. *In* Lespérance, P. J. (ed.): *Subcommission on Silurian Stratigraphy, Ordovician–Silurian Boundary Working Group. Field Meeting, Anticosti–Gaspé, Quebec 1981. Vo. 2, Stratigraphy and Paleontology*, 257–291.

Pollock, C. A., Rexroad, C. B. & Nicoll, R. S. 1970: Lower Silurian conodonts from Northern Michigan and Ontario. *J. Paleont. 44*, 743–764.

Rexroad, C. B. 1967: Stratigraphy and conodont paleontology of the Brassfield (Silurian) in the Cincinnati Arch area. *Indiana Geol. Survey Bull. 36*. 70 pp.

Urvantzev, N. N. (Урванцев, Н. Н.) 1933: Северная Земля. Краткий очерк исследования. [Severnaya Zemlya. A short essay of the study.] *Изд. Аркт. Инст. Ленинград.*

Walliser, O. H. 1964: Conodonten des Silurs. *Abh. Hess. Landesamtes Bodenforsch. 41*. 106 pp.

Ziegler, W. 1960: Conodonten aus dem Rheinischen Unterdevon (Gedinnium) des Remscheider Sattels (Rheinisches Schiefergebirge). *Paläont. Z. 34*, 169–201.

Silurian conodont faunas from Gotland

LENNART JEPPSSON

FOSSILS AND STRATA

ECOS III

A contribution to the Third European
Conodont Symposium, Lund, 1982

Jeppsson, Lennart 1983 12 15: Silurian conodont faunas from Gotland. *Fossils and Strata*, No. 15.
pp. 121–144. Oslo. ISSN 0300-9491. ISBN 82-0006737-8.

The faunal succession of conodonts on Gotland is briefly described, and some taxonomic
comments are presented. As presently known, the faunas are stratigraphically important for
correlations with areas outside Gotland, at several levels. The *Pterospathodus amorphognathoides*
Zone occurs in the Lower Visby Beds, indicating that the base of the Wenlockian is within that
unit. *Kockelella walliseri* occurs, often together with *Monograptus priodon* and *Rhipidium tenuistriatum*
and other characteristic fossils, in Slite unit f and other Slite units. These occurrences are here
interpreted as representing a single horizon which is identifiable world-wide. A variety of *K.*
variabilis occurs in the Hemse Marl NW part and dates at least parts of that unit as Late?
Bringewoodian and earliest Early Leintwardinian [late? late Gorstian and earliest early early
(sic!) Ludfordian]. The *Polygnathoides siluricus* Zone is dated as Late? Leintwardinian (late? early
Ludfordian) and traced from the coast of the Näs peninsula in the southwest to the easternmost
Östergarn peninsula, improving local correlations. The zone thus includes parts of the Hemse
Marl SE part, which unit is identified in a wider area than earlier known, and parts of the Millklint
Limestone (uppermost Hemse limestone unit). *H. snajdri crispa* appears in the lower Burgsvik
Beds and the oldest *H. steinhornensis* with alternating denticles yet found on Gotland occurs in the
Hamra unit b. In the Silurian type area in Britain such populations are only known from the latest
Whitcliffian (late late Ludfordian). □*Conodonta, biostratigraphy, Silurian, Gotland.*

Lennart Jeppsson, Department of Historical Geology and Palaeontology, Sölvegatan 13, S-223 62 Lund,
Sweden; 3rd November, 1982.

The early history of the stratigraphical research of Gotland
was described by Hede (1921). The 13 subdivisions (now
called Beds) introduced by him (Hede 1921, 1925 a) have
remained the stratigraphic frame work used both in the
geological map descriptions and in most subsequent work,
even though many represent a long time interval. In most of
the map descriptions these Beds are further subdivided, but as
most of these units have lacked names or symbols, they have
been little used. Laufeld (1974 a) extended the lettering of the
units begun by Hede, and that system is followed here. The
short remarks on the lithologic character of each unit are taken
from Hede's map descriptions, from Hede (1960) and from
Laufeld (1974 a). Hede's Beds and some of his minor units
were largely faunistic subdivisions; thus many of the Beds
include the whole range of lithologies common on Gotland.
Hede (1919, 1942) based his correlations with other areas
chiefly on the sparse occurrences of graptolites in the least
calcareous parts of some of the Beds. Martinsson (1967), using
the ostracodes, improved correlations both within Gotland
and with areas outside the island.

Several publications have dealt with, mentioned or illustrated conodonts from Gotland (e.g., Lindström 1964; Martinsson 1967; Fåhraeus 1968; Jeppsson 1969, 1972, 1976,
1979 a; Barnes *et al.* 1973), but few have discussed the faunas.
However, Fåhraeus (1969) listed the faunas in 22 samples,
most of them producing less than 25 elements each. I have
earlier commented upon the Hemse and younger conodont
faunas (1974) and described those in the Lower Visby, Upper
Visby and Högklint Beds in Vattenfallet (1979 c).

Recently (Holland 1980), a formal subdivision of the
Wenlockian and the Ludlovian has been decided upon. Each

series is now divided into two stages. Thus we have an
unambiguous base for the convenient terms early and late,
useful in the corresponding parts of the chronology. In the
Ludlovian a greater precision was potentially possible – and
partly achieved (cf. Holland 1980; Holland *et al.* 1980) –
following the introduction of four stages with nine substages
(Holland *et al.* 1963). In order not to lose this greater precision
in correlations, I use also these older units.

Collections

This preliminary review is based on over 600 samples (about
3–6 kg each), the majority of which was collected in the late
sixties and early seventies. (Regarding localities, see appendix.) Thereafter collecting was largely postponed due to lack of
funds for extracting the conodont elements. Research grants
from the Swedish National Research Council during the latter
part of the seventies have now made it possible to dissolve 0.5
kg of each sample and another 0.5–4 kg of some of the richest
ones. Nearly all of them have produced at least some conodont
elements, and it seems probable that no unit is barren. Most of
the collections are small, and it is evident that from several
stratigraphical units adequate faunas can be obtained only
from samples weighing tens of kilograms. One example of
what can be obtained from large samples is the fauna from
Möllbos 1. The first 0.5 kg dissolved yielded only 4 specimens,
and therefore nothing more was dissolved for some time.
However, the sample also showed that the strata contained
silicified fossils. I have organized the study of such fossils on
Gotland and have been able to get 'unemployment grants' for

the preparation of the samples, with the result that more than 700 kg have now been dissolved in acetic acid. One of the samples from Möllbos 1, G 77–28 (78.32 kg), has been picked for conodonts, and has produced about 577 elements representing an extremely well preserved conodont fauna, one of the very best I have from Gotland. Many other important conodont faunas deriving from that project are mentioned in the text. Their sample numbers are chosen so that they do not duplicate those of my other samples. However, in the unabbreviated sample designation they are separated by the letters PSSFG and LJ respectively. I have received many samples and collections from Claes Bergman, Stig M. Bergström, Kent Larsson, Sven Laufeld, Anders Martinsson, and Carl Pleijel. The numbers of these samples are here followed by the initials of the collector, but in the designation of my own samples I have omitted LJ.

State of preservation. – Many limestones on Gotland consist of more or less worn skeletal debris. The conodont elements, too, are more or less worn in these limestones. Where possible, samples have been taken from lithologies in which the elements should show the least wear, but not uncommonly the only conodont remains found are a few rounded fragments of mature, robust elements of the most robust species (usually the sp elements of *Hindeodella confluens* and the *H. steinhornensis* group). Being translucent, they can, however, be identified on internal structures. Larger samples might possibly increase the length of the species list, but the original relative frequencies of different species would remain unknown. The same is true of the numbers of each element within an apparatus (if only known from such localities), and the distribution of different age groups within a population.

In most samples the state of preservation is better, but usually all the bar elements are in pieces and all denticles broken except the blunt ones on sp and oz elements. This might well have been ascribed to careless laboratory treatment, but now and then a sample treated in the same way as the others, at the same time and by the same people, has produced perfect specimens. Thus, the inferior state of preservation is a result of breakage during the formation of the sediment or fracturing afterwards (see below).

In many samples the elements are internally fractured, often with many fractures per millimetre. In most elements these fractures are perpendicular to the denticles and to the process, respectively. That this fracturing has happened during or after diagenesis is evident from the fact that very often all the pieces of large or otherwise individually identifiable specimens can be found in the same sample. Further, many fractures can often be seen in the recovered pieces. Thus, the fracturing was probably caused by strain in the beds.

In many samples the elements are recrystallised. The surface of the elements has become irregular by overgrowth and by corrosion. The denticles change shape and may even show crystal facets. The white matter becomes less distinct. If recrystallisation occurred after fracturing, the pieces are often fused again, and thus more complete specimens may be recovered. When thermally affected, the organic carbon in the elements darkens (see Epstein *et al.* 1977), but any fusing material lacks organic carbon and remains light (Jeppsson 1976:109). Recrystallisation seems to be promoted by heating.

The beds on Gotland are not thermally affected, and thus complete bar elements are much rarer there than they are in many samples from slightly metamorphosed areas. In a few samples the elements were not only fractured, but the pieces were displaced before recrystallisation occurred, and thus the elements look strongly malformed. Both kinds of recrystallisation have been interpreted as evidence for the healing of fractured elements by the animal. That this is not the case is evident from the fact that they are related to locality and not to taxa (Jeppsson 1976).

As noted above, none of the collections from Gotland is thermally affected. On the contrary, the light colour is so well preserved that taxonomically important differences in colour can be used in reconstructing apparatuses and in the routine identification of fractured specimens. Perhaps collections of this kind should be separated as a zero group in the scale of conodont alteration indexes introduced by Epstein *et al.* (1977).

However, the colour may also be affected by other factors. In darker beds the elements are often slightly darkened, probably by very finely disseminated pyrite or by organic matter. In red beds oxidization may bleach the elements and disseminated haematite may stain them.

Less than one out of ten samples from Gotland produces a rich and well preserved fauna. However, some samples produce the very best collections possible.

Annotated list of conodont taxa reported here from Gotland

Correct taxonomy and nomenclature of conodonts require much more work per taxon than most other well-known groups of fossils. As yet, only a few of the taxa recognized on Gotland have been adequately studied. Therefore, some of the names are used herein with less confidence than others, and it is certain that future revision of the scope and nomenclature of some of the taxa will have to be undertaken.

There are conflicting opinions, or none at all, regarding the apparatuses of most of the Silurian conodont taxa with platform elements. Therefore, presence of these taxa has been registered only on the occurrence of the platform element.

Except for *Panderodus* and *Pseudooneotodus*, taxa with only simple cones have only been identified to the genus. Confident identification of these to species level awaits major taxonomic and nomenclatural revisions.

For most of the taxa I also outline their known distribution on Gotland. In most cases, more details are given with the descriptions of the faunas.

Apsidognathus walmsleyi Aldridge 1974. Aldridge noted that also elements identified as from *Ambalodus galerus* may belong here, and that receives some support from my collections. Lower Visby Beds.

Aulacognathus bullatus (Nicoll & Rexroad [1969]). Lowermost? Lower Visby Beds.

Belodella. A recent discussion was given by Barrick 1977. At least from low levels in the Hemse Beds into the Hamra Beds.

'*Carniodus*'. Walliser (1964) assigned many different small elements to *Carniodus* while he incorporated others that showed broad similarities in *Neoprioniodus* and *Roundya*. Some of these have been interpreted as elements of *Pterospathodus* (see Jeppsson 1979 c), while others may belong to *Hadrognathus* and *Apsidognathus*. Pending their proper identification, those not shown to belong to *Pterospathodus* are here referred to '*Carniodus*'. Lower Visby Beds.

Dapsilodus Cooper 1976. This genus was discussed by Barrick (1977). Slite to Hamra Beds, inclusive.

Decoriconus Cooper 1975. This genus was discussed by Barrick (1977). There seem to be at least two species on Gotland. Lower Visby Beds to Eke Beds, inclusive.

Distomodus dubius (Rhodes 1953), sensu Jeppsson 1972. Upper Hemse Beds and Hamra Beds.

Distomodus kentuckyensis Branson & Branson 1947, sensu Klapper & Murphy 1975. Lower and Upper Visby Beds.

Hadrognathus staurognathoides Walliser 1964, sensu Walliser 1964. Lower Visby Beds and lower part of Upper Visby Beds (redeposited in the latter unit?).

Hindeodella confluens (Branson & Mehl 1933), sensu Jeppsson 1969. Högklint to Sundre Beds, inclusive.

Hindeodella excavata (Branson & Mehl 1933), sensu Jeppsson 1969. Lower Visby to Sundre Beds, inclusive.

Hindeodella gulletensis (Aldridge 1972), sensu Jeppsson 1979 c. Högklint Beds, unit a, to lower Slite Beds, inclusive.

Hindeodella polinclinata (Nicoll & Rexroad [1969]), sensu Cooper 1977. One of many synonymous names is *Spathognathodus tauchionensis* Saladžius, 1975. Aldridge (1979) discussed this taxon and the differences between its elements and those of some related taxa. There seem to be minor differences between collections from different areas (Aldridge 1979); however, I think that those from Gotland do belong in this species. Lower Visby Beds.

Hindeodella sagitta rhenana (Walliser 1964), sensu Aldridge 1975a. Högklint to lower part of Slite Beds, inclusive.

'*Ozarkodina serrata*' Helfrich 1975?. In the Klintberg c beds at Botvaldevik 1 in sample G 75-36 there is a very large, complete, strange oz element. As the element is unique, it is difficult to exclude the possibility that it is a strongly malformed gerontic specimen of *H. confluens*. However, it is close to what Helfrich (1975: appendix 1:32) described as *Ozarkodina serrata*. That name is probably a junior homonym to *Astacoderma serratum* Harley 1861. The latter name has not been published in any combination with the generic name *Ozarkodina*, but it has been shown to be a senior unused synonym to *H. confluens*, often placed in *Ozarkodina* (cf. Rhodes 1953; Jeppsson 1969, 1974; Klapper & Philip 1971; ICZN article 57). In some of my other samples from the Klinteberg Beds

there are worn fragmentary specimens of possible elements of the same species. These do not deviate as much from *H. confluens* as the single oz element. In the Klinteberg a beds at Hällinge 2 in sample G 71-94 there is a very large tr element that at present cannot be placed in an apparatus. It somewhat resembles '*Trichonodella*' sp. A of Uyeno (1980, Pl. 10:17). It is possible that all these Klinteberg specimens represent the same taxon, and that it is the same as Helfrich's. Whether it is distinct from *H. confluens* at the species or subspecies level is unclear. Klinteberg Beds. Fig. 1 F.

Hindeodella snajdri (Walliser 1964). The apparatus-based concept of this taxon and its subdivision in subspecies in the late Ludlovian will be described by Schönlaub *et al.* Another subspecies occurs on Gotland in the late Wenlockian. Its relationship to *H. snajdri* s. str. is evident: the fusing of the cusp and adjacent denticles of the sp element to an edge very early in ontogeny strongly differs from the gerontic smoothing out of details seen in other taxa of *Hindeodella*. Also, the rest of the apparatus is so similar to the late Ludlovian representatives that the late Wenlockian taxon clearly should be separated at a subspecific level only. Uppermost Slite Beds to Sundre Beds, inclusive.

Hindeodella steinhornensis (Ziegler 1956). For a discussion, see Jeppsson (1974). The lineage can now be traced from *H. gulletensis* via scattered specimens in the late Wenlockian and early Ludlovian to more regularly occurring populations in the late Ludlovian.

Hindeodella wimani Jeppsson [1975]. Two subspecies are known (Jeppsson 1974). Burgsvik and Hamra Beds.

Hindeodella sp. m. This taxon is probably closely related to *H. confluens*. Characteristic is the fact that only the core of the cusp is transformed into white matter. The bases of the denticles are similarly constructed. Klinteberg and Hemse Beds.

Johnognathus huddlei Mashkova 1977. Lower Visby Beds.

Kockelella absidata Barrick & Klapper 1976. This is the third generic and second specific name on Walliser's (1964) '*fundamentata*'. The species is very variable. Thus, specimens from the Mulde and Klinteberg Beds at Loggarve 1 (see p. 132) differ in having fewer denticles.

In the Hemse Marl NW part at Kullands 1 in sample G 77-38 PSSFG there are four specimens of the sp element of what obviously is a well defined, distinct population. These have a denticle on the inner basal cavity lip. Similar specimens have also been found at Gerumskanalen 1 in sample G 77-37 PSSFG, at Gardsby 1 in sample G 82-27, and at Lilla Hallvards 1 in sample G 71-143. Walliser (1964) included similar specimens in what is now known as *K. absidata*. On the other hand Barrick & Klapper (1976) described the sp element of *K. stauros* as having one to two processes, the inner one with one to two denticles. However, at present I consider my specimens to be a variety of *K. absidata*, which is a very variable taxon (Walliser 1964, Pl. 23). The variation is evident also in my few specimens from Gotland, but at present it is

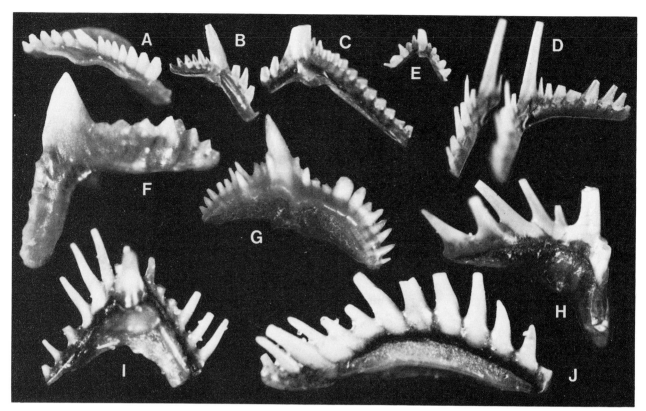

Fig. 1. □A–E. *Polygnathoides siluricus* Branson & Mehl 1933 sensu novum. Sample G 71-148 from Vaktård 3. The sp and hi elements are from individuals that were rather young, the ne element from one that was slightly older and the oz and pl elements from juvenile individuals. No tr element is illustrated. ×40. □A. An sp element. The two anteriormost denticles are broken away. LO 5590. □B. An oz element. LO 5591. □C. An ne element. The tips of both processes are broken away. The basal filling below the long process is preserved. LO 5592. □D. A hi element. Focus is on the lateral process in the left, incomplete, picture. That process is straight and about as long as the posterior one. The latter is deflected slightly outwards. LO 5593. □E. A pl element. LO 5594. □F. *Ozarkodina serrata?* A very large oz element.

Note that the posterior process is malformed with an extra lateral process. ×32. Sample G 75-36 from Botvaldevik 1. LO 5595. □G–J. Gen. et sp. indet. □G. A 3.08 mm long oz element. ×18. Sample ES 137, collected at Millklint 1 by Anders Martinsson who described it as a 'thickly bedded limestone with strophomenaceans and *Atrypa*'. LO 5596. □H. A hi element. Note that the denticles of the posterior process point in different directions. ×40. Sample ES 202 collected by Anders Martinsson at Gogs 1. LO 5597. □I. A tr? element. ×40. Sample ES 202 (A.M.). LO 5598. □J. An unidentified 1.85 mm long element. The basal filling is preserved. I have not been able to identify the cusp in this element. ×43. Sample G 67-54 from Gogs 1. LO 5599.

difficult to use it in stratigraphy, as the outline changes strongly during ontogeny, and thus growth series must be illustrated; the white matter distribution is also crucial, particularly in mature specimens. The variety from Kullands 1 and Gerumskanalen 1 is closest to Walliser's (1964, Pl. 23) specimens from the middle part of his *A. ploeckensis* Zone. Regarding the other elements of the apparatus, my collections seem to confirm the reconstruction of Barrick & Klapper 1976, although non-sp elements very similar to those that occur with sp elements of *K. absidata* have a much wider distribution on Gotland than the sp elements. I also note that Barrick & Klapper (1976) referred 55 non-sp elements to *K. stauros* at their locality Ca 2, but only one sp element. Here I therefore identify non-sp elements occurring without sp elements as *Kockelella?* sp. Slite Beds to Hemse Beds, inclusive.

Kockelella? ranuliformis (Walliser 1964). The species is here identified on the sp element only. Upper part of the Lower Visby Beds to the Högklint Beds, inclusive.

Kockelella variabilis Walliser 1957, sensu Barrick & Klapper 1976. The species has been listed as present only if the sp element was found. Lower Hemse Beds.

Kockelella walliseri (Helfrich 1975). The variation in its platform element is large and my samples (like Viira's) include both narrow forms of the kind on which Helfrich (1975) based the name *Spathognathodus walliseri* and broad forms of the kind on which Viira (1975) based the name *Spathognathodus corpulentus*. The two names are here considered synonymous. It seems that Viira's description was published on 1975 08 15 but Helfrich's on 1975 06 09 (personal communication from Richard Aldridge, who has corresponded with the author and publisher, respectively, in the matter). Thus *walliseri* has priority. Another aspect that has to be considered is that Helfrich's description was given on microfiche which was not accepted as a publication by the Code in force in 1975 (compare Cooper 1980), but it seems probable that his names will be validated. Pending the outcome of this question, I will use the oldest name. Barrick & Klapper (1976) transferred '*S. walliseri*' to *Kockelella* and that seems to be well founded.

A collection of seven fairly small elements from the Slite d beds at Stora Myre 1 is intermediate between *K.? ranuliformis* and typical *K. walliseri* regarding the lateral process. Thus some sp elements lack a lateral process altogether, whereas a few have a process with up to a couple of denticles. The difference may depend on the elements being juvenile. The species has been listed as present only if the sp element was found. Slite Beds.

Kockélella sp. a. I illustrated a large ne element of this taxon in 1974 (Pl. 12:5). It is closely similar to that of *K. absidata*, and I refer to it as *Kockelella* sp. a. I have now found four such elements from four localities on Gotland but identified only one, poorly preserved, possible sp element. Upper Hemse Beds.

Kockelella? sp. See *K. absidata* above. Upper Hemse Beds.

Ligonodina confluens Jeppsson 1972. Only those specimens originally referred to *L. c. confluens* are included here. Högklint c to Sundre Beds, inclusive.

Ligonodina aff. *confluens*. Described and illustrated in Jeppsson 1972 as *L. confluens* n. ssp.. Hemse Beds.

Ligonodina elegans Walliser 1964, sensu Jeppsson 1969. Hamra and Sundre Beds.

Lignodina excavata (Branson & Mehl 1933), sensu Jeppsson 1972.

Ligondina excavata novoexcavata Jeppsson 1972. Burgsvik to Sundre Beds, inclusive.

L. confluens, L. aff. *confluens* and *L. elegans* can easily be separated, as is the case with typical *L. excavata* in the Ludlovian. Apart from the frequent occurrence of more or less fragmentary stray specimens that cannot be positively identified as yet, there are also some other distinct populations. There are also problems in separating *Ligonodina* from some of the elements found in apparatuses with platforms. With these reservations, the following two taxa may be delimited:

Ligonodina silurica Branson & Mehl 1933? The elements have closely similar colour to those of *H. excavata* and *L. excavata*. Rexroad & Craig's (1971) description of the hi element stresses the distinguishing characters. The ne element was illustrated by them (Pl 80:11) and described as ?*Neoprioniodus latidentatus*. These two elements are those that are the easiest to identify, while the other elements do not differ so much from those of *L. excavata*. An evolutionary divergence is evident, as stratigraphically older populations are less distinct. However, the lineage can easily be followed down into the middle Wenlockian, but with regard to older populations, I have to reserve judgement on whether they belong here or to *L. excavata*. Into Hemse Beds.

Ligonodina sp. d is best known in the Hemse Marl SE part. The elements are brownish like those of *H. confluens*. The ne element is close to those of *Kockelella* in general shape. Thus, the cross section of the cusp is compressed and edged. The cusp, but not the rest of the element, resembles that of the corresponding element of *L. elegans*. The other elements exhibit some similarities in the direction of the processes to the corresponding elements of *L. elegans*. For example, the antero-lateral process of the hi element is directed obliquely back and down,

but less so than in *L. elegans*. However, all the elements of *Ligonodina* sp. d are much more robust, and the robust denticulation is quite distinctive.

'Ozarkodina serrata' see *Hindeodella* above.

'Ozarkodina sp. nov.' of Aldridge *et al.* 1982. Halla Beds to Hemse Beds, inclusive.

Panderodus. I hope to complete a monograph on this genus during 1984. There are seven or eight species on Gotland, subdivisible into about ten subspecies. Each taxon has eight to ten kinds of elements, including two short, recurved ones (ne and tr elements). Pending completion of the necessary type studies six of these taxa are here called:

Panderodus equicostatus (Rhodes 1953). The elements consist largely of a base with a very short cusp. Upper Visby Beds to Sundre Beds, inclusive.

Panderodus gracilis (Branson & Mehl 1933). Pending a second study of the type, I am slightly unsure of applying this name but use it here because my taxon is close to one of the species that occurs in the type formation. In the Silurian subspecies the base is much shorter than the cusp. Upper Hemse Beds.

Panderodus langkawiensis (Igo & Koike 1967). (This name has priority over the name *P. spasovi*.) Lower Visby Beds.

Panderodus recurvatus (Rhodes 1953). There is a strong indication that the name will have to be replaced by an older synonym. One older and one younger subspecies occur on Gotland. Lower Visby Beds to Hemse Beds, inclusive.

Panderodus unicostatus (Branson & Mehl 1933). Two or three consecutive subspecies occur on Gotland. The cusp in the elements of this species is slightly shorter than the base. Often one of the elements is serrated (*P. serratus* is a younger synonym). Lower Visby Beds to Sundre Beds, inclusive.

Panderodus sp. g. Two very different subspecies are presently included here. The cusp is longer than the base, and the elements may grow very robust. Some existing names may be applicable to this taxon, but I am not sure enough to use any of them here. Lower Visby to Hemse Beds, inclusive.

Pedavis thorsteinssoni Uyeno [1981]? Eke Beds.

Pelekysgnathus dubius Jeppsson 1972. The specimens from the *P. siluricus* Zone on Gotland usually differ from those illustrated from Skåne (Jeppsson 1972), with the cusp being much larger than the single denticle. Broken and worn elements are thus difficult to separate from the ne element of *D. dubius*. Upper Hemse Beds.

Polygnathoides siluricus Branson & Mehl 1933 (an improved concept is now possible, see Fig. 1 A–E). Upper Hemse Beds.

Pseudooneotodus. In the lowermost beds some elements are three-tipped but most are one-tipped. As yet it is impossible to say if both kinds of elements derive from the same taxon. Distinctly two-tipped elements occur together with one-tipped as soon as the three-tipped have disappeared, and then the distinctness of the two tips gradually decreases during the Wenlockian. In some cases I use the names:

Pseudooneotodus tricornis Drygant 1974. Lower Visby Beds.

Pseudooneotodus beckmanni (Bischoff & Sannemann 1958). Lower Visby Beds to Hamra Beds, inclusive.

Pseudooneotodus beckmanni bicornis Drygant 1974, sensu Jeppsson 1979 c. Upper part of the Lower Visby Beds to Högklint Beds, inclusive.

Pterospathodus amorphognathoides Walliser 1964. The complete apparatus includes six or seven elements. Lower Visby Beds.

Pterospathodus pennatus procerus (Walliser 1964), for a discussion, see Jeppsson 1979c. Lower and Upper Visby Beds.

Walliserodus. The genus has recently been discussed by Barrick (1977). There are at least six kinds of elements in the apparatus. Lower Visby Beds to Slite Beds, inclusive, questionably in the Hamra Beds.

Gen. et sp. indet. In the Upper Hemse Beds two kinds of undescribed elements occur: a large sp element and large rough elements with up to three rows of denticles with rounded cross sections. A connection between these two kinds of elements is possible. Fig. 1 G–J.

Biostratigraphy

The Lower Visby Beds

The Lower Visby Beds crop out intermittently in a strip, about 55 km long, along the NW coast. The largest known exposed thickness above sea level is about 12 m. The beds consist of marlstone and highly argillaceous limestone. There is a gradual increase in the content of carbonate upwards through the unit which continues in the following Upper Visby Beds. The base of the Upper Visby Beds is drawn solely on the faunal change (but see below).

Except in the uppermost beds (see below), the conodont fauna seems to be uniform and is much more diverse than those of the rest of the sequence on Gotland. The total number of species cannot be given as yet, as there are many elements which have not been combined into reconstructed apparatuses. However, it is well above 20 and may even be above 30. It is also characteristic that most of the lineages which are more or less ubiquitous in younger faunas are rare or absent. Thus, *Hindeodella confluens*, the *H. steinhornensis* group and *Panderodus equicostatus* are absent, and *H. excavata* is rare. *Belodella* and *Dapsilodus* are also absent.

Most abundant are *Pterospathodus amorphognathoides*, *Hindeodella polinclinata*, *Panderodus unicostatus*, and '*Carniodus*'. Other regular constituents are *Panderodus* sp. g. ssp. v., *P. recurvatus*, *P. langkawiensis*, *Walliserodus*, *Decoriconus*, *Pseudooneotodus* (one-tipped elements dominate, but three-tipped

also occur), and *Distomodus*. Of slightly less regular occurrence are *Apsidognathus walmsleyi*, *Hadrognathus staurognathoides*, *Johnognathus huddlei*, *Ligonodina excavata*, and a number of other species of *Hindeodella* and *Ligonodina*. the rarest taxon identified as yet is *Aulacognathus bullatus* (one good specimen). This taxon occurs at a locality which appears to represent the lowermost part of the Lower Visby Beds. Another rare taxon which may be restricted to that part is the ostracode *Barymetopon infantile* Martinsson 1964.

P. amorphognathoides is regularly present in collections from the part of the Lower Visby Beds here discussed; therefore the upper boundary for the *P. amorphognathoides* Zone is here drawn immediately above the last proven occurrence of that taxon. In Britain, at the type locality for the base of the Wenlockian, *P. amorphognathoides* occurs both in the uppermost Llandoverian and in the basal Wenlockian, as do all forms considered characteristic of the *P. amorphognathoides* Zone (Aldridge 1975b:613). There, the faunal change concurs with a lithological change from fine blue grey mudstone to calcareous siltstone (Aldridge 1975 b). Thus the *P. amorphognathoides* Biochron may well range still higher there (Aldridge 1975 b). On Gotland there is no described lithologic change between the Lower and Upper Visby Beds, only the faunistic one (cf. Martinsson 1967:359). However, the two units weather differently; the Lower Visby Beds produce a sticky clay which is not easily washed away from the section, while the clay in the Upper Visby Beds weathers to dust. In any event, the *P. amorphognathoides* Range-Zone does extend into the Wenlockian and that is an indication that the base of the Wenlockian is at least a couple of metres down in the Lower Visby Beds. A very large change in the conodont faunas seems to occur worldwide at or close to the end of the *P. amorphognathoides* Biochron. The details of the changes on Gotland will be discussed separately.

The top part of the Lower Visby Beds. – By definition, the top of the Lower Visby Beds is to be drawn at the abrupt faunal change. However, in practice it has often been drawn lower at many localities (Martinsson 1967:358). Thus, the beds at Nygårdsbäckprofilen 1 was by Hedström (1910) referred to the unit which now is called Upper Visby Beds, but Hede (1940) identified 2.5 m of Lower Visby Beds there. Neither of them found Lower Visby Beds at Vattenfallet from where they were reported by Jaanusson (1979). Similarly, at Rönnklint 1 Hedström (1910) reported that the Lower Visby Beds reach about 9 m above sea level and Hede (1940) gave 8 m, but Jaanusson (oral information 1981 08 24) draws the boundary some metres higher up. Samples from the top 2 to 3 m of the Lower Visby Beds lack one or more of *Pterospathodus amorphognathoides*, *Pseudooneotodus tricornis* and *H. polinclinata* but have instead one or more of *Pterospathodus pennatus procerus* and *Pseudooneotodus beckmanni bicornis*. The rest of the conodont fauna consists both of some surviving taxa from the older fauna and some new ones that continue in the Upper Visby Beds, like *Kockelella? ranuliformis*. Collections now available have both a much lower frequency and much lower diversity than older ones. The low conodont frequency seems to indicate a rapid rate of sedimentation and, thus, the interval between the *P. amorphognathoides* Zone and the Upper Visby Beds was very short. The change in frequency makes it difficult to establish the upper range especially of the many rare taxa present in older parts of the Lower Visby Beds. This problem is further complicated by reworking, but reworked specimens usually can be distinguished by their darker colour, especially of the basal filling and adjacent parts.

The fauna described previously from the Lower Visby Beds at Vattenfallet (Jeppsson 1979 c) belongs here and not to the *P. amorphognathoides* Zone. To that list can now be added that three species of *Panderodus* have been identified in both samples, *P. equicostatus*, *P. unicostatus* and *P. langkawiensis*.

The Upper Visby Beds

The early Wenlockian Upper Visby Beds crop out in the coastal cliffs along the NW coast for about 60 km. They consist of between some and 16 m of marlstone, argillaceous limestone and tabulate-dominated mounds.

I have earlier described the Upper Visby conodont faunas in the Vattenfallet section (Jeppsson 1979 c). To that description can now be added that *Panderodus* sp. a is *P. equicostatus*, *P.* sp. b is *P. unicostatus*, a taxon which also occurs in sample G 70-22, and questionably also in most other samples from this unit, and that 'gen. et sp. indet.' is *Hadrognathus* (as yet it cannot be excluded that the two specimens may be reworked). Also, in Fig. 70, read *H. gulletensis* for *H. s.* aff. *scanica* and *H.* aff. *confluens* for *H.* aff. *gulletensis*. My collections from other localities do not differ significantly from those from Vattenfallet.

The Högklint Beds

The early Wenlockian Högklint Beds crop out in an area about 80 km long by 1 km wide along the coast in the northwest. They consist of between less than 20 m and 35 m of bioherms and other kinds of limestones. The Beds are subdivided into four lettered units.

I have earlier (Jeppsson 1979 c) described the sequence of conodont faunas through the Högklint Beds in the Vattenfallet section. To that description can be added that the species identified as *Panderodus* sp. a is *P. equicostatus*. Regarding corrections in 1979c, Fig. 70, see above. Two of the conodont taxa found there – *H. sagitta* and *K.? ranuliformis* – have been used as zonal fossils (Walliser 1964; Barrick & Klapper 1976, respectively). In Vattenfallet *K.? ranuliformis* is not known above unit b, but that may well be due to the rarity of the taxon and the small samples available from the now inaccessible upper part of the section. *H. sagitta rhenana* occurs only in the upper part of unit c and in unit d. *H. s. rhenana* further occurs in the Högklint c at Galgberget 1.

In the area around the bay Kappelshamnsviken *H. s. rhenana* occurs in the Högklint b at Nymånetorp 1 in sample G 81-65 from 0.05–0.13 m above the reference level, and – together with *K.? ranuliformis* – in the Högklint b at Svarven 1 in sample ES 177 (A.M.) and in Högklint c in sample G 78-3 CB from 0.30 m above the reference level. A few specimens of *K.? ranuliformis* also occur in samples ES 113 (A.M.) and ES 178 (A.M.) from the same unit and locality. Both taxa occur at Västös Klint 1 in sample G 77-34 PSSFG.

On Fårö, *H. s. rhenana* occurs in the lower–middle Högklint Beds at Lautershornsvik 2 in sample G 73-73, in the slightly younger (Hede 1936:14) Högklint Beds at Lautershornsvik 3

in sample G 73-72, and, together with *K.? ranuliformis*, in the same unit at Aursviken 1 in sample G 77-25 CB. *K.? ranuliformis* occurs also in sample G 77-26 CB from Langhammarshammar 1.

The incoming of *H. s. rhenana* may be stratigraphically useful in a small area like Gotland, especially as the strike of the beds is close to an original depth contour (regarding larger regions, see Jeppsson 1979 c). If so, it would indicate that the 'pre-*H. s. rhenana*' part of the Högklint Beds thins out NE of Lickershamn and that the Högklint b lithology persists longer there. Alternatively the appearance of *H. s. rhenana* is progressively younger in the southwest direction. The only other significant event in the faunas is the appearance of *H. confluens* which occurs in the Högklint b at Svarven 1, in sample ES 178 (A.M.).

The Tofta Beds

The early Wenlockian Tofta Beds crop out in an area about 1–2 km wide and 50 km long. They chiefly consist of limestones rich in *Spongiostroma* and stromatoporoids. The Beds are up to 8 m thick but thin out to the NE and probably also to the SW.

The conodont faunas at hand from the Tofta Beds are few and small, but the diversity seems not to be extremely low.

At Vale 1 in sample G 77-14 *H. confluens* morphotype γ occurs; the elements are small but not juvenile. That morphotype was earlier only reported from Ludlovian and younger beds. There are further some fragmentary specimens that resemble those from the Högklint d, which I (1979c) have identified as resembling morphotype ε of *H. confluens*. The fauna also contains *Ligonodina* sp.

At Klockaremyr 1 in sample G 71-73, *H. sagitta rhenana* occurs together with *H.* cf. *confluens*.

The lower parts of the Slite Beds

The outcrops of the Wenlockian Slite Beds form an area up to 20 km wide and about 115 km long. Limestones occupy the northwestern part, except in the extreme southwest, and occur also in outliers across the belt in the northeastern half of the area. Marlstone forms a belt widest in the soutwest and narrowing in NE direction. However, marl is known all the way up to easternmost Fårö.

The lower part of the Slite limestones is divided into 5 lettered units, of these a, b and d occur only locally. Together the five units are up to 20 m thick in the SW and up to 10 m on Fårö (Hede 1936). Only in the extreme southwest are marly equivalents of these limestones known.

All conodont faunas extracted to date are small, and very little can yet be said about these beds (three faunas will be discussed together with the Slite f). Some of the small collections now at hand do not differ much from a Högklint fauna, e.g., a collection from the Slite unit b beds at Kluvstajn 1 (sample G 73-16) includes *H. gulletensis*, *H. excavata* and *P. equicostatus*. In a fauna from the Slite d beds at Martille 7 (sample G 73-18) *H. sagitta rhenana* occurs together with *H. confluens*, *Ligonodina*, *P. equicostatus* and *P.* sp.

Other collections are more like those from younger beds; thus, sample G 79-206 from the Slite c beds at Hällagrund 1, 0.05–0.10 m above the contact to the Högklint Beds, contains *H. excavata*, *H. confluens*, *P. equicostatus*, and *P. unicostatus*.

Slite f and contemporaneous beds

The Slite f (Laufeld 1974 a) is a distinct unit consisting of up to
2 m of fine-grained argillaceous limestone, in places interca-
lated with thin layers of marlstone (Hede 1960; Laufeld
1974 a). The unit has also been called the *Rhipidium tenuistriatum* beds after the pentamerid that in places is abundant in
the unit. It is known both in the southwest where the outcrops
form a narrow strip ending about 1.8 km ESE of Västerhejde
church (Hede 1940), and in the northeast, in southwestern
Fårö (Hede 1936, 1960) and adjacent parts of the main
Gotland (Hede 1933:52). Probably this kind of sediments,
formed on Gotland only during a short interval of time, was
laid down only in a narrow belt, while Slite Marl was deposited
southeast of it and different limestones northwest of it. For over
50 km along the strike of the beds, only such lithologies crop
out, and time-stratigraphic equivalents have to be identified
using fossils. This remains to be done, except for Hede's
(1933:43) report of *R. tenuistriatum?* from the Klinthagen area.

Another characteristic fossil for this unit is *Monograptus
priodon* (see Hede 1927 a:28). This taxon has a much more
wide-spread occurrence on Gotland than *R. tenuistriatum*. In
the southwest, it thus occurs, e.g., also in the limestones at
Hallbro Slott 6 (Hede 1942 loc. 19, 20) west of the strip with *R.
tenuistriatum*, and in the Slite Marl east of it. Considering the
slight general dip of the strata in the area and the topographical levels of these localities, they may well be outliers and the

---

Fig. 2. A stratigraphical – geographical diagram of the strata exposed
on Gotland. The diagram shows two of the dimensions of the beds,
time and place, when projected onto a N 45°E plane *i.e.* a plane
approximately parallel to the strike.

The SW-NE co-ordinates used for a locality are easily derived by
adding its S-N and its W-E UTM co-ordinates, in the CJ area. In the
CK area (*i.e.* northern Gotland) 100 must be added and on
easternmost Fårö, 200. Localities on Stora Karlsö have to be
measured. These co-ordinate lines are shown on the large inset map.
In order to facilitate identification of the SW-NE co-ordinates for the
marked localities, and to aid the addition of other localities, most of the
vertical lines in the pattern designating limestone areas are placed at
40, 50, 60, and so on, and at 45, 55, 65, and so on. Other co-ordinate
systems can, of course, be used equally easily, and NW-SE diagrams
for selected areas can be constructed in a similar way.

The age (time co-ordinate) is still rather uncertain for most
localities, as it is based on conodonts only. Therefore, all localities in
the same unit are placed on the same horizontal line if there is no
evidence of different ages.

Where the dip of the beds is normal and the topographic effects are
negligible the diagram is easy to construct and to interpret. In other
intervals, different lithologies and units may be superimposed on each
other. This is the case with the Slite Marl, which occurs SE of the
limestones, and with the Slite f which occurs southeast of some of the
other Slite limestones, chiefly basal Slite g. Similarly, the Mulde Beds
are superimposed on the Halla Beds of Lilla Karlsö. The diagram of
the upper part of the Hemse Beds is as complicated as the Slite Beds.

The outline of the figure is strongly simplified and also wider than
the exact boundaries in order to avoid interference with the symbols
for shore localities and other marginal localities.

The names of the different Beds are given adjacent to the figure, as
are the letters of many of the subunits. The narrow stratigraphic
column to the right on this side facilitates the identification of the
others.

Most of the sampled localities are marked, but names are included
only for those mentioned in the text. Some of the names are
abbreviated, but these may be identified using the locality list.

The distribution of selected taxa is given, both as range lines and
with marks directly on the locality symbol.

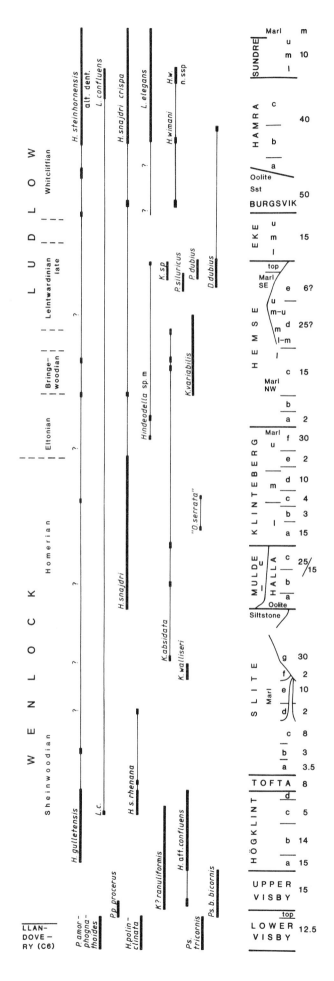

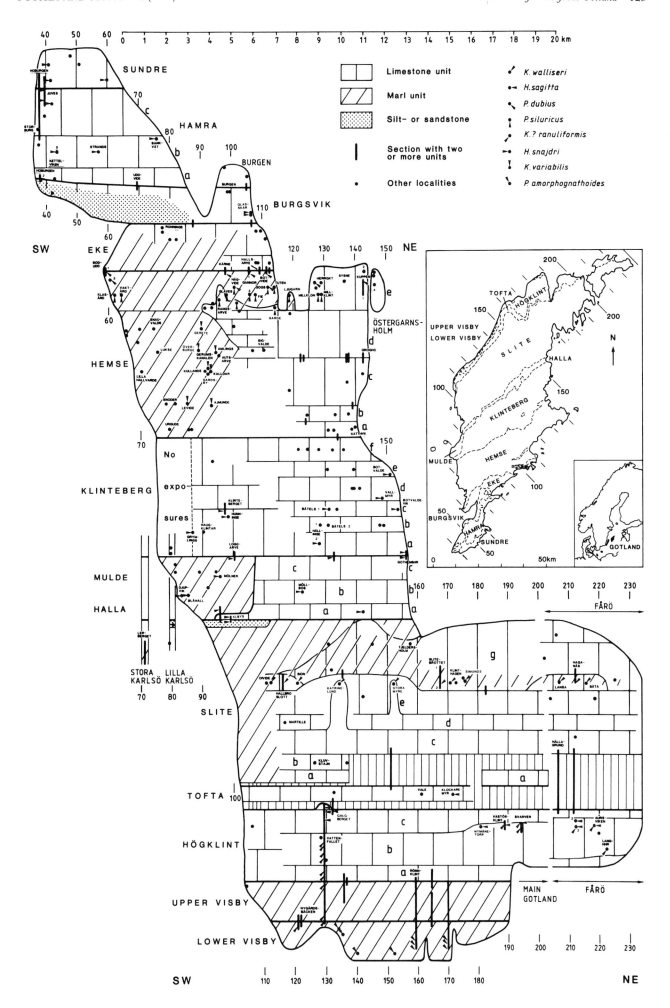

continuation, respectively, of this same horizon. *M. priodon* also occurs at Klinthagen, at least 9 m below the unit with *R. tenuistriatum?* (Hede 1933:43). In the northeast, in the areas of the Slite g expansion, *M. priodon* also occurs in the Slite Marl.

Small conodont faunas have been extracted from some localities with Slite f and contemporaneous beds. A typical fauna is that from the *R. tenuistriatum* beds in the shallow quarry Oivide 1, sample G 79-44 from 0.5 m below top. Stratigraphically most important is *K. walliseri*. In addition the fauna includes a few specimens of each of *Hindeodella excavata*, *H. confluens*, *Pseudooneotodus beckmanni*, *Panderodus equicostatus*, and *Ligonodina excavata?*. Sample ES 158 (A.M.) from the same locality has produced no *K. walliseri*, but instead two specimens of *Dapsilodus,* a genus known only from a few scattered localities on Gotland.

K. walliseri further occurs in the *R. tenuistriatum* beds (Slite f) at Sojvide 1, in sample G 81-58, and at Sion 1, in sample G 72-3.

At Hallbro Slott 6, *K. walliseri* occurs in sample G 73-12, 2.45–2.55 m below the '*Leperditia*-shale' described by Hedström (1910). Three other samples from higher levels in the 5 m section failed to produce this taxon. Hede (1942) reported *Monograptus priodon* from 5–5.5 m above the 2 m thick '*Leperditia*-shale' at a nearby locality. The lettered units have not been extended to this area, e.g., even Hede (1942) used Hedström's terminology. However, Hede's (1940) lithological descriptions, which are the base for the lettered units, are so similar to Hedström's (1910) descriptions that it is fairly safe to conclude that the '*Leperditia*-shale' is Slite e and the beds below it, Slite c (Slite d is missing here, Hede 1940:49, line 11). Hede (1940, 1960) did not extend the lettering above e (that was done by Laufeld 1974a) and he described the *R. tenuistriatum* beds only as the lower part of 'the next unit'. However, the lithologic character of the beds above the '*Leperditia*-shale' are those of Slite g. Thus, *K. walliseri* is found here 4.45–4.55 m below the base of Slite g, in the Slite c.

Further to the NE in the area where no *R. tenuistriatum* Beds are known, *K. walliseri* has also been found in the unit c beds at Katrinelund 1 in sample G 69-32 from about 1 m below the uppermost exposed beds. The species probably occurs in the Slite d beds at Stora Myre 1 (see p. 124).

South of the large bay Kappelshamnsviken, *K. walliseri* has been found in the lower part of Slite g, both at Simunds 1 in sample G 81-68 and at Klinthagen 2 in sample G 73-1. I have not been able to identify the section with both *R. tenuistriatum* and *M. priodon* described by Hede (1933:45) from the Klinthagen area as there are many quarries 1.8 km E of Stora Banne. However, his section was also from the lower part of Slite g.

In the quarry Slitebrottet 2 *K. walliseri* occurs in sample G 73-50 from about 36 to 36.5 m below sea level and in sample G 73-49 from 2.0 m above G 73-50. Dr. Hermann Jaeger has told me that in 1960 he found *M. priodon* in Slitebrottet 1, from about 1 m above the floor, that is about 2.5 m above my sample G 73-49.

On Fårö *K. walliseri* occurs in the *R. tenuistriatum* beds at Lansa 2 in sample G 73-70, and at Båta 1 in sample G 77-27 CB. It also occurs about 1 m below the upper boundary (Hede 1960:71) of the Slite Marl at Haganäs 1, in sample G 69-10. From the same 1.0 m thick unit of Slite Marl, Hede (1942, Loc. 35; 1960, Loc. 25) reported *M. priodon*. At this locality the Slite Marl is overlain by Slite g. Three other samples from higher

beds – including one from the bed just above that from which G 69-10 was taken – failed to produce *K. walliseri*.

The absolute conodont frequencies at these localities indicate that the rate of sedimentation was very different; 10 m at Hallbro Slott 6 and at Klinthagen 1 may correspond to only a few decimetres or less at Haganäs 1 and a few metres at Slitebrottet 2 (Jeppsson, in preparation). Thus, the sediments here discussed at the three former localities correspond to a comparable length of time.

Correlations within Gotland. – On Gotland *K. walliseri* occurs in all varied common lithologies on Gotland except reef limestone, and its distribution, like that of *M. priodon* is not correlated with lithology. One interpretation would be that both taxa had patchy distributions during a rather long time interval during which the Slite c to lower Slite g were deposited, one after the other. Alternatively, the interval containing *K. walliseri* on Gotland was short and these units were deposited side by side. This would mean that the base of the Slite g is oldest in the Klinthagen area – where also older Slite beds are developed as limestones – and progressively younger in a southeast direction where the Slite g expanded out over the Slite Marl. If so, the regional picture in *K. walliseri* time may have been one of an unchanged(?) high rate of sedimentation of skeletal carbonate sand, with still coarser carbonates in the northwest, locally with *Megalomus* beds. A narrow zone of *Rhipidium*-rich marl and marly limestone was locally developed in the former marginal carbonate sedimentation area, and a low rate of the continuous grain-by-grain sedimentation in some areas outside this zone would have permitted the formation of hardgrounds. In Slitebrottet 2 at least the beds 4 m above sample G 73-50 are graded and crossbedded in a way that strongly resembles Bouma's units c, d (and e?), and Bergman (1981, 1982) has found that a mudcarrying current deposited one of the lowermost accessible strata at Haganäs 1, just above a hardground. (Whether these sediments are to be described as flysch or instead as a related kind of 'bulk-deposited' sediments are outside the scope of this publication. 'Bulk-deposited' sediments seem, however, to be very important in the strata on Gotland.) A slightly raised sea level during the deposition of Slite f and contemporaneous beds may have caused marl to be deposited in the former marginal limestone area and produced starvation further out. When normal sea level was restored again, the area of carbonate sedimentation rapidly expanded seawards outside the old margin. The occurrence of graptolites in Slite f and contemporaneous beds agrees with a raised sea level, as they are usually interpreted as indicative of a more open marine environment. Berry & Boucot (1972) – citing Hede's (1942) data from Gotland – described the *M. priodon* group of graptolites as that which occurs in the most shallow areas; it occurs in the *Eocoelia* marine benthic life zone, which is the second one from the shore. On Gotland, *K. walliseri* occurs with *Panderodus equicostatus*, a species which seems to be characteristic of shallow water sediments there, while, e.g., *P. recurvatus* is not found.

Correlations outside Gotland. – In Estonia, in the middle part of the Paramaja Member of the Jaani 'stage', *K. walliseri* also occurs with graptolites, but here they have been identified as

Monograptus ex. gr. *flemingi* and *M.f.* cf. *primus* (see Viira 1975). '*M. flemingi*' is a descendent of *M. priodon* and treated as only subspecifically distinct by Jaeger & Schönlaub (1977). A comparison of the Estonian and Gotland specimens to see if the nomenclatural difference corresponds to a taxonomic one seems worthwhile. Viira found 5 sp elements of *K. walliseri* in an interval 37.5 m thick of the Ohesaare core, about 100 m below beds with *H. sagitta rhenana*. On Gotland the latter taxon occurs in the upper parts of the Högklint Beds, the basal Tofta Beds, and in Slite d, thus only below *K. walliseri*.

Another specimen of *K. walliseri* was reported from the Ukmärge core from Lithuania (Viira 1975).

Aldridge (*in* Bassett 1974:757) found two specimens of this taxon from within 3 m of the top of the Barr Limestone in central England, and later (1975 b) reported that it is common in this Limestone. The Barr Limestone is 9 or 10 m thick and was correlated with the upper *C. centrifugus*, the *C. murchisoni* and the lower *M. riccartonensis* Zones by Cocks *et al.* (1971, Fig. 4), and tentatively with the whole or part of the *M. riccartonensis* Zone by Bassett (1974). The lower part of this range corresponds to the Upper Visby Beds and the upper part to the Högklint a and b. However, the Barr faunal list also includes *Dicoelosia biloba* (see Bassett 1974:757), a brachiopod which has been found only in the Slite Beds and in younger units on Gotland, according to Bassett & Cocks (1974:11, 42). In the Upper Visby and in the Högklint Beds they found *D. verneuiliana*. Thus, both *K. walliseri* and *D. biloba* indicate a closer similarity with the Slite Beds than with older beds, and the correlation of the Barr Limestone with the graptolite zones may have to be restudied.

Aldridge (1975b:614) also recorded one specimen from the Dolyhir Limestone in Wales, co-occurring with *H. s. rhenana*. The Dolyhir Limestone was tentatively correlated with the top *C. centrifugus*, *C. murchisoni* and basal *M. riccartonensis* Zones by Bassett (1974). Aldridge *et al.* (1982) reported *K. walliseri* from the Brinkmarsh Formation.

Walliser (1964:88, 101) reported 1 sp element from the upper part of the *Ostracodenkalk* in the Rheinisches Schiefergebirge, where it occurs with *H. s. rhenana*.

Helfrich (1975) found a total of 14 sp elements of *K. walliseri* in a horizon up to about 2.1 m thick of the Cosner Gap Member of the Mifflinton Formation, at three localities in Virginia and West Virginia. At two of these it occurs in the lower local range of specimens identified by Helfrich as *H. s. bohemica*, at the third locality it is present just below the lowest occurrence of *H. s. bohemica*.

Barrick & Klapper (1976) identified 7 sp elements of *K. walliseri* from the middle part of the Fitzhugh Member of the Clarita Formation at two localities in Oklahoma. There it occurs in the uppermost samples with *K.? ranuliformis* and slightly above, in a horizon less than 1.0 m thick. At one locality it occurs with *H. s. rhenana*. On Gotland, *K.? ranuliformis* is not known above the middle part of the Högklint Beds, i.e. about 10 to about 40 m below Slite f. In Britain *K. walliseri* and *K.? ranuliformis* are similarly separated (Aldridge 1975b). In the sections described by Barrick & Klapper (1976) the Wenlockian is strongly condensed – it is not more than 4.5–6.5 m at the most in the two sections with *K. walliseri* – and evidence of stratigraphical leaks and reworking were reported from other levels in their sections. Therefore it may be wise to leave the question open whether the co-occurrence of the two taxa in their samples indicates a contemporaneity or not.

Nicoll & Rexroad (1968:60) reported that some of their specimens of *K.? ranuliformis* had an extra denticle on the cup. Cooper (1980:221) quoted them as having described this as a rudimentary extra process and evidently interpreted this to be the beginning of a lateral process (although the term 'rudimentary' would not imply this). Such specimens also occur on Gotland (Jeppsson 1979c:244, Fig. 72:2) and are within the range in variation of *K.? ranuliformis*. They should not be invoked in a discussion of the age of *K. walliseri*.

From Australia, Bischoff (1975:8) noted the presence of this taxon in 'higher horizons in the Borenore section'. Judging from the rest of the publication (Talent *et al.* 1975), this section is in New South Wales.

To sum up, although only slightly over 30 sp elements of *K. walliseri* have previously been reported, they have been found at 12 localities in 8 areas on three continents. Until now, I have found another 11 or 12 localities on Gotland with 28 sp elements apart from those from Stora Myre 1 (see above). At the present, beds with *K. walliseri* are assigned to various levels in the Wenlockian. This may indicate that the total range of the taxon includes a large part of the Wenlockian. If so, we would expect it to occur more than once in some of the areas or to exhibit some strong ecologic restrictions. However, in most areas the species was only found in a very thin interval. In different areas, this interval may be below, in, or above beds with *H. s. rhenana*, or in beds with '*H. s. bohemica*'. The picture is similarly diffuse in relation to graptolite ranges: above, in or below the *M. riccartonensis* Zone, and both with *M. priodon* and with its descendent *M.p. flemingi*. To suggest that *K. walliseri* may be useful for stratigraphy would thus challenge the supremacy in stratigraphy awarded to several other taxa. The variable distribution with respect to *H. sagitta* presents few problems, for as noted on p. 137, *H. sagitta* and *H. snajdri* are among the most facies-dependent Silurian conodonts. In most cases, it must be borne in mind that correlation with the type area or with graptolite zones is indirect, and errors may have accumulated in the process. There is now also so much evidence for ecological restriction of many graptolites.

In conclusion, we should consider the hypothesis that *K. walliseri* was distributed world-wide for a short interval in the early Wenlockian, probably in the late Sheinwoodian.

Slite g, Slite Marl and Slite Siltstone

The Wenlockian Slite g is up to 30 m thick and is the highest Slite limestone. Bioherms are abundant, as are more or less coarse-grained limestones. During the time of the deposition of Slite g the limestone belt expanded many kilometres, in the northeast over 10 km, out over the Slite Marl. The contact is exposed in many places. The oldest exposed Slite Marl is only slightly older than the oldest local Slite g; according to the conodont faunas the oldest Marl exposed is contemporaneous with the Slite f (see above). Most of the exposed Slite Marl is a lateral equivalent of the Slite g to the northwest or is younger, except in the Bara area and on Karlsöarna (the islands in the SW) where Halla limestone follows directly on the Slite limestone. The upper part of the Slite Marl has been subdivided into Lerberget Marl and the *Pentamerus gothlandicus* Beds (Hede 1927 b) in Stora Karlsö. The latter unit can be identified across Gotland to the east side of the island. The Slite Siltstone – up to 4.5 m – occurs only in the southwest (Hede 1921; Sivhed 1976; Bergman 1980), but the highest Slite Beds in the NE, the *Atrypa reticularis* Beds, may be developed as a sandy limestone (Hede 1928:40).

The conodont faunas available from the Slite g are small and include only long-ranging taxa, e.g., *Hindeodella excavata, H. confluens, Panderodus equicostatus, P. unicostatus* and *Ligonodina*, probably *L. excavata*.

The yield from the Slite Marl is more varied; beds with hardgrounds may produce large faunas while other levels give small but rather diversified faunas. As an example, many samples from sections in the Lerberget Marl at the type locality on Stora Karlsö have produced the following taxa: *Hindeodella excavata, H. confluens* (less frequent than *H. excavata*), *L. excavata, Panderodus unicostatus, P. recurvatus*, a third species of *Panderodus*, probably *P.* sp. g., *Pseudooneotodus* (broad-tipped), *Walliserodus*, and *Decoriconus*. None of these permit detailed correlations. In other faunas *P. equicostatus* occurs, usually together with *P. unicostatus*, i.e. the same panderodontid fauna as in the Slite g. However, *P. recurvatus* does occur in the upper Slite Marl also in the eastern part of the island. I have found it at Tjeldersholm 1, in sample G 73-8 from 0.35–0.40 m below the reference level, i.e. below the *Pentamerus gothlandicus* Beds there, too. Thus there is also some evidence for slightly deeper water in the eastern part of the area.

The samples from Slitebrottet 1 and 2, above the sample with *K. walliseri* (see above) include *H. excavata, H. confluens, L. silurica, P. equicostatus, P. unicostatus, P. recurvatus, P.* sp., *Pseudooneotodus beckmanni*, and *Kockelella absidata*. An sp element of *K. absidata* occurs in sample G 73-44, from about 17.2 m above the highest one with *K. walliseri* and another one in sample G 73-43 from 1.75 m above G 73-44.

A fauna from the topmost Slite Siltstone at Klintebys 1, in sample G 70-23, includes *Hindeodella snajdri* and *Panderodus equicostatus*.

The Halla, Mulde, and Klinteberg Beds

The late Wenlockian Halla Beds consist of about 15 m of oolite and argillaceous limestone in the northeast where they are subdivided into 3 units (Hede 1928:45, 1960), now lettered a to c (Laufeld 1974 a). The Halla Beds thin out toward the SW and disappear south of Klintehamn except for a thick pile of limestones with reefs on Karlsöarna, the islands west of the main Gotland (Hede 1927b:41–42, 47–50).

The late Wenlockian Mulde Beds consist of about 25 m of marlstone in the SW but they thin out and disappear in the area where the Halla Beds are still less than one metre thick. In a NW direction it can be noted that no Mulde Beds were reported by Hede (1927b) from Lilla Karlsö, below the Klinteberg Beds.

The late Wenlockian and early Ludlovian Klinteberg Beds, as presently known, crop out in an area about 12 km wide and 50 km long. There is an outlier on Lilla Karlsö. The beds consist of up to 70 m of limestone in the east where they have been subdivided into 6 units (Hede 1929, 1960), now lettered a to f (Laufeld 1974a). In the west, the facies is partly different, and there they are presently subdivided informally into lower, middle and upper (Laufeld 1974a).

Conodont faunas from the major part of these beds typically include a member of the *H. snajdri* lineage. In this interval *H. sagitta bohemica* is often reported (e.g., Fåhraeus 1969), but at least the specimens from Gotland should be referred to *H. snajdri* (see p. 123).

The conodont faunas in the Halla Beds may be illustrated with the following species lists.

In the Halla a – the Bara oolite – at Bara 1 in sample G 70-13 there are a few specimens of *H. snajdri, H. excavata, P. beckmanni*, and *Panderodus*.

In some faunas, e.g., one extremely well preserved from the Halla b at Möllbos 1 from sample G 77-28 PSSFG (see also p. 122), *H. confluens* is the dominating *Hindeodella* species, while *H. excavata* is very rare. *P. equicostatus* occurs in good numbers as do *L. excavata, H. snajdri*, and *P. beckmanni*. The taxon illustrated as *Ozarkodina* sp. nov. by Aldridge *et al.* 1982 is represented by juvenile elements.

In other faunas *P. equicostatus* is lacking, as are all other species of *Panderodus*. This is so in many samples both from Halla c and the basal Klinteberg Beds at Gothemshammar 1, 2 and 9, where *H. confluens* and *H. snajdri* seem to be equally frequent.

In the west, in the Mulde Beds, slightly different faunas have been found. Thus, at Mölner 1 in sample G 71-6 from the top bed in the quarry, *H. excavata* dominates while *H. confluens* is rare. *H. snajdri* is even less frequent, but does occur. Both *P. equicostatus* and *P. unicostatus* occur. *Decoriconus* has not been identified in this sample but occurs in sample G 71-7, 1.70 m below G 71-6. In sample ES 166 (A.M.) an oz element of *Kockelella* occurs, probably of *K. absidata*. Faunas of the same general aspect have also been found in several samples from Djupvik 2 and Blåhäll 1. At Djupvik 2, *K. absidata* is present in samples G 70–26 and G 70–28 from 0.95–1.5 m below and 2.24 m above the reference level, respectively. Small collections from other localities in the Mulde Beds give the same picture of *H. excavata* as the dominating *Hindeodella* species while *H. snajdri* and *H. confluens* are rare; a co-occurrence of *P. equicostatus* and *P. unicostatus*, and sporadic presence of *Pseudooneotodus* and *Decoriconus*.

The contact between the Mulde and the Klinteberg Beds is exposed at Loggarve 1. Sample G 77-21 from the topmost Mulde Beds (0–0.05 m below the boundary) has produced a very well preserved fauna dominated by *H. excavata*, but also with *H. snajdri, P. equicostatus, Pseudooneotodus, Decoriconus*, and '*Ozarkodina* sp. nov.' of Aldridge *et al.* 1982. An ne element, which probably belongs to *K. absidata*, has also been found. An sp element of *K. absidata* occurs in the next sample (G 77-22) from 0–0.05 m above the boundary, i.e. in the basal Klinteberg Beds (see also p. 123).

With few exceptions the collections at hand from the Klinteberg Beds are very small, and thus the present account is a very preliminary one. *H. snajdri* continues through most of the Klinteberg Beds both in the southwest and in the east. Thus it occurs *inter alia* in the basal Klinteberg Beds at Gothemshammar 2 and 6 (see above), in the lower–middle Klinteberg Beds at Hunninge 1 in sample G 71-3 from about 5.5 m below the top of the quarry and about 2.3–3 m above the floor of the quarry; questionably in unit a at Hällinge 2; in unit c at Båtels 1 in samples G 71-90 and G 71-91 from 0.90 and 2.10 m above the lowermost exposed beds, respectively; and in unit d at Vallmyr 1 in sample ES 138 (A.M.). In the east the highest find is from unit e at Botvalde 1, in sample G 75-37. The highest find of *H. snajdri* to date in the southwest is from Klinteberget 1 in sample G 67-29 from about 8 m below the top of the section. In Britain similar forms occur in the late Homerian (Aldridge 1975b). With the exception of two questionable specimens from the Hemse Beds *H. snajdri* is absent from younger beds on Gotland until another subspecies appears in the Burgsvik Beds.

In the Klinteberg c beds at Botvaldevik 1 in sample G 75-36 '*O. serrata*'? occurs. The species may also occur in other samples (see p. 123).

The rest of the known faunas may be illustrated by the following species lists.

From Grymlings 1, the most southwesterly locality with Klinteberg Beds on main Gotland, the following species have been identified in sample G 71-1: *H. excavata, H. confluens, H. snajdri* and *P. equicostatus*.

In the lower Klinteberg Beds at Haugklintar 2, in sample G 70-31, a much less abundant but otherwise similar fauna occurs except that *H. snajdri* has not been found, and there is a *Ligonodina* species, perhaps *L. excavata*, included.

In the Klinteberg Beds unit f at Sutarve 2 in sample G 77-31 PSSFG, *H. confluens, H. excavata, P. unicostatus, P. recurvatus?, Ligonodina silurica?* and *Hindeodella* sp. m. occur.

The Hemse Beds

The outcrops of the Ludlovian Hemse Beds form a diagonal belt across Gotland about 15 km wide and 65 km long. The maximum thickness is usually given as about 100 m (Hede 1921, 1960) but is probably less. In the NE the Hemse Beds consist of about 50 m of limestone subdivided into 5 units (Hede 1929:25–54, 1958:142–144, 1960:47–48) which now are lettered a to e (Laufeld 1974:11–12). The thin unit a may contain bioherms as do unit d and unit e. These two units together are 20 m thick according to Hede 1929 but up to 25 m thick and up to 6 m thick, respectively, according to Hede 1960. The limestones continue into the 'north-central' area where these subdivisions have not been mapped. However, as the limestones consist largely of bioherms and their lateral equivalents it is likely that they are mostly representative of unit d (compare Laufeld 1974a:12) or younger, i.e. unit e and even Eke Beds (compare Martinsson 1967). Laufeld (1974a) separated his localities as undiff. L-M, M-U and U.

Marlstones are exposed to the south and west of the limestones. They are here stratigraphically divided into three parts. Hede (1972b:24) subdivided the central area of marl into an older part in the northwest and a younger one in the southeast, and, following Laufeld (1974a:12), they are here separated as NW and SE Marl. Hede did not extend this subdivision to the western coast, but judging from the conodont faunas the boundary continues from Kvinngårde (compare Hede 1927b:24) southwestward to the inner part of Linviken or slightly further to the south; Vaktård 3 and Klasård 1 have typical Hemse Marl SE conodont faunas. On the other side of the island, faunas of similar age occur in the fissure filling (mapped as thin-bedded crinoidal limestone by Watkins 1975, Figs. 3 and 4) in the Hemse d? reef at Ljugarn 1. This filling, consisting of limestone and marl, may thus be considered an eastern outlier of the Hemse Marl SE.

The topmost Hemse Marl is usually separated as a distinct unit (Munthe 1910; Laufeld 1974 a, b; Larsson 1979).

On average, the Hemse Beds are more productive of conodonts than the Slite and Klinteberg Beds. Also the known diversity is larger; altogether over 20 species have now been delimited. However, only a few localities, with Hemse Marl SE, have ten or more species and only 5 to 7 species regularly occur. The three species *H. excavata, H. confluens* and *P. unicostatus* probably together make up about 75% or more of the number of elements in most samples. About half of the number of species occur only as stray specimens, many of them are represented by fewer than ten specimens each in my Hemse collections.

The Hemse Beds expose the very pronounced ecologic constraint of some species. Thus, *Panderodus recurvatus* occurs practically only in the areas where marl was laid down, where it is present throughout the sequence, while *Panderodus* sp. g. has been found throughout the limestone sequence but is restricted to that. *P. unicostatus* is present everywhere and *P. equicostatus* nowhere. Other taxa seem to be largely stratigraphically restricted. Thus, *Polygnathoides siluricus* occurs across the island independent of lithology. Others combine both restrictions, thus *Panderodus gracilis* occurs only in the distal parts of the Hemse Marl SE and in the topmost Hemse Beds.

Hemse Marl NW. – The Hemse Marl NW probably represents a rather long time span. Further studies will probably result in several stratigraphically distinctive conodont faunas being identified in it.

A few platform elements of *Kockelella variabilis* have been found at Levide 3 in sample G 79-8, at Gerumskanalen 1 in samples G 71-39 and G 77-37 PSSFG, at Överburge 1 in sample G 79-5, at Källdar 2 in sample G 71-35, at Amlings 1 in sample G 77-39b PSSFG, and at Gerete 1 in sample G 79-6. At Snoder 3 in sample G 71-10 from the bottom of the small ditch there is a juvenile specimen which either belongs here or to *Ancoradella*. *K. variabilis* also occurs at Ajmunde 1 in sample G 71-40. That locality is within the area which Hede (1927b) mapped as Klinteberg Beds. However, the locality is only 100–200 m NW of the approximate (Hede 1927b:23) boundary to the Hemse Beds, and the whole known fauna is the same as that in the Hemse Marl NW. Thus, it seems to be better to consider the locality an outlier or an extension of the same beds as those exposed at the other localities with *K. variabilis*.

Kockelella variabilis is rather long-ranging, but it changes strongly with time as illustrated by Walliser (1964:40, Pl. 16) from Cellon. The sequence is closely similar elsewhere (cf. e.g. Rexroad & Craig 1971; Klapper & Murphy 1974; Barrick & Klapper 1976), and it may be possible to subdivide the species into subspecies of stratigraphical importance. The specimens from Gotland are closest to those from the middle part of the *A. ploeckensis* Zone as defined at Cellon but strongly different from those in older strata and from those in the *P. siluricus* Zone. Similar specimens occur in Britain in the Upper Bringewood Formation (Rhodes & Newall 1963) and also in the lowermost lower Leintwardine Formation at the type locality for the base of the Ludfordian = the base of the Leintwardinian (Aldridge 1975b: 615, Pl. 1:19).

Sp elements of *Kockelella absidata* occur in collections from the Hemse Marl NW at Lukse 1 in sample G 71-12 and at Amlings 1 in sample G 77-39c PSSFG. *K.* cf. *absidata* occurs at Autsarve 1 in sample G 71-43. A distinct population occurs at Kullands 1 in sample G 77-38 PSSFG, at Gerumskanalen 1 in sample G 77-37 PSSFG, at Gardsby 1 in sample G 82-27, and at Lilla Hallvards 1 in sample G 71-143 (see p. 123). At Cellon such specimens occur in the lower and middle *A. ploeckensis* Zone (Walliser 1964). The locality Lilla Hallvards 1 has a *M. chimaera* Zone graptolite fauna (Jaeger 1981). In the type area for the Ludlovian, *M. chimaera* occurs in the Upper Elton Formation (Holland *et al.* 1963:101), but its upper range there is not known because the Lower Bringewood Formation there has not yielded any graptolites.

Fåhraeus (1969) dissolved two samples from the Hemse Marl NW, reported *Ancoradella* from both, and illustrated the oral surface of a fragmentary specimen. In small specimens the upper surface is indistinguishable from that of specimens of *Kockelella* (Walliser 1964:29). I have recollected both localities but not yet found *Ancoradella*, although *K. variabilis* occurs at Källdar 2.

The rest of the conodont fauna can be illustrated by the following faunal lists in stratigraphically ascending order:

Urgude 2, in sample G 79-10, a rather small collection comprises *H. excavata, P. unicostatus?, Kockelella?, Dapsilodus.*

Snoder 3, in sample G 71-10, from the bottom of the small ditch: *H. excavata* (dominant), *H. confluens, P. unicostatus, P. recurvatus, Kockelella?, Belodella?, H. steinhornensis?* (an unquestionable specimen of this species occurs in sample G 71-43 from Autsarve 1, in a closely similar fauna).

Lukse 1, in sample G 71-12: *H. confluens, H. excavata, H. steinhornensis, P. unicostatus, P. recurvatus, K. absidata, Decoriconus,* and *Belodella.*

Snauvalds 1, in sample G 71-14 (a small fauna): *H. excavata, H. confluens, P. unicostatus, P. recurvatus.*

The conodont faunas in the older Hemse limestones. – The conodont faunas in the Hemse a, b, c, d, undiff. L to M-U limestones are rich but similar, and as yet it has mostly been difficult to use them for correlation of the limestone units with the marls. Most of the faunas are dominated by large robust fragmentary elements of *H. confluens* and a distinct subspecies of *Panderodus* sp. g. In most samples there are also robust fragments of a species of *Ligonodina.* These fragments cannot be distinguished from *L. excavata* but they are too poorly preserved for a positive identification. That taxon can be positively identified in some samples. *H. excavata* is subordinate in frequency and *P. unicostatus* occurs in most samples. In the easternmost area there are a few specimens of *Hindeodella* sp. m. Scattered occurrences of some other taxa complete the picture. Some faunal lists can illustrate the faunas.

In the Hemse a at Katthammarsvik 1, sample G 67-30, *H. confluens* dominates over *H. excavata,* and *Panderodus* sp. g. over *P. unicostatus.* Fragments of *Ligonodina,* probably *L. excavata,* and of *Hindeodella* sp. m. also occur.

In the Hemse c at Grogarnshuvud 1 occur *H. confluens* (which dominates), *H. excavata, H.* sp. m., *Panderodus* sp. g., *P. unicostatus, L. excavata, L.* aff. *confluens,* and *H. steinhornensis.*

In the lower–middle Hemse limestone at Källdar 2, in sample G 71-35, *H. confluens* (dominates), *H. excavata, P. unicostatus, P.* sp. g., *P. beckmanni* and *K. variabilis* occur.

In the lower–middle Hemse limestone at Sigvalde 2, in sample G 71-115 from 0–0.1 m below the reference level and below the reference point, *H. confluens, H. excavata* (rare), *Panderodus* sp. g., *P. unicostatus* and *L.* aff. *confluens* occur.

The POLYGNATHOIDES SILURICUS *Zone.* – The Leintwardinian Hemse Marl SE, and at least parts of unit e, the Millklint Limestone, contain *P. siluricus.* The species is usually rare, the frequency of the sp element often being about one specimen per thousand. Fortunately, this horizon is also characterized by a very high absolute frequency of conodont elements. The faunas further show a very high diversity. Another still rarer taxon is *Kockelella* sp. a. It occurs in samples with *P. siluricus* and in the succeeding fauna.

The very well-preserved fauna from Vaktård 3 in sample G 71-148 is typical for the most westerly localities. *H. confluens* strongly dominates over *H. excavata; P. unicostatus, P. recurvatus, P. gracilis, D. dubius* and *P. siluricus* occur in frequencies between 100 and 10 specimens per kilogram. *P. beckmanni* is slightly less frequent in my collections, but it is probable that a finer screen would retain more than 10 elements per kilogram. In addition, a few elements of *Ligonodina* are found. Evidence combined from several localities indicates that both *L. excavata* and another species may occur. A fragmentary specimen of a three-branched element completes the fauna list. Similar faunas with *P. siluricus* occur in the Hemse Marl at Klasård 1 in sample G 71-150. There *Decoriconus* and a single juvenile specimen of *Belodella* have also been found.

Further to the northeast, the faunas are similar except that *Panderodus gracilis* does not occur at this level, but *Ligonodina* sp. d does. Such faunas occur in the Hemse Marl SE at Gläves 1, in sample G 75-30, at Hägvide 1 in G 71-129 (both with *Pelekysgnathus dubius*), at Gannor 3 in G 71-125, and at Fie 3 in G 71-128.

P. siluricus is also found, at Rangsarve 1, to the west of these localities, within the area where the Hemse Marl NW is overlain by reef limestone. Here it is at a topographically higher level than the closest Hemse Marl NW localities. The fauna is slightly different from that at Gläves 1, Hägvide 1, Gannor 3, and Fie 3, principally in the presence of *L. silurica?,* which is the most abundant species of *Ligonodina* at Rangsarve, and in the rarity of *D. dubius.* This locality may be at a slightly lower level than my others with *P. siluricus.* The fauna deviates strongly from those in the Hemse Marl NW, and considering the local topography, I find it likely that the beds correspond in age to the lower part of the Hemse Marl SE. Within the same limestone area, but 13 km to the ENE, in the large quarry Garde 1, *P. siluricus* occurs in the three samples G 82-35, G 71-224, and G 82-36. They are from the uppermost 1.64 m of the exposed sequence. The collections presently available are small, but seem to agree best with those from Rangsarve 1. Thus, *D. dubius* remains to be found at Garde 1. Fåhraeus' (1969) report of a single specimen of *P. siluricus* from Ekese (about 5 km ENE of Garde 1) may indicate a similar fauna there.

In the northeasternmost part of the area with Hemse Marl SE, at Gogs 1, the fauna is known from very large collections, many of which have been prepared by Anders Martinsson. Samples, collections, and specimens have also been received from Kent Larsson, Carl Pleijel and Björn Sundquist. The fauna includes a number of very rare species, some of which are poorly known or undescribed. The following taxa have been recognized to date: *H. confluens*, *H. excavata*, *P. unicostatus*, and *D. dubius* are regularly present and *P. recurvatus* occurs in good numbers in many samples. Less frequent, and absent from some half-kilogram samples, are *P. siluricus* and both kinds of undescribed elements included under gen. et sp. indet. Very rare are *Belodella*, *Decoriconus*, *Pelekysgnathus dubius* and *Pseudooneotodus*. There are also at least two, but probably several, species of *Ligonodina* and/or another genus with similar denticulation.

A similar fauna occurs at Tuten 1, in sample G 81-51. The first 0.5 kg have produced *H. confluens*, *H. excavata*, *P. unicostatus*, *P. siluricus*, *L. excavata*, *L. silurica?*, gen. et sp. indet., a few specimens that are questionably identified as *Panderodus* sp. g., and *Belodella*.

Closely similar, but badly fragmented, collections have also been found further to the NE, in the fissure filling of Hemse Marl SE at Ljugarn 1 in sample G 79-50, in the Millklint Limestone at Millklint 1 in sample ES 137 (A.M.) and at Millklint 3 in sample G 69-32. Like those from Gogs 1 and Tuten 1, all three faunas include gen. et sp. indet. There is at least one real difference from the Gogs fauna: *Panderodus* sp. g occurs, as in older Hemse limestone units, and *P. recurvatus* is absent except for stray specimens at Ljugarn 1.

In the Hemse beds at Kuppen 1, in sample G 81-44 from 0–0.11 m above the reference level, there is an identical fauna. The fauna differs markedly from studied Hemse d faunas, in which *D. dubius* and gen. et sp. indet. are absent. Hede (1929:38) described what is now (Laufeld 1974a, cf. Hede 1960) separated as d and e, as one stratigraphic unit, and described the Millklint Limestone only as the highest lithologic unit within it (Hede 1929:49). His description of the beds in question (1929:40) at Kuppen 1, is closely similar to that of the strata at Millklint. However, he did not introduce the term Millklint Limestone until some pages after he had described both Kuppen 1 and the type area. Anyhow, the similarities in the conodont faunas are so large that I here refer the beds above the reference level to the Hemse e. Similarly, the marls and limestones at Rangsarve 1 should be referred to the Hemse Marl SE and unit e respectively.

My correlations agree closely with those based on ostracodes by Martinsson (1967). For example, ostracodes and conodonts both indicate that beds at Rangsarve (Martinsson's Linde area) and in the east are to be correlated with the Hemse Marl SE. As is to be expected when more fossils are put in stratigraphic use it is possible to refine some biostratigraphical divisions slightly; in particular, the *P. siluricus* Zone on Gotland is very thin and does not include the top of the Hemse Marl, which can thus be separated.

The age of the P. SILURICUS Zone. – Hede (1919, 1942:20, loc. 1 c, 2 c) reported *Monograptus bohemicus* from Vaktård 3 and Klasård 1. Some of Hede's other finds of *M. bohemicus* on Gotland are at localities which he mapped as Hemse Marl SE, and at least one (Bodudd 3) – as far as can be judged – has a *Pelekysgnathus dubius* fauna, and thus represents the uppermost Hemse Marl. In Great Britain, at that time, *M. bohemicus* was only known in strata that were referred to the *M. nilssoni* Zone (Hede 1919), the lowermost Ludlovian graptolite Zone.

However, it does occur higher; Jaeger (1975) reported it from Cellon in the upper part of the *Cardiola-Niveau*, i.e. in the upper part of the *P. siluricus* Zone, an occurrence he dated as younger than *M. leintwardinensis*. He has also concluded (1981:12) that a graptolite fauna from the Hemse Marl NW beds at Lilla Hallvards 1 belongs in the *M. chimaera* Zone, which succeeds the *M. nilssoni* Zone. Thus the Hemse Marl SE cannot belong to the *M. nilssoni* Zone. Walliser (1964:97) found the *P. siluricus* Zone between *M. fritschi linearis* and *M. ultimus* at Hviždalka in Bohemia, and Křiž & Schönlaub (1980) showed that the basal bed of the *P. siluricus* Zone at Můslovka in the same area contains *M. fritschi linearis* (*M. bohemicus* occurs just below the *P. siluricus* Zone there).

Thorsteinsson & Uyeno (1980 [1981]:23) also discussed the age of the *P. siluricus* Zone. They referred to Klapper & Murphy (1974) as concluding 'that the lower range of *P. siluricus* overlaps the Zone of *M. chimaera* in Nevada'. Apparently their reference is to Figs. 6 and 8 in Klapper & Murphy's publication, where *M. chimaera* is shown to range about 120 and 80 feet, respectively, above the interval with *P. siluricus*. However, Klapper & Murphy did not themselves refer to this graptolite range as the *M. chimaera* Zone, a term which they used for an older interval with the *A. ploeckensis* Zone (Klapper & Murphy 1974:12, Fig. 7). That correlation agrees well with that discussed above for parts of the Hemse Marl NW.

The *P. siluricus* Zone is known from many other areas, but not yet from the type area for the Silurian in Great Britain. However, Martinsson (1967) reported an ostracode fauna with *Neobeyrichia scissa* and *N. lauensis* from several of the Hemse Marl SE localities with *P. siluricus*, and using this, he could correlate them (and the top of the Hemse Marl) with the late Leintwardinian. This agrees closely with the European graptolite and conodont evidence dicussed above, and it is thus the most likely age of the *P. siluricus* Zone.

The uppermost part of the Hemse Marl. – The conodont faunas in the top of the Hemse Marl are similar to those in the Hemse Marl SE except that *P. siluricus* and some other of the very rare taxa have not been found, while at least *P. dubius* is slightly more frequent. The faunas show similar changes from the west to the east. At Bodudd 1 in sample G 81-30, from the lowermost beds exposed, there is a very well preserved fauna. It includes *P. gracilis* (about 20 elements per kg) together with *H. excavata*, *H. confluens*, *P. unicostatus*, *D. dubius*, *P. dubius*, *P. beckmanni*, and at least one unidentified species. At Bodudd 3 in a much larger collection (sample G 71-151) *Decoriconus*, *P. recurvatus*, and *Belodella* also occur.

Stray specimens of *P. gracilis* occur also further to the east, at Kärne 3 (S of Burgen) at least up to about 0.75 m below the top of the Hemse Beds in sample G 71-196, and at Hallsarve 1 in samples G 69-26 and G 69-27 from 0.15–0.25 m and 0–0.05 m below the reference level (=the base of the Eke Beds), respectively. At the former locality *P. dubius* also occurs; the collections at hand from Hallsarve are much smaller. *P. dubius* also occurs at Gannor 1 in sample G 71-120 from 0.65 m below the reference level and at Botvide 1, less than 1 m below the boundary. At that locality in sample G 66-248 SL from 0.50–0.55 m below the boundary, *Hindeodella* sp. m has been found.

The youngest Hemse limestones. – A very large conodont collection from the Millklint Limestone at Millklintdalen 2, sample G 77-44 PSSFG, contains *H. confluens, H. excavata, P. unicostatus, P.* sp. g, *P. recurvatus, D. dubius, Ligonodina* sp., *Kockelella* sp. a, and questionably also *P. dubius.* A much smaller collection from Herrgårdsklint 1 (sample G 71-80) contains *H. confluens, H. excavata, L. excavata, P. unicostatus,* and questionably also *P. dubius.* At Kuppen 1, in sample G 81-45, from 0.11–0.14 m above the reference level, *H. confluens, H. excavata, D. dubius, P. dubius* (many specimens per kg), *Pseudooneotodus, P. unicostatus, P.* sp. g, *Ligonodina, Kockelella?* sp., and *Hindeodella* sp. m occur. This fauna differs from that in G 81-44 (see above), inter alia in the greater frequency of *P. dubius* and the absence of gen. et sp. indet. Two smaller collections from Kuppen 2, G 81-46 from 0.03–0.08 m and G 81-47 from about 2.5 m above the reference level there, have only produced *H. confluens, H. excavata?, P.* sp. g, *P. unicostatus?, Ligonodina,* and *Kockelella?* sp. Samples from Östergarnsholm 1, 2, and 3 (G 78-19 CB, G 78-18 CB, and G 78-20 CB, respectively) and sysne 1 (sample G 81-49) contain similar faunas; *D. dubius* occurs in the samples from Östergarnsholm 3 and Sysne 1 (the collections available from the other two localities are too small for any conclusion), and a possible specimen of *P. dubius* and another of '*Ozarkodina* sp. nov.' of Aldridge *et al.* 1982 in the collection from Östergarnsholm 3. It is possible that *D. dubius* occurs below the *P. siluricus* Zone on Gotland; thus, the beds on Östergarnsholm and at Sysne may be older than that zone. However, other samples must be younger; thus they indicate that the Hemse limestones reach above the *P. siluricus* Zone (cf. Martinsson 1967). As yet I have no good faunistic characteristic to distinguish any possible outliers of the Eke Beds (see below), but at least the collections with *P. dubius* and/or *D. dubius* agree closest with the uppermost Hemse Marl.

The Eke Beds

The Late Leintwardinian to early Whitcliffian Eke Beds crop out in an area some kilometres wide and 28 km long. Outliers in the area of the Hemse Marl SE mapped by Hede (1925b) extend the area by another 10 km. Martinsson (1967:373) identified outliers also in the Hemse limestone area, extending the distribution a further 18 km to the NE. Regarding evidence of the former extension, see also Jeppsson 1982. In most of the area, the beds consist of about 10 m of calcareous mudstone and argillaceous limestone abundant in calcareous algae. In the NE the lower levels are to a large extent developed as crinoidal limestone with small bioherms.

The lower few metres of the Eke Beds are exposed in many places, and I have many fairly large faunas from them. The yield is lower than in the Hemse Beds, and the diversity of the faunas is much lower. Only *H. excavata, H. confluens,* and *Panderodus unicostatus* are regular, and *Pseudooneotodus beckmanni* occurs in some samples. Thus, the faunal constituents are those that dominate the upper Hemse faunas, with the difference that all the 'exotic' taxa have disappeared.

Higher in the Eke Beds this fauna is replaced by a distinctly different one. The meagre faunas are dominated by *Panderodus equicostatus,* with *Pseudooneotodus beckmanni* and *Decoriconus* in most collections. Two specimens of *Dapsilodus* have been found, one at Ronnings 1 in sample G 79-47 and another at Ronehamn 3 in sample G 69-51. Taxa with bar elements are

equally rare; one sp element of *H. steinhornensis* has been found in G 69-51, and the oldest find on Gotland of *L.* cf. *elegans,* one fragmentary hi element, derives from Ronnings in sample G 71-190. Another species of *Ligonodina* – probably *L. excavata* – also occurs at this level, represented by a fragment in sample G 69-51. Most interesting, however, is a single fragmentary specimen of an icriodontid in sample G 71-190. The specimen is difficult to identify, but is closest to *Pedavis thorsteinssoni* Uyeno 1981, and may well belong there.

The Burgsvik, Hamra and Sundre Beds

The Whitcliffian Burgsvik Beds crop out in a belt about 50 km long and largely less than 1 km wide. In the Burgsvik area, the beds consist of up to 45 m of sandstone and claystone overlain by up to 2 m of mixed beds, including oolites, pisolites and sandstone (see Stel & de Coo 1977). In the south, the lithologies are similar, but in the extreme NE, at Burgen, the Burgsvik Beds are only 7–8 m thick and oolites and biohermal limestones are the most important lithologies.

The latest Ludlovian Hamra Beds crop out in an area 30 km long. There are also some outliers 15–20 km further to the NE. In the southwest the Hamra Beds are mostly 20–25 m thick, but locally reach 40 m. They are subdivided into three units, designated a, b, and c, the lower one consisting of algal nodules and the other two of more or less argillaceous limestone and biohermal and associated detrital limestones. In the outliers to the NE there is less than 10 m of biohermal limestone remaining.

The uppermost stratigraphical unit on Gotland is the latest Ludlovian Sundre Beds. They crop out in an area up to 18 km long along the coast in the southeast and reach a maximum of 10 m of crinoidal and biohermal limestones.

Faunas from the Burgsvik Beds. – On Närsholmen, the northeasternmost land-area with Burgsvik Beds, conodonts have been found in the oolithic lower Burgsvik Beds at Glasskär 1 in sample G 72-18. As yet only a few tens of specimens have been extracted, but the fauna is diverse and includes *H. confluens, H. wimani, H. snajdri, L. excavata novoexcavata* and *P. equicostatus.* In Skåne *H. wimani* appears in beds containing *H. s. scanica.* Thus, it is possible that the Bjärsjölagård Limestone in Skåne is to be correlated with parts of the Burgsvik Beds on Gotland and the white sandstone (unit 4 of Eichstädt 1888, unit 2 of Grönwall 1897) in the Öved–Ramsåsa Group in Skåne may be correlated with the sandstones in the upper part of the Burgsvik Beds.

Faunas from the top Burgsvik Beds and the Hamra a. – Relatively large samples have been dissolved from the uppermost Burgsvik Beds and lowermost Hamra Beds, but the faunas recovered are very small. At Uddvide 2, nine specimens of *H. steinhornensis* have been found in samples from four levels in the Burgsvik Beds and four more in one sample from the basal Hamra Beds. In the latter sample *H. confluens* and *L. excavata novoexcavata* also occur. At Hoburgen 2 the meagre faunas are dominated by *H. excavata,* while *H. steinhornensis* has not been identified. The faunas there also include *P. equicostatus, H. confluens,* and *L. excavata novoexcavata.* The specimens of *H. steinhornensis* from Uddvide 2 are very similar to the youngest *H. s. scanica* from Klinta in Skåne, but none of the relevant elements are well enough preserved to say whether the denticles alternate in size or not. I have described the interval at Klinta with these elements as that of the younger *H. s. scanica* fauna (1974). That fauna is characterized by the absence of other species of *Hindeodella, Ligonodina,* and *Distomodus.* The collections at hand from the Burgsvik Beds at Uddvide thus agree also in this aspect. The differences between Uddvide 2 and Hoburgen 2 may either indicate a biofacies difference or more probably, an age difference. If so, the beds at Hoburgen 2 are probably the younger.

Faunas from Hamra b. – At Bankvät 1, in sample G 81-39, *H. confluens, H. snajdri crispa,* and a population of *H. steinhornensis* with alternating denticulation occur in good numbers. *H. wimani* and *L. elegans* are rare. It is notable that neither *H. excavata* nor *Panderodus* has been identified among the several hundred elements extracted. One specimen of the latter taxon occurs in sample G 72-2 from Bankvät 1. A closely similar fauna occurs at Strands 1 in sample G 75-14, which has *H. confluens, H. steinhornensis, H. snajdri crispa, H. wimani,* and *L. elegans.*

During the Whitcliffian and earliest 'post-Whitcliffian' much happened in the evolution of the conodonts, e.g., *H. steinhornensis* became more abundant and evolved alternating denticulation, *H. s. snajdri* evolved into *H. s. crispa* (Schönlaub *et al.,* in prep.), the lineage of *H. wimani* appeared and evolved rapidly, and that of *L. elegans* is again commonly represented in collections. *H. wimani* appears in the lower Burgsvik Beds (see above). *H. steinhornensis* with alternating denticulation is rare in the collection from Strands, dominates in those from Bankvät, and is not found in the other Hamra b collections; thus, Bankvät seems to represent the very youngest Hamra b and Strands a slightly older level. Also the first definite *L. elegans* appears at Strands. Further taxonomic studies are necessary to identify the levels defined by the evolutionary events.

In the type area for the Ludlovian in Britan, *H. steinhornensis* with alternating denticulation occurs in the very latest Whitcliffian, but the other more or less contemporaneous developments in other taxa have not been recognized (Aldridge, written communication). This may indicate that the uppermost units on Gotland are younger than the Upper Whitcliffe Formation in the type area. On the other hand Martinsson (1967) has shown that low levels in the Downton Castle Sandstone Formation are younger than the Sundre Beds. Thus several tens of metres of sediments on Gotland may have been formed during an interval poorly represented in the type area of the Ludlovian.

The faunas from Strands and Bankvät are very similar to the Wenlockian ones from Gothemshammar in which a different subspecies of *H. snajdri* and *H. confluens* dominate, while *H. excavata* and *Panderodus* are rare or absent. The fauna may also be compared with those from the upper parts of the Högklint Beds in Vattenfallet (Jeppsson 1979c). There *H. confluens* appears slightly before *H. sagitta,* a close relative of *H. snajdri, H. excavata* is much rarer than in the lower Högklint Beds, and *Panderodus* is very rare. Another close relative of *H. snajdri, H. steinhornensis,* also often occurs in faunas in which *Panderodus* and *H. excavata* are rare or absent, e.g., in Skåne (Jeppsson 1974) and in the *Beyrichienkalk* (Jeppsson 1981). None of these

three closely related taxa is limited to collections in which *Panderodus* and *H. excavata* are rare or absent, but they seem to be more rare in other collections. As is the case in the Halla Beds (see p. 132), *H. snajdri* is rarer, and the faunas more normal further to the southwest. Thus, *H. snajdri* occurs in sample G 77-33 PSSFG (3.99 kg) from Kättelviken 5 but is not found in five 0.5 kg samples higher up in the section. Sample G 77-33, which is from less than 5 m above the base of the Hamra Beds, has also yielded *H. excavata, H. confluens, H. steinhornensis, P. equicostatus, Belodella, Pseudooneotodus, Dapsilodus?,* and *L. excavata?.*.

Faunas from the Hamra c. – Some of the beds at Juves have produced rich faunas, which I plan to describe in another publication. The following taxa have been found to date: *H. confluens, H. excavata, H. snajdri crispa, H. steinhornensis, H. wimani* n. ssp., *L. confluens, L. elegans, L. excavata novoexcavata, P. equicostatus, P. unicostatus, P. beckmanni,* and questionably *Belodella.* Of these *H. wimani* and *Belodella* are very rare, and several of the other taxa are not met with in the smaller collections. Other localities do not seem to add much to the picture of the fauna, although *P. unicostatus* has not been found at Hoburgen 3.

In Anders Martinsson's sample Hoburgen I from the upper Hamra Beds at Storburg 1 there is an element of *Distomodus dubius.* In Britain this taxon continues into the basal Downton Beds (Aldridge 1975b), but on Gotland this is the only find to date from post-Leintwardinian beds. The species is regular in the Hemse Marl SE, e.g., at Gogs, and it may be significant that Anders Martinsson designated the Hoburgen I sample as 'Gogs lithology'.

Sundre faunas. – In the Sundre Beds at Juves *H. excavata, H. confluens, H. steinhornensis, L. confluens, L. excavata novoexcavata, Panderodus equicostatus,* and *P. unicostatus* occur. Most or all of the differences between this list and that from the Hamra c at the same locality can be accounted for by the smaller sizes of the collections from the Sundre Beds. Other Sundre faunas available contain a similar set of species.

Acknowledgements. – The work reported here has been undertaken at the Department of Historical Geology and Palaeontology, University of Lund, where working facilities have been provided by Gerhard Regnéll and Anita Löfgren.

My first field work on Gotland took place with Stig M. Bergström in 1967. My wife has assisted me on nearly all of the subsequent field work there, much of it together with Anders Martinsson or Sven Laufeld. These two, like Stig M. Bergström, enthusiastically shared their knowledge of the island and its geology. Much of the field work was done with other associates within the Baltic-Scanian Silurian Project and Project Ecostratigraphy, the projects which provided the funds for the field work. Regarding samples and specimens and their collectors, see p. 122.

Some of the conodont identifications and correlations have benefitted from discussions with Richard J. Aldridge and Hans Peter Schönlaub, and the manuscript has been improved by comments from Richard J. Aldridge, Sven Laufeld, Anita Löfgren, Anders Martinsson, and Hans P. Schönlaub. Richard Aldridge also took care of the linguistic revision (I have changed a few things subsequently and might have introduced new items of poor language), and Hans P. Schönlaub was the editor's referee.

Christin Andreasson drafted fig. 1, Ingrid Lineke typed the manuscript, and Sven Stridsberg took the photographs.

Grants from the Swedish National Science Research Council,

Kungliga Fysiografiska Sällskapet i Lund, Lunds Geologiska Fältklubb, Hierta-Retzius Stiftelse and the University of Lund have paid the many students who have extracted most of the conodont elements.

The manuscript was written during the tenure of research position granted by the Swedish Natural Science Research Council, who also payed the printing costs.

To all, my sincere thanks.

I dedicate this study to the memory of Anders Martinsson.

References

Aldridge, R. J. 1972: Llandovery conodonts from the Welsh Borderland. *Bulletin of the British Museum (Natural History) Geology 22*, 125–231.

Aldridge, R. J. 1974: An *amorphognathoides* Zone conodont fauna from the Silurian of the Ringerike area, south Norway. *Norsk Geologisk Tidskrift 54*, 295–303.

Aldridge, R. J. 1975a: The Silurian conodont *Ozarkodina sagitta* and its value in correlation. *Palaeontology 18*, 323–332.

Aldridge, R. J. 1975b: The stratigraphic distribution of conodonts in the British Silurian. *Journal of the Geological Society (London) 131*, 607–618.

Aldridge, R. J. 1979: An upper Llandovery conodont fauna from Peary Land, eastern North Greenland. *Grønlands Geologiske Undersøgelse Rapport 91*, 7–23.

Aldridge, R. J., Dorning, K. J. & Siveter, D. J. 1982: Distribution of microfossil groups across the Wenlock Shelf of the Welsh Basin. *In* Neale, J. W. & Brasier, M. D. (eds.): *Microfossils from Recent and Fossil Shelf Seas*, 18–30. British Micropalaeontological Society Series.

Barnes, C. R., Sass, D. B. & Poplawski, M. L. S. 1973: Conodont ultrastructure: The Family Panderodontidae. *Life Sciences Contribution Royal Ontario Museum 90*. 36 pp.

Barrick, J. E. 1977: Multielement simple-cone conodonts from the Clarita Formation (Silurian), Arbuckle Mountains, Oklahoma. *Geologica et Palaeontologica 11*, 47–68.

Barrick, J. E. & Klapper, G. 1976: Multielement Silurian (late Llandoverian–Wenlockian) conodonts of the Clarita Formation, Arbuckle Mountains, Oklahoma, and phylogeny of *Kockelella*. *Geologica et Palaeontologica 10*, 59–99.

Bassett, M. G. 1974: Review of the stratigraphy of the Wenlock Series in the Welsh Borderland and South Wales. *Palaeontology 17*, 745–777.

Bassett, M. G. & Cocks, L. R. M. 1974: A review of Silurian brachiopods from Gotland. *Fossils and Strata 3*, 56 pp.

Bergman, C. F. 1979: Ripple marks in the Silurian of Gotland, Sweden. *Geologiska Föreningens i Stockholm Förhandlingar 101*, 217–222.

Bergman, C. F. 1980: Macrofossils of the Wenlockian Slite Siltstone of Gotland. *Geologiska Föreningens i Stockholm Förhandlingar 102*, 13–25.

Bergman, C. F. 1981: Palaeocurrents, wave marks and reefs; a palaeogeographical instrument applied to the Silurian of Gotland. – The whole environment – Slite Beds, Gotland. *In* Laufeld, S. (ed.): Proceedings of Project Ecostratigraphy Plenary Meeting, Gotland, 1981. *Sveriges Geologiska Undersökning, Rapporter och meddelanden 25*, 6.

Bergman, C. F. 1983: Palaeocurrents in the Silurian Slite Marl of Fårö, Sweden. *Geologiska Föreningens i Stockholm Förhandlingar 105:2* (in press).

Berry, W. B. N. & Boucot, A. J. 1972: Silurian graptolite depth zonation. *24th International Geological Congress Montreal 7*, 59–65.

Bischoff, G. C. O. 1975: Conodonts. *In* Talent, J. A., Berry, W. B. N. & Boucot, A. J.: Correlation of the Silurian Rocks of Australia, New Zealand, and New Guinea. *The Geological Society of America, Special Paper 150*, 6–9.

Bischoff, G. & Sannemann, D. 1958: Unterdevonische Conodonten aus dem Frankenwald. *Notizblatt des Hessischen Landesamtes für Bodenforschung zu Wiesbaden 86*, 87–110.

Branson, E. B. & Branson, C. C. 1947: Lower Silurian conodonts from Kentucky. *Journal of Paleontology 21*, 549–556.

Branson, E. B. & Mehl, M. G. 1933–1934: Conodont studies. *Missouri University Studies 8*. 349 pp.

Brood, K. 1982: *Gotländska fossil*. 95 pp. P. A. Nordstedt & Söners Förlag. Stockholm.

Cherns, L. 1982: Palaeokarst, tidal erosion surfaces and stromatolites in the Silurian Eke Formation of Gotland, Sweden. *Sedimentology 29*, 819–833.

Claesson, Charlotte 1979: Early Palaeozoic geomagnetism of Gotland. *Geologiska Föreningens i Stockholm Förhandlingar 101*, 149–155.

Cocks, L. R. M., Holland, C. H., Rickards, R. B. & Strachan, I. 1971: A correlation of Silurian rocks in the British Isles. *Journal of the Geological Society, London 127*, 103–136.

Cooper, B. J. 1975: Multielement conodonts from the Brassfield Limestone (Silurian) of southern Ohio. *Journal of Paleontology 49*, 984–1008.

Cooper, B. J. 1976: Multielement conodonts from the St. Clair Limestone (Silurian) of southern Illinois. *Journal of Paleontology 50*, 205–217.

Cooper, B. J. 1977: Toward a familial classification of Silurian conodonts. *Journal of Paleontology 51*, 1057–1071.

Cooper, B. J. 1980: Toward an improved Silurian conodont biostratigraphy. *Lethaia 13*, 209–227.

Drygant, D. M. 1974: Simple conodonts of the Silurian and lowermost Devonian of the Volyn-Podolian area. *Палеонтологический сборник 10*, 64–70. In Russian, with an English summary.

Eichstädt, F. 1888: Anteckningar om de yngsta öfversiluriska aflagringarna i Skåne. *Geologiska Föreningens i Stockholm Förhandlingar 10*, 132–156.

Eisenack, A. 1975: Beiträge zur Anneliden-Forschung. I. *Neues Jahrbuch für Geologie und Paläontologie Abhandlungen 150*, 227–252.

Epstein, A. G., Epstein, J. B. & Harris, L. D. 1977: Conodont color alteration – an index to organic metamorphism. *United States Geological Survey Professional Paper 995*. 27 pp.

Fåhraeus, L. E. 1968: Conodont zones in the Ludlovian of Gotland, Sweden. *The Geological Society of America, North-Central Section. Abstracts with Programs, 1968 Annual Meeting*, 43–44.

Fåhraeus, L. E. 1969: Conodont zones in the Ludlovian of Gotland and a correlation with Great Britain. *Sveriges Geologiska Undersökning C 639*. 33 pp.

Franzén, C. 1977: Crinoid holdfasts from the Silurian of Gotland. *Lethaia 10*, 219–234.

Grönvall, K. A. 1897: Öfversikt af Skånes yngre öfversiluriska bildningar. *Geologiska Föreningens i Stockholm Förhandlingar 19*, 188–244. Also in *Sveriges Geologiska Undersökning C 170*. 59 pp.

Harley, J. 1861: On the Ludlow Bone-Bed and its Crustacean remains. *The Quarterly Journal of the Geological Society of London 17*, 542–552.

Harper, C. W. 1969: Rib branching patterns in the brachiopod *Atrypa reticularis* from the Silurian of Gotland, Sweden. *Journal of Paleontology 43*, 183–188.

Hede, J. E. 1919: Om några nya fynd av graptoliter inom Gottlands silur och deras betydelse för stratigrafin. *Sveriges Geologiska Undersökning C: 291*. 31 pp.

Hede, J. E. 1921: Gottlands silurstratigrafi. *Sveriges Geologiska Undersökning C: 305*. 100 pp.

Hede, J. E. 1925a: Beskrivning av Gotlands silurlager. *In* Munthe, H., Hede, J. E. & v. Post, L.: Gotlands geologi, en översikt. *Sveriges Geologiska Undersökning C: 331*, 13–30.

Hede, J. E. 1925b: Berggrunden (Silursystemet). *In* Munthe, H., Hede, J. E. & v. Post, L.: Beskrivning till Kartbladet Ronehamn. *Sveriges Geologiska Undersökning Aa: 156*, 14–51.

Hede, J. E. 1927a: Berggrunden (Silursystemet). *In* Munthe, H., Hede, J. E. & Lundqvist, G.: Beskrivning till kartbladet Klintehamn. *Sveriges Geologiska Undersökning Aa: 160*, 12–54.

Hede, J. E. 1927b: Berggrunden (Silursystemet). *In* Munthe, H., Hede, J. E. & v. Post, L.: Beskrivning till kartbladet Hemse. *Sveriges Geologiska Undersökning Aa: 164*, 15–56.

Hede, J. E. 1928: Berggrunden (Silursystemet). *In* Munthe, H., Hede, J. E. & Lundqvist, G.: Beskrivning till kartbladet Slite. *Sveriges Geologiska Undersökning Aa: 169*, 13–65.

Hede, J. E. 1929: Berggrunden (Silursystemet). *In* Munthe, H., Hede, J. E. & Lundqvist, G.: Beskrivning till kartbladet Katthammarsvik. *Sveriges Geologiska Undersökning Aa: 170*, 14–57.

Hede, J. E. 1933: Berggrunden (Silursystemet). *In* Munthe, H., Hede, J. E. & Lundqvist, G.: Beskrivning till kartbladet Kappelshamn. *Sveriges Geologiska Undersökning Aa: 171*, 10–59.

Hede, J. E. 1936: Berggrunden. *In* Munthe, H., Hede, J. E. & Lundqvist, G.: Beskrivning till kartbladet Fårö. *Sveriges Geologiska Undersökning Aa: 180*, 11–42.

Hede, J. E. 1940: Berggrunden. *In* Lundqvist, G., Hede, J. E. & Sundius, N.: Beskrivning till kartbladen Visby och Lummelunda. *Sveriges Geologiska Undersökning Aa: 183*, 9–68.

Hede, J. E. 1942: On the correlation of the Silurian of Gotland. *Lunds Geologiska Fältklubb 1892–1942. Also in Meddelanden från Lunds Geologisk-Mineralogiska Institution 101*, 25 pp.

Hede, J. E. 1958: [Silurian entries in] *Lexique stratigraphique international 1, Europe 2c Suède-Sweden-Sverige.* 498 pp.

Hede, J. E. 1960: The Silurian of Gotland. *In* Regnéll, G. & Hede, J. E.: The Lower Paleozoic of Scania. The Silurian of Gotland. *International Geological Congress XXI Session Norden* 1960, Guidebook Sweden d. Stockholm. Also in *Publications from the Institutes of Mineralogy, Palaeontology and Quaternary Geology, University of Lund 91*, 44–89.

Hedström, H. 1904: Detaljprofil från skorpionfyndorten – *Pterygotus*-lagret – i siluren strax S om Visby. *Geologiska Föreningens i Stockholm Förhandlingar 26*, 93–96.

Hedström, H. 1910: The stratigraphy of the Silurian strata of the Visby district. *Geologiska Föreningens i Stockholm Förhandlingar 32*, 1455–1484.

Hedström, H. 1923b: Remarks on some fossils from the diamond boring of the Visby cement factory. *Sveriges Geologiska Undersökning C 314*, 26 pp.

Helfrich, C. T. 1975: Silurian conodonts from Wills Mountain Anticline, Virginia, West Virginia, and Maryland. *The Geological Society of America Special Paper 161.* 82 + 86 pp.

Holland, C. H. 1980: Silurian series and stages: decisions concerning chronostratigraphy. *Lethaia 13*, 238.

Holland, C. H., Lawson, J. D. & Walmsley, V. G. 1963: The Silurian rocks of the Ludlow district, Shropshire. *Bulletin of the British Museum (Natural History) Geology 8*, 93–171.

Holland, C. H., Lawson, J. D., Walmsley, V. G. & White, D. E. 1980: Ludlow stages. *Lethaia 13*, 268.

Igo, H. & Koike, T. 1967: Ordovician and Silurian Conodonts from the Langkawi Islands, Malaya, Part 1. *Geology and Palaeontology of Southeast Asia 3*, 1–29.

Jaanusson, V. 1979: Stratigraphical and environmental background. *In* Jaanusson, V., Laufeld, S. & Skoglund, R. (eds.): Lower Wenlock faunal and floral dynamics – Vattenfallet section, Gotland. *Sveriges Geologiska Undersökning C 762*, 11–38.

Jaanusson, V., Laufeld, S. & Skoglund, R. 1979 (eds.): Lower Wenlock faunal and floral dynamics – Vattenfallet Section, Gotland. *Sveriges Geologiska Undersökning C 762.* 294 pp.

Jaeger, H. 1975: Die Graptolithenführung im Silur/Devon des Cellon-Profils (Karnische Alpen). *Carinthia 2*, 111–126.

Jaeger, H. 1981: Comments on the graptolite chronology of Gotland. *In* Laufeld, S. (ed.): Proceedings of Project Ecostratigraphy Plenary Meeting, Gotland 1981. *Sveriges Geologiska Undersökning, Rapporter och meddelanden 25*, 12.

Jaeger, H. & Schönlaub, H. P. 1977: Das Ordoviz/Silur-Profil im Nölblinggraben (Karnische Alpen, Österreich). *Verhandlungen der Geologischen Bundesanstalt 1977*, 349–359.

Janvier, Ph. 1978: On the oldest known teleostome fish *Andreolepis hedei* Gross (Ludlow of Gotland), and the systematic position of the lophosteids. *Eesti NSV Teaduste Akadeemia Toimetised 27. Köide Geoloogia 3, (1978)*, 88–95.

Jeppsson, L. 1969: Notes on some Upper Silurian multielement conodonts. *Geologiska Föreningens i Stockholm Förhandlingar 91*, 12–24.

Jeppsson, L 1972: Some Silurian conodont apparatuses and possible conodont dimorphism. *Geologica et Palaeontologica 6*, 51–69.

Jeppsson, L. 1974 [1975]: Aspects of Late Silurian conodonts. *Fossils and Strata 6.* 79 pp.

Jeppsson, L. 1976: Autecology of Late Silurian conodonts. *In* Barnes, C. R. (ed.): Conodont Palaeoecology. *The Geological Association of Canada Special Paper 15*, 105–118.

Jeppsson, L. 1979a: Conodont element function. *Lethaia 12*, 153–171.

Jeppsson, L. 1979b: Growth, element arrangement, taxonomy and ecology of selected conodonts. *Publications from the Institutes of Mineralogy, Paleontology, and Quaternary Geology, University of Lund, Sweden, 218.* 42 pp.

Jeppsson, L. 1979c: Conodonts. *In* Jaanusson, V., Laufeld, S. & Skoglund, R. (eds.): Lower Wenlock faunal and floral dynamics – Vattenfallet Section, Gotland. *Sveriges Geologiska Undersökning C 762*, 225–248.

Jeppsson, L. 1981: The conodont faunas in the *Beyrichienkalk. In* Laufeld, S. (ed.): Proceedings of Project Ecostratigraphy Plenary Meeting. Gotland 1981. *Sveriges Geologiska Undersökning, Rapporter och meddelanden 25*, 13–14.

Jeppsson, L. 1982: Third European Conodont Symposium (ECOS III). Guide to excursion. *Publications from the Institutes of Mineralogy, Paleontology and Quaternary Geology, University of Lund, Sweden, 239.* 32 pp.

Kershaw, S. 1981: Stromatoporoid growth form and taxonomy in a Silurian biostrome, Gotland. *Journal of Paleontology 55*, 1284–1295.

Kershaw, S. & Riding, R. 1978: Parameterization of stromatoporoid shape. *Lethaia 11*, 233–242.

Klapper, G. & Murphy, M. A. 1974 [1975]: Silurian–Lower Devonian conodont sequence in the Roberts Mountains Formation of central Nevada. *University of California Publications in Geological Sciences 111.* 87 pp.

Klapper, G. & Philip, G. M. 1971: Devonian conodont apparatuses and their vicarious skeletal elements. *Lethaia 4*, 429–452.

Kříž, J. & Schönlaub, H. P. 1980: Stop 1. Daleje Valley, Mušlovka Quarry section. *In* Schönlaub, H. P. (ed.): Second European conodont symposium (ECOS II) Guidebook Abstracts. *Abhandlungen der Geologischen Bundesanstalt 35*, 153–157.

Larsson, K. 1975: Clastic dikes from the Burgsvik Beds of Gotland. *Geologiska Föreningens i Stockholm Förhandlingar 97*, 125–134.

Larsson, K. 1979: Silurian tentaculitids from Gotland and Scania. *Fossils and Strata 11.* 180 pp.

Laufeld, S. 1974a: Silurian Chitinozoa from Gotland. *Fossils and Strata 5.* 130 pp.

Laufeld, S. 1974b: Reference localities for palaeontology and geology in the Silurian of Gotland. *Sveriges Geologiska Undersökning C 705.* 172 pp.

Laufeld, S. & Jeppsson, L. 1976: Silicification and bentonites in the Silurian of Gotland. *Geologiska Föreningens i Stockholm Förhandlingar 98*, 31–44.

Laufeld, S. & Martinsson, A. 1981: *Reefs and Ultrashallow Environments. Guidebook to the Field Excursions in the Silurian of Gotland, Project Ecostratigraphy Plenary Meeting 22nd–28th August, 1981.* 24 pp. Museum Department, Geological Survey of Sweden.

Liljedahl, L. 1981: Silicified bivalves from the Silurian of Gotland. *In* Laufeld, S. (ed.): Proceedings of Project Ecostratigraphy Plenary Meeting, Gotland, 1981. *Sveriges Geologiska Undersökning, Rapporter och meddelanden 25*, 22.

Liljedahl, L. 1983: Two silicified Silurian bivalves from Gotland. *Sveriges Geologiska Undersökning C 799.* 51 pp.

Lindström, M. 1964: *Conodonts.* 196 pp. Elsevier. Amsterdam, London, New York.

Martinsson, A. 1962: Ostracodes of the family Beyrichiidae from the Silurian of Gotland. *Bulletin of the Geological Institutions of the University of Uppsala 41.* 369 pp.

Martinsson, A. 1967: The succession and correlation of ostracode faunas in the Silurian of Gotland. *Geologiska Föreningens i Stockholm Förhandlingar 89*, 350–386.

Martinsson, A. 1972: Review of Manten, A. A.: Silurian reefs of Gotland. *Geologiska Föreningens i Stockholm Förhandlingar 94*, 128–129.

Mashkova, T. V. (Машкова, Т. В.) 1977: Новые конодонты зоны *Amorphognathoides* из нижнего силура Подолии. [New conodonts of the *Amorphognathoides* Zone from the Lower Silurian of Podolia.] *Палеонтологический Журнал 4*, 127–131.

Mori, K. 1970: Stromatoporoids from the Silurian of Gotland. Part 2. *Stockholm Contributions in Geology 22.* 152 pp.

Munthe, H. 1910: On the sequence of strata within southern Gotland. *Geologiska Föreningens i Stockholm Förhandlingar 32*, 1397–1453.

Nicoll, R. S. & Rexroad, C. B. 1968 [1969]: Stratigraphy and conodont paleontology of the Salamonie Dolomite and Lee Creek Member of the Brassfield Limestone (Silurian) in southeastern

Indiana and adjacent Kentucky. *Indiana Geological Survey Bulletin 40.* 73 pp.

Poulsen, K. D., Saxov, S., Balling, N. & Kristiansen, J. I. 1982: Thermal conductivity measurements of Silurian limestones from the Island of Gotland, Sweden. *Geologiska Föreningens i Stockholm Förhandlingar 103*, 349–356.

Rexroad, C. B. & Craig, W. W. 1971: Restudy of conodonts from the Bainbridge Formation (Silurian) at Lithium, Missouri. *Journal of Paleontology 45*, 684–703.

Rhodes, F. H. T. 1953: Some British Lower Palaeozoic conodont faunas. *Philosophical Transactions of the Royal Society of London, B 237*, 261–334.

Rhodes, F. H. T. & Newall, G. 1963: Occurrence of *Kockelella variabilis* Walliser in the Aymestry Limestone of Shropshire. *Nature 199*, 166–167.

Saladžius, V. 1975: Conodonts of the Llandoverian (Lower Silurian) deposits of Lithuania. *In:* Фауна и стратиграфия палеозоя и мезозоя Прибалтики и Белоруссии. 219–226. Vilnius. (In Russian, with an English summary.)

Sivhed, U. 1976: Sedimentological studies of the Wenlockian Slite Siltstone on Gotland. *Geologiska Föreningens i Stockholm Förhandlingar 98*, 59–64.

Stel, J. H. & de Coo, J. C. M. 1977: The Silurian upper Burgsvik and lower Hamra–Sundre beds, Gotland. *Scripta Geologica 44*, 1–43.

Stridsberg, S. 1981a: Apertural constrictions in some oncocerid cephalopods. *Lethaia 14*, 269–276.

Stridsberg, S. 1981b: Silurian oncocerid nautiloids from Gotland. *In* Laufeld, S. (ed.): Proceedings of Project Ecostratigraphy Plenary Meeting, Gotland, 1981. *Sveriges Geologiska Undersökning, Rapporter och meddelanden 25*, 32.

Sundquist, B. 1981: The whole environment – Silurian Slite Beds, Gotland. – Petrography. *In* Laufeld, S. (ed.): Proceedings of Project Ecostratigraphy Plenary Meeting, Gotland, 1981. *Sveriges Geologiska Undersökning, Rapporter och meddelanden 25*, 33.

Sundquist, B. 1982a: Wackestone petrography and bipolar orientation of cephalopods as indicators of littoral sedimentation in the Ludlovian of Gotland. *Geologiska Föreningens i Stockholm Förhandlingar 104*, 81–90.

Sundquist, B. 1982b: Carbonate petrography of the Wenlockian Slite Beds at Haganäs, Gotland. *Sveriges Geologiska Undersökning C 796.* 79 pp.

Talent, J. A., Berry, W. B. N. & Boucot, A. J. 1975: Correlation of the Silurian Rocks of Australia, New Zealand, and New Guinea. *The Geological Society of America, Special Paper 150.* 108 pp.

Thorsteinsson, R. & Uyeno, T. T. 1980 [1981]: Biostratigraphy. *In* Thorsteinsson, R.: Stratigraphy and conodonts of Upper Silurian and Lower Devonian rocks in the environs of the Boothia Uplift, Canadian Arctic Archipelago. Part 1. Contributions to stratigraphy. *Geological Survey of Canada Bulletin 292*, 21–31.

Uyeno, T. T. 1980 [1981]: Stratigraphy and conodonts of Upper Silurian and Lower Devonian rocks in the environs of the Boothia Uplift, Canadian Arctic Archipelago. Part 2. Systematic study of conodonts. *Geological Survey of Canada Bulletin 292*, 39–75.

Viira, V. 1975: A new species of *Spathognathodus* from the Jaani Stage of the East Baltic. *Eesti NSV Teaduste Akadeemia Toimetised 24*, 233–236. (In Russian, with an English summary.)

Walliser, O. H. 1957: Conodonten aus dem oberen Gotlandium Deutschlands und der Karnischen Alpen. *Notizblatt des Hessischen Landesamtes für Bodenforschung zu Wiesbaden 85*, 28–52.

Walliser, O. H. 1964: Conodonten des Silurs. *Abhandlungen des Hessischen Landesamtes für Bodenforschung zu Wiesbaden 41*, 106 pp.

Walliser, O. H. 1972: Conodont apparatuses in the Silurian. *Geologica et Palaeontologica SB1*, 75–79.

Walmsley, V. G. 1965: *Isorthis* and *Salopina* (Brachiopoda) in the Ludlovian of the Welsh Borderland. *Palaeontology 8*, 454–477.

Walmsley, V. G. & Boucot, A. J. 1975: The phylogeny, taxonomy and biogeography of Silurian and Early to Mid Devonian Isorthinae (Brachiopoda). *Palaeontographica Abteilung A 148*, 34–108.

Watkins, R. 1975: Silurian brachiopods in a stromatoporoid bioherm. *Lethaia 8*, 53–61.

Ziegler, W. 1956: Unterdevonische Conodonten, insbesondere aus dem Schönauer und dem Zorgensis-Kalk. *Notizblatt des Hessischen Landesamtes für Bodenforschung zu Wiesbaden 84*, 93–106.

Appendix: Localities

In the list of references to each locality, an asterisk marks the publication in which a full description of the locality is found. For localities described in Laufeld 1974b, Laufeld & Jeppsson 1976 and Larsson 1979, earlier literature is not cited. Grid references refer to the Swedish National Grid system and (within parentheses) the Universal Transverse Mercator (UTM) system.

AJMUNDE 1. Earlier referred to as Klinteberg Beds, Klinteberg Marl top. In my opinion, the same unit as other localities referred to Hemse Marl NW. Beds with *Kockelella variabilis*. *Age:* Ludlovian; probably Bringewoodian.
References: Laufeld 1974a, b*; Laufeld & Jeppsson 1976; Larsson 1979.

AMLINGS 1. Hemse Beds, Hemse Marl NW part. Beds with *Kockelella variabilis*. *Age:* Ludlovian; probably Late Bringewoodian, possibly earliest Leintwardinian.
References: Laufeld 1974a, b*; Laufeld & Jeppsson 1976; Larsson 1979.

AURSVIKEN 1. Höglint Beds, the middle of three Höglint units separated by Hede (1936), probably high in that unit; lower–middle part (Laufeld 1974b). Beds with *Kockelella? ranuliformis* and *Hindeodella sagitta rhenana*. *Age:* Early Wenlockian.
References: Laufeld 1974a, b*.

AUTSARVE 1. Hemse Beds, Hemse Marl NW part. *Age:* Ludlovian; probably Bringewoodian.
References: Laufeld 1974a, b*.

BANKVÄT 1. Hamra Beds, unit b. Beds with *Hindeodella snajdri crispa*. *Age:* Latest Ludlovian.
References: Laufeld 1974a, b*; Larsson 1979; Jeppsson 1982.

BARA 1. Halla Beds, unit a, i.e. Bara Oolite. Beds with *Hindeodella snajdri*. *Age:* Late Wenlockian.
References: Laufeld 1974a, b*; Larsson 1979; Jeppsson 1982.

BÅTA 1. Slite Beds, unit f. i.e. *R. tenuistriatum* Beds. Beds with *Kockelella walliseri*. *Age:* Early Wenlockian.
References: Laufeld 1974a, b*; Larsson 1979.

BÅTELS 1. Klinteberg Beds, unit c. Beds with *Hindeodella snajdri*. *Age:* Late Wenlockian.
References: Laufeld 1974a, b*.

BLÅHÄLL 1. Mulde Beds, lower part. *Age:* Late Wenlockian.
References: Laufeld 1974a, b*; Larsson 1979; Claesson 1979; Poulsen et al. 1982.

BODUDD 1. Hemse Beds, Hemse Marl uppermost part and Eke Beds, basal part. The Hemse Beds contain the *Pelekysgnathus dubius* fauna. *Age:* Ludlovian; Late Leintwardinian.
References: Laufeld 1974a, b*; Larsson 1979; Jeppsson 1982.

BODUDD 3. 633050 164550 (CJ 3065 2972), ca 5000 m SW of Näs church. Topographical map sheet 5I Hoburgen NO & 5J Hemse NV. Geological map sheet Aa 152 Burgsvik.
Shore exposure on the northern side of the bulge in the coast line, with five bushy service-trees *(Sorbus)*, only three of them alive and standing.
Hemse Beds, Hemse Marl uppermost part. Strata with the *Pelekysgnathus dubius* fauna. *Age:* Ludlovian; Late Leintwardinian.
References: Hede 1919:21, Loc. 9; Hede 1942:20, Loc. 1C.

BOTVALDE 1. 638860 167638 (CJ 6588 8526), ca 1800 m NE of Gothem church. Topographical map sheet 6J Roma NV & NO. Geological map sheet Aa 169 Slite.
Shallow 'quarry', about 40 m SW of the road and a few metres E of the edge of the pine forest, 450 m NW of the road intersection at Tummungs.

Klinteberg Beds, unit e. Beds with *Hindeodella snajdri. Age:* Wenlockian; Late Homerian.

Reference: Hede 1928:61, lines 5–10.

Note: Hede's description of the beds and of their stratigraphical position (Hede 1928:61) is closely similar to his description of the beds that are now referred to unit e (Hede 1929:20), and the beds can be considered to belong to unit e. Botvalde 1 is thus the northernmost known locality with that unit.

BOTVALDEVIK 1. Klinteberg Beds, unit c. *Age:* Late Wenlockian.
Reference: Larsson 1979*.

BOTVIDE 1. Hemse Beds, Hemse Marl uppermost part and Eke Beds lowermost part. The Hemse Beds contain the *Pelekysgnathus dubius* fauna. *Age:* Ludlovian; Late Leintwardinian.
References: Laufeld 1974a, b*; Jeppsson 1974:10; Larsson 1979; Laufeld & Martinsson 1981; Jeppsson 1982; Cherns 1982.

DJUPVIK 2. 635556 164074 (CJ 2776 5504), ca 4450 m WNW of Eksta church. Topographical map sheet 6I Visby SO. Geological map sheet Aa 164 Hemse.

Cliff section just SW of the disturbed beds below the place where there are some trees NW of the road, about 675 m SW of the harbour of Djauvik.

Reference level: There are four deeply caved marl beds (bentonites?) in the section and some less distinct ones. The middle two are about 2 m above the base of the section and about 0.2 m apart. The lower of these is selected as the reference level. The remaining two distinct marl beds occur 1.5 m below and 1.10 m above the reference level; *Tussilago farfara* grows in the uppermost bed.

Mulde Beds, lower part. Beds with *Hindeodella snajdri. Age:* Late Wenlockian.

Reference: Hede 1927b (to the whole 1.1 km long section).

Note: The thicknesses between the marl beds are so similar to those described by Laufeld 1974a from Djupvik 1, that they are almost certainly the same horizons. Thus the same reference level can be used at the two localities.

FIE 3. 635294 167097 (CJ 5770 5005), ca 1700 m NE of När church. Topographical map sheet 6J Roma SV. Geological map sheet Aa 156 Ronehamn.

Ditch exposure along the field road (not marked on the topographical map) about 10 m E of the road between Kauparve and Hägdarve.

Hemse Beds, Hemse Marl, southeastern part. Beds with *Polygnathoides siluricus. Age:* Ludlovian; Leintwardinian.

GALGBERGET 1. Högklint Beds, unit c and Tofta Beds, basal part. The Högklint Beds contain *Hindeodella sagitta rhenana. Age:* Early Wenlockian.

References: Laufeld 1974a, b*; Larsson 1979; Claesson 1979; Laufeld & Martinsson 1981; Jeppsson 1982.

GANNOR 1. Hemse Beds, Hemse Marl, uppermost part and Eke Beds, basal part. The Hemse Beds contain the *Pelekysgnathus dubius* fauna. *Age:* Ludlovian; Late Leintwardinian.

References: Laufeld 1974a, b*; Eisenack 1975, Fig. 30; Laufeld & Jeppsson 1976; Larsson 1979; Claesson 1979.

GANNOR 3. Hemse Beds, Hemse Marl SE part. *Polygnathoides siluricus* Zone. *Age:* Ludlovian; Leintwardinian.

References: Laufeld 1974a, b*.

GARDE 1. 636109 166785 (CJ 5513 5853), ca 3100 m N12°E of Garde church. Topographical map sheet 6J Roma SV. Geological map sheet Aa 170 Katthammarsvik.

Abandoned, large quarry (marked on both maps) 650 m SE of point 53,46. My section was measured about 50 m west of the eastern end of the quarry.

Reference level: The lower surface of the oldest bed with large (up to 100 mm) stromatoporoids, the boundary between the two units in Hede 1929.

Hemse Beds, unit e. At least the beds from 0.2 m below the reference level belong in the *Polygnathoides siluricus* Zone. *Age:* Ludlovian; Leintwardinian.

Reference: Hede 1929:47, line 39–p. 48, line 19.

GARDSBY 1. 635366 165328 (CJ 4015 5222), ca 2200 m N (and slightly E) of Fardhem church. Topographical map sheet 6J Roma SV. Geological map sheet Aa 164 Hemse.

Exposure 5 m N of the road in the small ditch, perpendicular to the road, 25 m NW of the solitary house 350 m ESE Gardsby.

Hemse Beds, Hemse Marl NW part. *Age:* Ludlovian; Bringewoodian or Leintwardinian.

GERETE 1. Hemse Beds, Hemse Marl NW part. Beds with *Kockelella variabilis. Age:* Ludlovian; Bringewoodian or Leintwardinian.
References: Laufeld 1974a, b*; Franzén 1977; Larsson 1979.

GERUMSKANALEN 1. Hemse Beds, Hemse Marl NW part. Beds with *Kockelella variabilis. Age:* Ludlovian; probably Late Bringewoodian, possibly earliest Early Leintwardinian.
References: Laufeld 1974a, b*; Laufeld & Jeppsson 1976; Jeppsson 1982.

GLASSKÄR 1. Burgsvik Beds, lower part. Fauna with *Hindeodella wimani* and *H. snajdri. Age:* Ludlovian; Whitcliffian.
References: Laufeld 1974a, b*; Larsson 1979.

GLÄVES 1. Hemse Beds, Hemse Marl SE part. The locality was unavailable in 1971 but in 1975 the ditch along the northern side of the main road and along the western side of the approach road to the farm had recently been deepened. My sample is from the ditch along the approach road 5 m from the main road. *Polygnathoides siluricus* Zone. *Age:* Ludlovian; Leintwardinian.
References: Laufeld 1974b*; Jeppsson 1974:7; Larsson 1979.

GOGS 1. Hemse Beds, Hemse Marl SE part. *Polygnathoides siluricus* Zone. *Age:* Ludlovian; Leintwardinian.
References: Laufeld 1974a, b*; Jeppsson 1972:63, Fig. 1B, 1974:7, 12, 1976:108, 109, 113–117; Janvier 1978; Larsson 1979; Jeppsson 1982; Brood 1982:24–25.

GOTHEMSHAMMAR 1. Halla Beds, unit c and Klinteberg Beds, unit a. Beds with *Hindeodella snajdri. Age:* Late Wenlockian.
References: Laufeld 1974a, b*; Larsson 1979.

GOTHEMSHAMMAR 2. Halla Beds, unit c and Klinteberg Beds, unit a. Beds with *Hindeodella snajdri. Age:* Late Wenlockian.
References: Laufeld 1974a, b*; Larsson 1979; Claesson 1979.

GOTHEMSHAMMAR 3. Halla Beds, unit c and Klinteberg Beds, unit a. Beds with *Hindeodella snajdri. Age:* Late Wenlockian.

Low section on the shore, partly covered by scree from the raised beach ridge. Drive 3.3 km from road 146 to where the road splits into three parallel ones close to each other. The locality is immediately NW of a large *Salix* shrub to the left of the northern road.
References: Laufeld 1974b*; Larsson 1979; Poulsen *et al.* 1982.

GOTHEMSHAMMAR 6. Halla Beds, unit c. Beds with *Hindeodella snajdri. Age:* Late Wenlockian.
References: Laufeld 1974b*; Claesson 1979.

GROGARNSHUVUD 1. Hemse Beds, unit c. *Age:* Early Ludlovian.
References: Laufeld 1974a, b*; Larsson 1979; Claesson 1979; Laufeld & Martinsson 1981; Jeppsson 1982; Sundquist 1982a.

GRYMLINGS 1. 635566 164505 (CJ 3210 5485), ca 1970 m NNE of Eksta church. Topographical map sheet 6I Visby SO. Geological map sheet Aa 164 Hemse.

Exposure 1 m W of the road at the intersection of the road side ditch and the drainage ditch that crosses the road 750 m WSW of Grymlings.

Klinteberg Beds, lower–middle part. Beds with *Hindeodella snajdri. Age:* Late Wenlockian.

Note: This is the most southwesterly exposure available for the Klinteberg Beds except for Karlsöarna.

HAGANÄS 1. Slite Beds, Slite Marl (the part which is contemporaneous with unit f) and unit g. The lowermost exposed Slite Marl contains *Kockelella walliseri. Age:* Early Wenlockian.

References: Laufeld 1974a, b*; Larsson 1979; Bergman 1981, 1982; Sundquist 1981, 1982b.

HÄGVIDE 1. Hemse Beds, Hemse Marl SE part. *Polygnathoides siluricus* Zone. *Age*: Ludlovian; Leintwardinian.
 References: Harper 1969:188 (USNM Loc. 10026) *not* Fig. 3; Laufeld 1974a, b*; Larsson 1979.

HÄLLAGRUND 1. Högklint Beds, unit c and Slite Beds, unit c. *Age*: Early Wenlockian.
 References: Laufeld 1974a, b*.

HALLBRO SLOTT 6. 63860 16417 (CJ 3104 8530), ca 4500 m W and slightly S of Västerhejde church. Topographical map sheet 6I Visby NO. Geological map sheet Aa 183 Visby & Lummelunda.
 Section 300 metres E of Allhagemyr. For a measured section, see Hedström 1910, Pl. 55.
 Reference level: The lower boundary of Hedström's (1910) 'Leperditia shale', about 1.8 m thick, and consisting of marly limestone and marl.
 Slite Beds, units c, e, and g. *Samples*: G 73–12, 2.55–2.45 m below the reference level, strata with *Kockelella walliseri*; G 73–13, 0.05–0.15 m above the reference level; G 73–14, 0.80–0.90 m above the reference level; G 73–15, 2.50 m above the reference level. *Age*: Early Wenlockian.
 Reference: Hedström 1910, Fig. 5, Pl. 55.
 Note: To avoid confusion, I recommend that the localities in the Hallbro Slott area numbered by Hedström 1910p. 1479, should keep their numbers; therefore this locality is numbered 6 even though it is the first Hallbro Slott locality to be described in the formal system.

HÄLLINGE 2. Klinteberg Beds, unit a. Beds questionably with *Hindeodella snajdri*. *Age*: Late Wenlockian.
 References: Laufeld 1974a, b*.

HALLSARVE 1. Hemse Beds, Hemse Marl uppermost part, and Eke Beds, lowermost part. The Hemse Beds probably contain the *Pelekysgnathus dubius* fauna. However, *P. dubius* itself remains to be found. *Age*: Ludlovian; Late Leintwardinian.
 References: Laufeld 1974a, b*; Jeppsson 1974:10; Larsson 1979; Cherns 1982.

HAUGKLINTAR 2. Klinteberg Beds, lower part. *Age*: Late Wenlockian.
 References: Laufeld 1974a, b*.

HERRGÅRDSKLINT 1. Hemse Beds, unit e. Beds probably with the *Pelekysgnathus dubius* fauna. *Age*: Ludlovian; Late Leintwardinian.
 References: Laufeld 1974a, b*; Larsson 1979.

HOBURGEN 2. Burgsvik Beds, top part and Hamra Beds, unit a. *Age*: Ludlovian, Whitcliffian.
 References: Laufeld 1974a, b*; Jeppsson 1974:13; Larsson 1979; Claesson 1979; Laufeld & Martinsson 1981; Jeppsson 1982.

HUNNINGE 1. Klinteberg Beds, lower-middle part. Beds with *Hindeodella snajdri*. *Age*: Late Wenlockian.
 References; Laufeld 1974a, b*; Larsson 1979.

JUVES 2, 3, 4, 5. Hamra Beds unit c and Sundre Beds, lower part. Beds with *Hindeodella snajdri crispa*. *Age*: Latest Ludlovian.
 References: Jeppsson 1972:59, 60; Barnes *et al.* 1973; Laufeld 1974a, b*; Jeppsson 1974:11, 45, 79; Laufeld & Jeppsson 1976*; Larsson 1979; Claesson 1979; Jeppsson 1982.

KÄLLDAR 2. 635145 165412 (CJ 4095 5172), ca 1150 m W of Linde church. Topographical map sheet 6J Roma SV. Geological map sheet Aa 164 Hemse.
 About 1.5 m high, north-facing section in the abandoned quarry just N of Kälder.
 Hemse Beds, lower–middle part. Beds with *Kockelella variabilis*. *Age*: Ludlovian; Bringewoodian or possibly earliest Leintwardinian.
 References: Hede 1927b:33 line 3–12; Fåhraeus 1969:12.

KÄRNE 3. 644770 166630 (CJ 5266 4525), ca 3800 m ESE of Burs church. Topographical map sheet 5I Hoburgen NO & 5J Hemse NV. Geological map sheet Aa 156 Ronehamn.

Exposures (fresh in 1971) in a N-S running ditch 150 m SE Kärne (that Kärne which is on the south slope of Burgen). West of the ditch there is an E-W boundary between two fields. The contact is found 30 m north of this boundary.
 Hemse Beds, Hemse Marl uppermost part and *Eke Beds*, basal part. The Hemse Beds contain the *Pelekysgnathus dubius* fauna. *Age*: Ludlovian; Late Leintwardinian.

KATRINELUND 1. Slite Beds, unit c. Beds with *Kockelella walliseri*. *Age*: Wenlockian.
 References: Laufeld 1974a, b*.

KÄTTELVIKEN 5. Hamra Beds, unit b. Beds with *Hindeodella snajdri*. *Age*: Ludlovian; Late Whitcliffian or slightly younger.
 References: Laufeld & Jeppsson 1976*; Larsson 1979.

KATTHAMMARSVIK 1. Hemse Beds, unit a. *Age*: Early Ludlovian.
 References: Laufeld 1974a, b*; Laufeld & Martinsson 1981.

KLASÅRD 1. 633190 164580 (CJ 3105 3107), ca 3800 m SW of Näs church. Topographical map sheet 5I Hoburgen NO & 5J Hemse NV. Geological map sheet Aa 152 Burgsvik.
 Shore exposure on the bulge of the shore with three shrubby service-trees (*Sorbus*) at the end of the field road. Sample G 71-150 from 10 m N of these trees.
 Hemse Beds, Hemse Marl SE part. *Polygnathoides siluricus* Zone. *Age*: Ludlovian; Leintwardinian.
 References: Hede 1919: 20 line 24–p. 21 line 8, loc. 8; Hede 1942:20, loc. 2c.

KLINTEBERGET 1. Klinteberg Beds, lower-middle parts. Beds with *Hindeodella snajdri*. *Age*: Late Wenlockian.
 References: Laufeld 1974a, b*; Larsson 1979; Claesson 1979: Laufeld & Martinsson 1981; Jeppsson 1982.

KLINTEBYS 1. Slite Beds, Slite Siltstone (top) and Halla Beds. Both Beds contain *Hindeodella snajdri Age*: Late Wenlockian.
 References: Laufeld 1974a, b*; Sivhed 1976:60; Larsson 1979; Bergman 1979, 1980; Laufeld & Martinsson 1981.

KLINTHAGEN 1. 641340 167870 (CK 7000 0975), ca 2700 m NNE of Lärbro church. Topographical map sheet 7J Fårösund SO o. NO. Geological map sheet Aa 171 Kappelshamn.
 Old quarry, ca 650 m SSE St. Vikers.
 Slite Beds, unit g. Beds with *Kockelella walliseri*. *Age*: Early Wenlockian.
 Reference: Hede 1933:43, lines 5–28.

KLOCKAREMYR 1. Tofta Beds. Beds with *Hindeodella sagitta rhenana*. *Age*: Early Wenlockian.
 References: Laufeld 1974a, b*; Larsson 1979.

KLUVSTAJN 1. Slite Beds, unit b. *Age*: Early Wenlockian.
 References: Laufeld 1974a, b*.

KULLANDS 1. Hemse Beds, Hemse Marl NW part. *Age*: Ludlovian; Bringewoodian or Early Leintwardinian.
 References: Laufeld 1974a, b*; Laufeld & Jeppsson 1976; Larsson 1979.

KUPPEN 1. Hemse Beds. The beds immediately above the reference level belong to unit e. A *Polygnathoides siluricus* Zone fauna occurs up to 0.11 m above the reference level, followed by beds with the *Pelekysgnathus dubius* fauna. *Age*: Ludlovian; Leintwardinian.
 References: Laufeld 1974a, b*; Larsson 1979; Jeppsson 1982.

KUPPEN 2. Hemse Beds. *Age*: Ludlovian; Leintwardinian.
 References: Laufeld 1974a, b*; Kershaw & Riding 1978, Figs. 5, 12; Laufeld & Martinsson 1981; Kershaw 1981; Jeppsson 1982.

LANGHAMMARSHAMMAR 1. Högklint Beds, lower–middle part. Beds with *Kockelella? ranuliformis*. *Age*: Early Wenlockian.
 References: Laufeld 1974a, b*.

LANSA 2. 642560 169423 (CK 8645 2075), ca 3000 m W of Fårö

church. Topographical map sheet 7J Fårösund SO & NO. Geological map sheet Aa 180 Fårö.

Ditch exposure 0–100 m S of the road, about 400 m SE the crossroads 650 m E of Lansa and 1100 m S of Marpes. Sample G 73–70 from 1 m S of the road.

Slite Beds, unit f, *Rhipidium tenuistriatum* beds. Beds with *Kockelella walliseri*. *Age*: Early Wenlockian.

Reference: Hede 1936, p. 28, line 30–p. 29 line 9.

LAUTERSHORNSVIK 2. Högklint Beds, the middle of the three Högklint units on Fårö described by Hede (1936); lower–middle part (Laufeld 1974b). Beds with *Hindeodella sagitta rhenana*. *Age*: Early Wenlockian.

References: Laufeld 1974a, b*.

LAUTERSHORNSVIK 3. Högklint Beds, the upper part of the middle of the three Högklint units on Fårö described by Hede (1936); lower–middle part (Larsson 1979). Beds with *Hindeodella sagitta rhenana*. *Age*: Early Wenlockian.

Reference: Larsson 1979*.

LERBERGET (I need to revisit the locality to be able to refer my locality to those described by Laufeld 1974b).

LEVIDE 3. 635246 164760 (CJ 3440 5143), ca 1200 m SW of Levide church. Topographical map sheet 6I Visby SO. Geological map sheet Aa 164 Hemse.

Exposure in the ditch along the western edge of the small forest, 47 m S of the private road that passes Hagalund, 350 m NE that farm.

Hemse Beds, Hemse Marl NW part. Beds with *Kockelella variabilis*. *Age*: Ludlovian; probably Bringewoodian.

LILLA HALLVARDS 1. Hemse Beds, Hemse Marl NW part. Beds with *Kockelella absidata*. *Age*: Early Ludlovian.

References: Laufeld 1974a, b*; Larsson 1979; Jaeger 1981; Laufeld & Martinsson 1981.

LJUGARN 1. Hemse Beds upper part, probably unit d, in that limestone there is a fissure filling of Hemse Marl, SE part (mapped as fine-grained limestone by Watkins 1975). The latter unit belongs to the *Polygnathoides siluricus* Zone. *Age*: Ludlovian; the fissure filling is Leintwardinian.

References: Laufeld 1974a, b*; Watkins 1975.

LOGGARVE 1. 636574 164813 (CJ 3597 6468), ca 2800 m NE of Klinte church. Topographical map sheet 6I Visby SO. Geological map sheet Aa 160 Klintehamn.

The most westerly roadside section south of the old road from Loggarve towards Hejde, about 400 m E of Loggarve, about 25 m E of the forest edge.

Reference level: The Mulde–Klinteberg boundary.

Mulde Beds, topmost part (about 0.1 m) and *Klinteberg Beds* (about 2.5 m), lowermost part. In the Mulde Beds occurs *Hindeodella snajdri*. *Age*: Late Wenlockian.

References: Hede 1927a:37 line 14 (to the area in general) and p. 38 lines 15–16 (to the area in general); Jeppsson 1982.

LUKSE 1. Hemse Beds, Hemse Marl NW part. *Age*: Ludlovian; probably Bringewoodian or Leintwardinian.

References: Laufeld 1974a, b*; Larsson 1979.

MARTILLE 7. 638472 164371 (CJ 3298 8390), ca 4150 m (W)NW of Stenkumla church. Topographical map sheet 6I Visby NO. Geological map sheet Aa 183 Visby & Lummelunda.

20 m E of road 140 and 20 m N of the private road to Martille; the bedrock occurs 0.5 m below the surface.

Slite Beds, unit d. Beds with *Hindeodella sagitta rhenana*. *Age*: Early Wenlockian.

Reference: Hede 1940:49, lines 9–10.

MILLKLINT 1. Hemse Beds, unit e, i.e. Millklint Limestone. *Polygnathoides siluricus* Zone. *Age*: Ludlovian; Leintwardinian.

References: Laufeld 1974b*; Jeppsson 1969, Figs. 1A–F, 2A–F, 1972, Fig. 1A(tr), B, Pl. 1:25–30, 1974; Larsson 1979.

MILLKLINT 3. Hemse Beds, unit e, i.e. Millklint Limestone. *Polygnathoides siluricus* Zone. *Age*: Ludlovian; Leintwardinian.

References: Laufeld 1974a,b*; Jeppsson 1974:7; Laufeld & Jeppsson 1976.

MILLKLINTDALEN 2. Hemse Beds, unit e, i.e. Millklint Limestone. *Age*: Ludlovian, Late Leintwardinian.

References: Laufeld 1974a, b*; Laufeld & Jeppsson 1976.

MÖLLBOS 1. Halla Beds, unit b. Beds with *Hindeodella snajdri*. *Age*: Late Wenlockian.

References: Laufeld 1974a, b*; Laufeld & Jeppsson 1976; Larsson 1979; Claesson 1979; Laufeld & Martinsson 1981; Stridsberg 1981a, b; Liljedahl 1981, 1983.

MÖLNER 1. Mulde Beds, upper part. Strata with *Hindeodella snajdri*. *Age*: Late Wenlockian.

References: Laufeld 1974a, b*; Larsson 1979.

NYGÅRDSBÄCKPROFILEN 1. 638880 164417 (CJ 3374 8794), ca 2970 m NW of Västerhejde church. Topographical map sheet 6I Visby NO. Geological map sheet Aa 183 Visby & Lummelunda.

Brook section at the mouth of Nygårdsbäcken (bäcken = the brook) and shore section SW of it. My samples are taken 5 to 10 m from the ravine.

Lower Visby Beds, upper part (at least 2.3 m exposed), and *Upper Visby Beds*. *Age*: Earliest Wenlockian.

References: Hedström 1910:1463, 1474; Hede 1940:13; Jeppsson 1982.

NYMÅNETORP 1. Högklint Beds, unit b, upper part. Beds with *Hindeodella sagitta rhenana*. *Age*: Early Wenlockian.

References: Laufeld 1974a, b*; Laufeld & Martinsson 1981:6.

OIVIDE 1. Slite Beds, unit f, i.e. *Rhipidium tenuistriatum* Beds. Strata with *Kockelella walliseri*. *Age*: Early Wenlockian.

References: Laufeld 1974a, b*; Larsson 1979; Jeppsson 1982.

ÖSTERGARNSHOLM 1. Hemse Beds, unit d?. *Age*: Ludlovian; Leintwardinian.

References: Laufeld 1974a, b*; Larsson 1979.

ÖSTERGARNSHOLM 2. Hemse Beds, unit d?. *Age*: Ludlovian; Leintwardinian.

References: Laufeld 1974a, b*; Larsson 1979.

ÖSTERGARNSHOLM 3. Hemse Beds, unit d?. *Age*: Ludlovian; Leintwardinian.

The locality will be described by Björn Sundquist.

ÖVERBURGE 1. 635215 165146 (CJ 3820 5085), ca 1680 m NW of Fardhem church. Topographical map sheet 6J Roma SV. Geological map sheet Aa 164 Hemse.

Exposure on the eastern side of the ditch, 10 m S of the ditch and road intersection 300 m NW Överburge.

Hemse Beds, Hemse Marl NW part. Beds with *Kockelella variabilis*. *Age*: Ludlovian; Bringewoodian or Leintwardinian.

RANGSARVE 1. Hemse Beds, upper part. Beds with *Polygnathoides siluricus*. *Age*: Ludlovian; Leintwardinian.

References: Laufeld 1974a, b*; Laufeld & Jeppson 1976; Larsson 1979; Claesson 1979; Laufeld & Martinsson 1981.

RONEHAMN 3. 634152 166181 (CJ 478 393), ca 4800 m SSE Ronehamn church. Topographical map sheet 5I Hoburgen NO & 5J Hemse NV. Geological map sheet Aa 156 Ronehamn.

Excavation 500 m WSW the harbour of Ronehamn.

Eke Beds, uppermost part. *Age*: Ludlovian; Whitcliffian.

Reference: Jeppson 1974:11, 12, 13.

RONNINGS 1. Eke Beds, upper part. *Age*: Ludlovian; Whitcliffian.

References: Laufeld 1974a, b*; Larsson 1979: Jeppsson 1982.

RÖNNKLINT 1. 641175 165700 (CK 4828 0984), ca 3750 m N (and slightly W) of Lummelunda church. Topographical map sheet 7J Fårösund SV & NV. Geological map sheet Aa 183 Visby & Lummelunda.

Cliff section in the channel down the cliff face just south of the reef that forms Rönnklint.

Reference level: The bentonite level about 6 m above base of section (= about 8 m above sealevel). There is another major bentonite 2.50 m higher up, and a thick limestone bed 3.5 m further up.

Lower Visby Beds, Upper Visby Beds, and *Högklint Beds*. The main part of the Lower Visby Beds contains the *Pterospathognathus amorphognathoides* Zone. *Age*: The age of the strata exposed is Llandoverian, late Telychian (C6), and Early Wenlockian. The Lower Visby Beds span the boundary between the Llandoverian and the Wenlockian.

References: Hede 1940:13, lines 16–17; Brood 1982:18–19, 24–25.

SIGVALDE 2. 636070 166462 (CJ 5195 5833), ca 2680 m ENE of Etelhem church. Topographical map sheet 6J Roma SV. Geological map sheet AA 156 Ronehamn.

Inland cliff section ca 470 m E of the eastern end of Sigvalde träsk and just south of the road. For a photograph of the locality, see Munthe 1910, Fig. 26 or Hede 1925b, Fig. 10.

Hemse Beds, lower-middle part (probably unit c and perhaps both that and unit d). *Age*: Ludlovian, probably early.

Reference level: The upper limit of the *Ilionia* limestone as drawn in Munthe 1910, Fig. 26 (= Hede 1925b, Fig. 10).

Reference point: The fissure visible 3 cm from the left margin of the photograph in Munthe 1910, Fig. 26 and Hede 1925b, Fig. 10.

References: Munthe 1910:1432, Fig. 26; Hede 1925b:16, line 21–31, Fig. 10; Martinsson 1962:56, the second locality mentioned under Sigvalde.

SIMUNDS 1. Slite Beds, unit g. The lowermost beds contain *Kockelella walliseri*. *Age*: Early Wenlockian.

Reference: Laufeld & Martinsson 1981:6*.

SION 1. Slite Beds, unit f, i.e. *Rhipidium tenuistriatum* Beds. Beds with *Kockelella walliseri*. *Age*: Early Wenlockian.

References: Laufeld 1974a, b*.

SLITEBROTTET 1. Slite Beds, Slite Marl and Slite g. *Age*: Wenlockian.

References: Walmsley 1965:469, Pl. 62:23–27, 33–35; Laufeld 1974a, b*; Walmsley & Boucot 1975:65, Pl. 3:9–11; Eisenack 1975, Fig. 18; Larsson 1979.

SLITEBROTTET 2. Slite Beds, Slite Marl, i.a. the parts corresponding to the Slite f. Beds with *Kockelella walliseri*. *Age*: Early Wenlockian.

References: Laufeld 1974a, b*; Claesson 1979.

SNAUVALDS 1. Hemse Beds, Hemse Marl NW part. *Age*: Ludlovian; Bringewoodian or, more probably, Early Leintwardinian.

References: Laufeld 1974a, b*; Larsson 1979.

SNODER 3. 634805 164490 (CJ 3135 4725), ca 2500 m NO of Silte church. Topographical map sheet 5I Hoburgen NO & 5J Hemse NV. Geological map sheet 164 Hemse.

Low (about 1.5 m) section from the bottom of a small ditch down to below the water level in the Snoder-a drainage ditch, 5 m S of the ditch and road intersection, on the private road between Snausarve and Snoder.

Hemse Beds, Hemse Marl NW. Beds with *Kockelella variabilis*? *Age*: Ludlovian; Bringewoodian or Early Leintwardinian.

References: Hede 1927b:26, line 29–p. 27, line 5 (list of over 50 taxa of macrofossils and ostracodes based on excavated material from this locality and others); Mori 1970:23, loc. 131 (excavated material from the ditch); Jeppsson 1982.

SOJVIDE 1. Slite Beds, unit f, i.e. *Rhipidium tenuistriatum* Beds. Beds with *Kockelella walliseri*. *Age*: Early Wenlockian.

Reference: Larsson 1979*.

STORA MYRE 1. Slite Beds, unit d. Beds with *Kockelella walliseri*? *Age*: Early Wenlockian.

References: Laufeld 1974a, b*; Larsson 1979.

STORBURG 1. Hamra and Sundre Beds. *Age*: Latest Ludlovian.

References: Laufeld 1974a, b*; Larsson 1979.

STRANDS 1. Hamra Beds, unit b. Beds with *Hindeodella snajdri crispa*. *Age*: Latest Ludlovian.

References: Laufeld 1974a, b*; Larsson 1979.

SUTARVE 2. Klinteberg Beds, unit f, top. *Age*: Early Ludlovian.

References: Laufeld 1974a, b*; Laufeld & Jeppsson 1976.

SVARVEN 1. Högklint Beds, unit b. Beds with *Hindeodella sagitta rhenana* and *Kockelella? ranuliformis*. *Age*: Early Wenlockian.

References: Laufeld 1974a, b*; Larsson 1979.

SYSNE 1. Hemse Beds, unit d? *Age*: Ludlovian; Late Leintwardinian.

Reference: Larsson 1979*.

TJELDERSHOLM 1. Slite Beds, *Pentamerus gothlandicus* Beds, and immediately younger beds (but not *Atrypa reticularis* Beds). *Age*: Late? Wenlockian.

References: Laufeld 1974a, b*; Larsson 1979.

TUTEN 1. 635702 167280 (CJ 5962 5408), ca 3900 m NE of Lau church. Topographical map sheet 6J Roma SV. Geological map sheet Aa 156 Ronehamn.

Blocks on the 'piers' of the 'harbour' at Tuten, excavated when it was deepened.

Hemse Beds, Hemse Marl, SE part. *Polygnathoides siluricus* Zone. *Age*: Ludlovian; Leintwardinian.

UDDVIDE 2. Burgsvik Beds, upper part and Hamra Beds, basal part. *Age*: Ludlovian; Whitcliffian.

References: Laufeld 1974a, b*; Jeppsson 1974:13; Larsson 1975:129.

URGUDE 2. 635021 164333 (CJ 2995 4943), ca 1800 m W of Sproge church. Topographical map sheet 6I Visby SO. Geological map sheet Aa 164 Hemse.

Ditch exposure, 47 m west of the road that runs NNW to St. Norrgårde, in the ditch that runs east-west south of Tjängdarve.

Hemse Beds, Hemse Marl NW part. *Age*: Early Ludlovian.

VAKTÅRD 3. 633370 164594 (CJ 3130 3285), ca 2900 m WSWW of Näs church. Topographical map sheet 5I Hoburgen NO & 5J Hemse NV. Geological map sheet Aa 152 Burgsvik.

Shore exposure about 100 metres N of the pier. There are four small former fishing houses at Vaktård, the two middle ones close to each other near the pier. The northern one looks very old. North and northwest of it there are two very small bulges on the shoreline and north of these a very small point. Sample G 71–148 is from the central of these three protusions.

Hemse Beds, Hemse Marl, SE part. *Polygnathoides siluricus* Zone. *Age*: Ludlovian; Leintwardinian.

References: Hede 1919:19 line 11–p. 20 line 23, loc. 7; Hede 1942:20, loc. 3c.

VALE 1. 641185 166046 (CK 5168 0967), ca 1975 m NW of Stenkyrka church. Topographical map sheet 7J Fårösund SV & NV. Geological map sheet Aa 183 Visby & Lummelunda.

Temporary exposure, immediately south of the road from road 149 to Vale, about 700 m from road 149, and slightly closer to the western than the eastern end of the straight part of the road.

Tofta Beds, probably lower part. *Age*: Early Wenlockian.

VALLMYR 1. Klinteberg Beds, unit d. Beds with *Hindeodella snajdri*. *Age*: Wenlockian, very close to the end.

Reference: Larsson 1979*.

VÄSTÖS KLINT 1. Högklint Beds, units b and c. Beds with *Kockelella? ranuliformis* and *Hindeodella sagitta rhenana*. *Age*: Early Wenlockian.

References: Laufeld 1974a, b*; Laufeld & Jeppsson 1976; Larsson 1979.

VATTENFALLSPROFILEN 1 (VATTENFALLET). Lower and Upper Visby Beds, and Högklint Beds. Beds with *Pterospathodus pennatus procerus*, *Kockelella? ranuliformis* and *Hindeodella sagitta rhenana*. *Age*: Early Wenlockian.

References: Hedström 1904:93, line 11–p. 96, line 17, 1923b:195, Fig. 2; Hede 1925:15, line 18 from below–p. 16, line 3; Martinsson 1972:128–129; Laufeld 1974b*; Bassett & Cocks 1974:5; Laufeld & Jeppsson 1976; Franzén 1977:223, 226; Larsson 1979; 43 papers *in* Jaanusson, Laufeld & Skoglund 1979 (eds.); Claesson 1979; Bengtson 1981; Jeppsson 1982; Brood 1982:48, Pl. 9:3.

New developments in conodont biostratigraphy of the Silurian of China

LIN BAO-YU

FOSSILS AND STRATA

ECOS III

A contribution to the Third European
Conodont Symposium, Lund, 1982

Lin Bao-yu 1983 12 15: New developments in conodont biostratigraphy of the Silurian of China. *Fossils and Strata*, No. 15, pp. 145–147. Oslo. ISSN 0300-9491. ISBN 82-0006737-8.

Abundant Silurian conodont faunas occur in China and, based on their stratigraphic distribution, ten zones can be recognized. In ascending stratigraphic order, the zones are: (1) *Spathognathodus obesus* Assemblage Zone; (2) *Spathognathodus parahassi − S. guizhouensis* Assemblage Zone; (3) *Pterosphathodus celloni* Zone; (4) *Pterospathodus amorphognathoides* Zone; (5) *Spathognathodus sagitta bohemicus* Zone; (6) *Kockelella variabilis* Assemblage Zone; (7) *Ancoradella ploeckensis* Zone; (8) *Polygnathoides siluricus* Zone; (9) *Spathognathodus crispus* Zone; (10) *Ozarkodina remscheidensis eosteinhornensis* Zone. Zones 1–3 are of Llandovery age, zones 4 and 5 are of Wenlock age, zones 6–9 are of Ludlow age and zone 10 is of Pridoli age. The Silurian conodont faunas of China are most closely related to those of Europe, and both may belong to an Atlantic biostratigraphic province. □*Conodonta, biostratigraphy, Silurian, China.*

Lin Bao-yu, Institute of Geology, Chinese Academy of Geological Sciences, Beijing, China; 21st May, 1982.

It has been shown that the importance of conodonts in the subdivision of Silurian strata is second only to that of graptolites. Conodonts are particularly important in the subdivision and correlation of Silurian strata of the shelly facies. The work of Walliser (1964, 1971), Aldridge (1972, 1975), Aldridge *et al.* (1979), and Nicoll & Rexroad (1969) on biostratigraphy of Silurian conodonts in Europe and North America has provided significant conodont zonations for these areas.

Silurian deposits, especially those of the shelly facies, are widely distributed in China (Lin 1979), thus providing excellent conditions for the study of Silurian conodonts. Research on Silurian conodonts in China has just been initiated and has thus lagged ten to fifteen years behind work in western Europe and North America. It has, however, developed rapidly in the last few years and research has now been carried out in Yunnan, Guizhou and Tibet by Wang (1980), Zhou *et al.* (1981), Qiu (1982) and Lin & Qiu (1983), respectively. A preliminary sequence of Silurian conodont zones has been established after comprehensive study of the data thus obtained. The conodont sequence will play a very important role in the subdivision of Silurian strata of the shelly facies in China. Based on the presently available data, the Silurian conodont sequence in China can be subdivided into ten zones. The main faunal elements and geographic occurrence of each of these zones are described below, in ascending stratigraphic order.

(1) *Spathognathodus obesus* Assemblage Zone. This zone is recognized in the Lower Silurian Xiangshuyan Formation in northeast Guizhou Province, China (Zhou *et al.* 1981). The major faunal elements are: *Spathognathodus obesus* Zhou, Zhai & Xian, *Paltodus* aff. *P. migratus* Rexroad, *Cyrtoniodus* sp., *Aphelognathus* sp., *Acodus curvatus* Branson & Branson and *Panderodus unicostatus* (Branson & Mehl).

(2) *Spathognathodus parahassi – S. guizhouensis* Assemblage Zone. This can be subdivided into two subzones. The lower subzone is called the *Spathognathodus parahassi* Subzone and it has been recovered from the Baisha Formation and the Rongxi Formation of Guizhou Province (Zhou *et al.* 1981). The major faunal elements are: *Spathognathodus parahassi* Zhou, Zhai & Xian, *Exochognathus brassfieldensis* (Branson & Branson), *Ozarkodina edithae* Walliser, *Microcoelodus egregius* (Walliser), and *Panderodus* sp.

The upper subzone is called the *Spathognathodus guizhouensis* Subzone and it is recognized in the lower part of the Xiushan Formation of Guizhou Province (Zhou *et al.* 1981). The main faunal elements are: *Spathognathodus guizhouensis* Zhou, Zhai & Xian, *Hadrognathus* cf. *H. staurognathoides* Walliser, *Exochognathus caudatus* (Walliser), *E. brassfieldensis* (Branson & Branson), *Ozarkodina hanoverensis* Nicoll & Rexroad, and *Hibbardella shiqianensis* Zhou, Zhai & Xian.

(3) *Pterospathodus celloni* Zone. Elements of this zone have been recovered from the upper part of the Xiushan Formation in Guizhou Province (Zhou *et al.* 1981) and the upper part of the Dewukaxia Formation in Xianza region, Tibet (Qiu 1982). The major elements are: *Dapsilodus obliquicostatus* (Branson & Mehl), *Ozarkodina excavata excavata* (Branson & Mehl), *Panderodus serratus* Rexroad, *Spathognathodus hassi* Pollock, Rexroad & Nicoll, *P. celloni* (Walliser), and *Apsidognathus tuberculatus* Walliser. In Guizhou, *P. celloni* is associated with *Sichuanoceras* suggesting a Wenlock age, whereas it seems to occur at a lower level (without *Sichuanoceras*) in the Xianza region, Tibet; in Malaysia it is associated with *P. amorphognathoides* (Igo & Koike 1968). These data suggest that the upper boundary of *P. celloni* range may be diachronous.

(4) *Pterospathodus amorphognathoides* Zone. This is known from the base of the Kede Formation in Pazhuo district, Tingri

	Western Europe (Walliser 1971)			Britain Aldridge et al. 1980		N. America Rexroad & Nicoll 1971	China This paper
	Graptol.-zones (Elles et Wood)	Conodont zones	Conodont stufen				
Gedin.	*M. uniformis*	*woschmidti*				*woschmidti*	*woschmidti*
Pridoli	*M. transgrediens* / *M. bouceki* / *M. ultimus*	*eosteinhorn-ensis*	*steinhornensis*	D	*O. r. eostein-hornensis*	*S. r. eostein-hornensis*	*O. r. eostein-hornensis*
U. Ludlow	36 *M. fritschi linearis*	*crispus*	*crisp.*	W		*S. snajdri*	*S. crispus*
		latialatus	*latial.*	L			
M. Lud.	33 *M. chimaera*	*siluricus*	*Kockelella*	B	*K. variabilis*	*Kockelella*	*P. siluricus*
L. Lud.	32 *M. nilssoni*	*ploeckensis*					*A. ploeckensis*
		crassa		E			*K. variabilis*
Wenlock	31	*sagitta*		H	*S. sagitta bohemicus*	*S. sagitta*	*S. sagitta bohemicus*
		patula				*P. amorphogna-thoides*	
	26	*amorphogna-thoides*	*Apsidognathus*	S	*P. amorpho-gnathoides*	*S. ranuliformis* A.Z.	*P. amorpho-gnathoides*
	25	*celloni*		T	*P. celloni*	*N. celloni* A.Z.	*P. celloni*
Llando-very				F	*H. stauro-gnathoides*		*S. parahassi-* *S. guizhou-ensis* A.Z.
				I	*I. discreta-* *I. deflecta*	*I. irregularis* A.Z.	*S. obesus* A.Z.
	16			*R			
Ord.							

Fig. 1. Correlation of Silurian conodont zones in China with those of other regions. *R-Rhuddanian, I-Idwian, F-Fronian, T-Telychian, S-Sheinwoodian, H-Homerian, E-Eltonian, B-Bringewoodian, L-Leintwardinian, W-Whitcliffian, D-Downtonian.

County, Tibet (Lin & Qiu 1983) and from the lower part of the Renheqiao Formation in the western part of Yunnan (Ni *et al.*1982). The major elements are: *Pterospathodus amorpho-gnathoides* Walliser, *Spathognathodus pennatus procerus* Walliser, and *Panderodus gracilis* (Branson & Mehl). In the western part of Yunnan, *S. pennatus procerus* is associated with *Monograptus flexilis* (Wang & Wang 1981). Wang & Wang (1981) believe that this zone is younger in China than it is in Europe and North America, possibly corresponding to graptolite zone 29 of the British sequence.

(5) *Spathognathodus sagitta bohemicus* Zone. At present, this species is only known from the middle and upper parts of the Kede Formation in Tingri County, Tibet (Lin & Qiu 1983).

(6) *Kockelella variabilis* Assemblage Zone. It is known from the Zhongcao Formation in the lower part of the Upper Silurian in Shidian, Yunnan Province (Ni *et al.* 1982) and from the Keya Formation in Nyalam County, Tibet (Lin & Qiu 1983). Typical faunal elements of the zone include: *Ozarkodina excavata excavata* (Branson & Mehl), *Panderodus liratus* Nowlan & Barnes, *P. gracilis* (Branson & Mehl), *Dapsilodus obliquicosta-tus* (Branson & Mehl) and *Kockelella variabilis* Walliser. *K. variabilis* is generally limited to the interval from the late Silurian *O. crassa* through *P. siluricus* zones.

(7) *Ancoradella ploeckensis* Zone. It has only been found in the

Keya Formation in the lower part of the Upper Silurian in Nyalam County, Tibet (Lin & Qiu 1983).

(8) *Polygnathoides siluricus* Zone. It is known only from the Keya Formation in Nyalam County, Tibet (Lin & Qiu 1983).

(9) *Spathognathodus crispus* Zone. It was first recognized in the upper part of the Silurian Bailongjiang Formation in Tewo County, Gansu Province (Wang & Wang 1981). It is now also known from the Miaogao formation in Qujing, Yunnan Province (Wang 1980). The main faunal elements of the zone are: *Spathognathodus crispus* Walliser, *Hindeodella priscilla* Stauf-fer. This zone has recently been reported from the Yulongsi Formation (Wang 1981).

(10) *Ozarkodina remscheidensis eosteinhornensis* Zone. The species assigned to this zone have been found in the *Camarocrinus asiaticus* beds in the upper part of the Renheqiao Formation in the west of Yunnan Province (Ni *et al.* 1982). The main components of the zone are: *Ozarkodina excavata excavata* (Branson & Mehl), *O. remscheidensis eosteinhornensis* (Walliser). The taxa assigned to this zone have also been recovered from the Pazhuo Formation in Tingri County, Tibet (Lin & Qiu 1983) and some elements have been found at the top of the Gayang Formation (Lin & Qiu 1983). In Tibet the main faunal constituents are: *Ozarkodina remscheidensis eosteinhornensis* (Walliser), *O. confluens* Branson & Mehl and *O. excavata*

excavata (Branson & Mehl). It is interesting to note that the mainly early Devonian *Icriodus woschmidti* Ziegler is also recovered with this fauna. This species is known from the latest Silurian but not in co-occurrence with *O. remscheidensis eosteinhornensis* (Walliser), therefore its recovery here raises some problems in the placement of the Silurian–Devonian boundary in the Pazhuo Formation.

The correlation of the Silurian conodont zonal sequence in China with that in Europe and North America is shown in Fig. 1. Based on data currently available it is apparent that Silurian conodont faunas of China are most similar to those of the Carnic Alps in Europe and differ considerably from those of North America. Thus China and Europe may belong to the same biogeographic province.

Acknowledgements. – The writer is indebted to Dr. Godfrey S. Nowlan, Geological Survey of Canada, Drs. Lennart Jeppsson and Anita Löfgren, University of Lund, and Dr. Carl B. Rexroad, Indiana Geol. Survey, for their careful reading and correcting of the English version of the manuscript and for many valuable instructions.

References

Aldridge, R. J. 1972: Llandovery conodonts from the Welsh Borderland. *Bulletin of the British Museum (Natural History) Geology 22*, 125–231.

Aldridge, R. J. 1975: The stratigraphic distribution of conodonts in the British Silurian. *Quarterly Journal of the Geological Society of London 131*, 607–618.

Aldridge, R. J., Dorning, K. J., Hill, P. J., Richardson, J. B. & Siveter, D. J. 1979: Microfossil distribution in the Silurian of Britain and Ireland. *In* Harris, A. L., Holland, C. H. & Leake, B. E. (eds.): *The Caledonides of the British Isles – reviewed.* Geological Society of London.

Barrick, J. E. 1977: Multielement simple-cone conodonts from the Clarita Formation (Silurian), Arbuckle Mountains, Oklahoma, *Geologica et Palaeontologica 11*, 47–68.

Barrick, J. E. & Klapper, G. 1976: Multielement Silurian (Late Llandoverian–Wenlockian) conodonts of the Clarita Formation, Arbuckle Mountains, Oklahoma, and phylogeny of *Kockelella.* *Geologica et Palaeontologica 10*, 59–100.

Cooper, B. J. 1975: Multielement conodonts from the Brassfield Limestone (Silurian) of southern Ohio. *Journal of Paleontology 49*, 984–1008.

Cooper, B. J. 1976: Multielement conodonts from the St. Clair Limestone (Silurian) of southern Illinois. *Journal of Paleontology 50*, 205–217.

Cooper, B. J. 1977: Toward a familial classification of Silurian conodonts. *Journal of Paleontology 51*, 1051–1071.

Cooper, B. J. 1980: Toward an improved Silurian conodont biostratigraphy. *Lethaia 13*, 209–227.

Igo, H. & Koike, T. 1968: Ordovician and Silurian conodonts from the Langkawi Islands, Malaya, Part II. *Geol. Palaeont. Southeast Asia 4*, 1–21.

Jeppsson, L. 1969: Notes on some Upper Silurian multielement conodonts. *Geologiska Föreningens i Stockholm Förhandlingar 91*, 12–24.

Jeppsson, L. 1972: Some Silurian conodont apparatuses and possible conodont dimorphism. *Geologica et Palaeontologica 6*, 51–59.

Jeppsson, L. 1974: Aspects of Late Silurian conodonts. *Fossils and Strata 6*, 1–54.

Lin Bao-yu 1979: [The Silurian system of China.] *Acta Geologica Sinica 3*, 171–191. (In Chinese with English summary.)

Lin Bao-yu & Qiu Hong-rong 1983: [The Silurian System in Xizang (Tibet).] *Contribution to the Geology of the Qinghai-Xizang (Tibet) Plateau 8*, 15–28. (In Chinese with English summary.)

McCracken, A. D. & Barnes, C. R. 1981: Conodont biostratigraphy and paleoecology of the Ellis Bay Formation, Anticosti Island, Quebec, with special reference to Late Ordovician – Early Silurian chronostratigraphy and the system boundary. *Geological Survey of Canada, Bulletin 329*, 51–134.

Nicoll, R. S. & Rexroad, C. B. 1968 (1969): Stratigraphy and conodont paleontology of the Salamonie Dolomite and Lee Creek Member of the Brassfield Limestone (Silurian) in Southeastern Indiana and adjacent Kentucky. *Indiana Geological Survey Bulletin 40*, 73 pp.

Ni Yu-nan, Chen Ting-en, Cai Chong-yang, Li Guo-hua, Duan Yan-xue & Wang Ju-de 1982: [The Silurian rocks in western Yunnan.] *Acta Palaeontologica Sinica 21:1*, 119–132 (In Chinese.)

Ni Shi-zhao 1980: Conodonts from the Lojoping Formation (Silurian) at Lojopin, Yichang, Western Hubei Province, China (Abstract). *Geological Society of America, Abstracts with Programs 12:5*.

Nowlan, G. S. & Barnes, C. R. 1981: Late Ordovician conodonts from the Vauréal Formation, Anticosti Island, Quebec. *Geological Survey of Canada, Bulletin 329*, 1–49.

Qiu Hong-rong 1982: The Silurian conodonts from Xizang (Tibet) (Abstract). *Recueil d'articles colloque Franco-Chinois sur la geologie de L'Himalaya, Quilin, Chine*, 15. (In English.)

Rexroad, C. B. & Nicoll, R. S. 1971: Summary of conodont biostratigraphy of the Silurian System of North America. *In* Sweet, W. C. & Bergström, S. M. (eds.): Symposium on conodont biostratigraphy. *Geological Society of America Memoir 127*, 207–227.

Walliser, O. H. 1964: Conodonten des Silurs. *Abhandlungen des Hessischen Landesamtes für Bodenforschung zu Wiesbaden 41*. 106 pp.

Walliser, O. H. 1971: Conodont biostratigraphy of the Silurian of Europe. *In* Sweet, W. C. & Bergström, S. M. (eds.): Symposium on conodont biostratigraphy. *Geological Society of America Memoir 127*, 195–206.

Wang Cheng-yuan 1980: [Upper Silurian conodonts from Qujing District, Yunnan.] *Acta Paleontologica Sinica 19:5*, 369–379. (In Chinese with English summary.)

Wang Cheng-yuan 1981: [New observation on the age of the Yulongsi Formation of Qujing, Yunnan.] *Journ. Stratigr. 5:3*, 240. (In Chinese.)

Wang Cheng-yuan & Wang Zhi-hao 1981: [Conodont sequence in China (Cambrian–Triassic System).] *12th Annual Conference, Palaeontological Society of China, Selected Papers* 105–115. (In Chinese.)

Zhou Xi-yun, Zhai Zhi-qiang & Xian Si-yuan 1981: [On the Silurian conodont biostratigraphy, new genera and species in Guizhou Province.] *Oil & Gas Geology 2:2*, 123–140. (In Chinese with English summary.)

Paedomorphosis in the conodont family Icriodontidae and the evolution of *Icriodus*

THOMAS W. BROADHEAD AND RONALD McCOMB

FOSSILS AND STRATA

ECOS III

A contribution to the Third European Conodont Symposium, Lund, 1982

Broadhead, Thomas W. & McComb, Ronald 1983 12 15: Paedomorphosis in the conodont family Icriodontidae and the evolution of *Icriodus*. Fossils and Strata, No. 15, pp. 149–154. Oslo ISSN 0300-9491. ISBN 82-0006737-8.

The genus *Icriodus* probably evolved from *Pedavis* by paedomorphic loss of anterior lateral processes through neoteny. Loss of the posterior spur characteristic of the *I. latericrescens* group also was a probable neotenic process that led to most species of Middle and Late Devonian *Icriodus*, although similarity between Early Devonian species lacking a posterior spur and coeval species of *Pelekysgnathus* does not eliminate the possibility of a polyphyletic origin for *Icriodus*. Two major increases in species diversity occurred, in the Early Devonian for *I. latericrescens* group species with posterior spurs, and in the late Early Devonian or early Middle Devonian for those lacking posterior spurs. □*Conodonta*, Icriodus, Pedavis, *evolution, paedomorphosis, Lower Devonian.*

Thomas W. Broadhead, Department of Geological Sciences, University of Tennessee, Knoxville, Tennessee 37996-1410, USA; Ronald McComb, Exxon Company USA, Houston, Texas 77001, USA; 15th October, 1982.

The genus *Icriodus* appears suddenly in the stratigraphic record (latest Silurian, Pridoli), and the initial species, *I. woschmidti*, rapidly achieved virtual worldwide distribution. The ease in correlating the *I. woschmidti* Zone on an intercontinental scale is biostratigraphically important because its base serves as an approximate correlative of the Silurian–Devonian boundary in many stratigraphic sequences that are devoid of graptolites.

Little attention has been devoted previously to examination of evolutionary trends and processes in conodonts excepting the recognition of presumed ancestor–descendant relationships. Many biostratigraphically important species such as *I. woschmidti* appear without obvious ancestral forms and have been regarded as cryptogens. Ancestor–descendant species or genera have been recognized by relatively high degrees of morphologic similarity in both apparatus composition and individual rapidly evolving elements (usually Pa and Pb). Sudden morphologic shifts that produced innovative forms must be evaluated both in terms of presumed closely related forms (reflected in the taxonomy) and in the context of ontogenetic development of both possible ancestor and descendant morphologies.

Evolutionary patterns and processes

Natural selection acts upon individuals throughout their life histories. Selective pressures that result in morphologic change of skeletal elements produce the evolutionary changes observed in the fossil record, although responsible selective pressures may not be discernible. Either increase in morphologic complexity or decrease in complexity may result from selection, and the latter commonly is expressed as a carry-over of juvenile features of proportions into fully mature adult stages (for a thorough discussion, see Gould 1977).

Morphologic elaboration during ontogeny is commonly recapitulative, wherein ontogenetic stages reflect the phylogenetic succession of adult ancestral morphologies. Maintenance of morphologic simplification (i.e. juvenile morphologies) during ontogeny is paedomorphosis. Both of these patterns are observable in conodonts, with paedomorphosis especially important in the development of progressively simplifying lineages.

Gould (1977:229) briefly summarized the processes responsible for and the morphologic results of the two kinds of paedomorphosis: progenesis and neoteny. In the first case, reproductive maturity is reached at an early age before complete development of somatic structures. Such progenetic individuals commonly appear small, sexually mature, but otherwise possessing juvenile characteristics which failed to elaborate much following reproductive maturity. Gould noted (1977:324) that many Recent progenetic species inhabit ephemeral environments where high fecundity is important for the survival of only a few individuals. Such conditions might influence evolution among conodonts living in otherwise unfavorable environments, such as those subject to rapid salinity fluctuations.

Neotenic development is the result of retarded elaboration, but commonly not retarded size increase, of morphologic features without an alteration of the timing of sexual maturity (Gould 1977:229). Resulting individuals tend to be of normal adult size, are sexually mature, but possess juvenile somatic morphologies. Gould suggested (1977:291) that such patterns may commonly occur within density-dependent communities that commonly lack broadly fluctuating environmental factors.

Origin of *Icriodus*

Previous ideas. – Klapper & Philip (1972:103) believed that species of *Pedavis* comprised 'a generic line completely sepa-

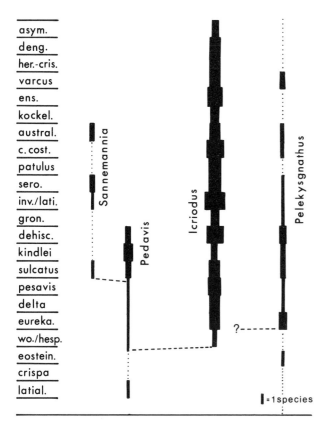

Fig. 1. Evolutionary relationships and species diversity of late Late Silurian (*Pedavis latialata* Zone) – early Late Devonian (top of lowermost *Polygnathus asymmetricus* Zone) genera of Icriodontidae. Zonation and species based upon Klapper & Johnson (1980) for Devonian and Walliser (1971) for Silurian.

rate from *Icriodus*'. This philosophy was based upon considerably greater complexity in all elements of the *Pedavis* apparatus (Klapper & Philip 1971) and greater similarity of Pa elements of *P. pesavis* and *P. latialata* (Klapper & Philip 1972:103).

Klapper & Murphy (1975:52) suggested the possible origin of *I. woschmidti* from *Pelekysgnathus index*, but noted that 'transitional specimens that would bridge the morphologic gap, especially between the respective I [=Pa] elements, are lacking'. Specimens of *P. index* illustrated by Klapper & Murphy (1975, Pl. 12) show a wide variation in breadth of denticles on the anterior process, but lack a posterior spur characteristic of *I. woschmidti* and other *I. latericrescens* group species. Nicoll (1982:209) maintained that *Pelekysgnathus dubius* (lower Ludlow) and not *P. index* (Pridoli) was the earliest species of Icriodontidae (excluding *Icriodella*). Thus it can be argued either that *Pelekysgnathus* (sensu Klapper & Murphy 1975) evolved paedomorphically from *Pedavis* or that *Pelekysgnathus* (sensu Nicoll 1982) may have been ancestral to the more morphologically complex *Pedavis* (Fig. 1).

Chatterton & Perry (1977:792) do not specifically comment on the origin of *Icriodus* but suggest a close relationship between it and *Pelekysgnathus*. Morphologic similarities between *Icriodus* and *Pelekysgnathus* tend to be in degree of denticulation which tends to be somewhat varible in each, requiring subjective judgements on generic assignment in some instances (e.g. *I. csakyi* of Chatterton & Perry 1977, regarded as a *Pelekysgnathus* by Klapper & Johnson 1980:451). This lends credibility to the suggestion by Chatterton & Perry (1977:792) that 'species that could be or have been assigned to

the genus *Pelekysgnathus*, as it is broadly conceived, may have been derived from other genera during four different evolutionary events (origin of "*P.*" *index* in the Late Silurian, origin of *I. hadnagyi* in the Lochkovian, origin of "*P.*" *furnishi* group [now assigned to *Sannemannia*, see Fig. 1] in the Early Devonian and origin of *P. communis* in the late Devonian)'.

Neotenic origin from PEDAVIS.–The Pa element of species of *Pedavis* characteristically possesses two anteriorly directed lateral processes. Commonly the outer of these is smaller or has reduced, irregular dentition, or both. The inner process is characteristically fully developed and denticulate and is generally aligned with the trend of the posterior spur (Fig. 2A–D). Anterior lateral processes occur only rarely in specimens of *Icriodus*, but when present (e.g. Klapper & Ziegler 1967, Pl. 8:5 for *I. beckmanni*; Carls 1975, Pl. 1:13 for *I. rectangularis lotzei*, Pl. 2:22 for *I. vinearum*, Pl. 3:41 for *I. r. rectangularis*; Klapper & Johnson 1980, Pl. 2:16 for *I. bilatericrescens*) invariably align with the posterior spur and are located on the inner side. In most specimens of *Icriodus*, the form of the posterior part of the basal cavity reflects this trend in species characterized by well-developed posterior spurs (Fig. 2E–G) and in those lacking a posterior spur (Fig. 2 H–L).

We suggest that the highly significant form of the basal cavity with respect to anterior lateral processes strongly supports the argument in favor of the origin of *Icriodus* from *Pedavis* and not from *Pelekysgnathus*. The probable mechanism would have been selection against the anterior lateral processes in early developmental stages of *Pedavis*. Conodont elements (other than simple cones) commonly add denticles throughout ontogeny so all processes grow by denticle addition and size increase of existing denticles (see Carls 1975, Pl. 1:14, 15 for comparison of juvenile and adult specimens of *P. pesavis*). Development of the anterior lateral processes would have been arrested early in ontogeny with the remaining features continuing to increase in size. In other words, juveniles of the earliest species of *Icriodus*, *I. woschmidti*, should resemble juveniles of the ancestral species of *Pedavis*, but at a larger size. A similar process probably reduced the denticulate S elements of *Pedavis* to nondenticulate S elements characteristic of *Icriodus*. Because Early Devonian *Icriodus* are characteristically as large as most specimens of *Pedavis*, a neotenic derivation is more likely than a progenetic one.

Although no direct ancestor–descendant relationship between a species of *Pedavis* and *Icriodus woschmidti* has been established, Pa elements of *P.* n.sp. E (Fig. 3A, B, E–G) from the Lower Devonian (*I. woschmidti* Zone) of west-central Tennessee, USA, are similar in some aspects to Pa elements of *I. woschmidti* (Fig. 3 H–J) from the same rocks. Denticulation patterns of the central anterior processes are remarkably similar (compare Fig. 3A, I). The form of the posterior spur is also similar, although the spur of *P.* n.sp. E lacks an elongate form with separate extension of the basal cavity seen in specimens of *P. pesavis* and *P. latialata*.

Evolutionary development of *Icriodus*

For the purpose of examining evolutionary trends within *Icriodus*, Early to early Late Devonian species were divided into

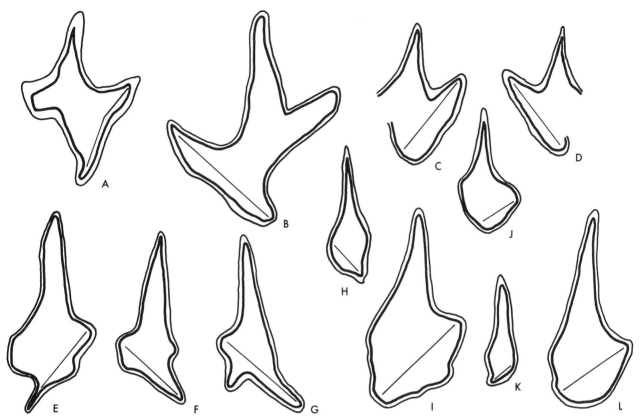

Fig. 2. Basal cavity form of species of *Pedavis* and *Icriodus* showing homology of lateral expansions in posterior region (indicated by thin, straight lines). □A. *P. latialata* (Klapper & Murphy 1975, Pl. 12:16). □B. *P. pesavis* (Carls 1969, Pl. 1:2B). □C, D. *P.* n. sp. E (outer lateral process not preserved in either). □E. *I. woschmidti* (Ziegler 1960, Pl. 15:18c). □F. *I. steinachensis* (Klapper & Johnson 1980, Pl.

2:21). □G. *I. latericrescens* n. subsp. A (Klapper & Ziegler 1967, Pl. 8:2b). □H. *I. angustoides angustoides* (Carls 1975, Pl. 3:48b).□I. *I. angustoides castilianus* (Carls 1969, Pl. 3:10c). □J. *I. expansus* (Klapper & Johnson 1980, Pl. 3:9). □K. *I. angustus* (Klapper & Johnson 1980, Pl. 3:6). □L. *I. corniger rectirostratus* (Bultynck 1970, Pl. 3:1). All specimens sketched to same scale, ×45.

three groups, each representing a morphologic grade in development of the posterior end of the platform (Fig. 4). Species that possess a well-developed denticulate posterior spur (Fig. 4A) show rapid diversification during the Early Devonian (*I. woschmidti* Zone – *P. dehiscens* Zone) but become exceedingly rare by the late Early Devonian. Species possessing a poorly defined, commonly nondenticulate spur (Fig. 4B) exhibit no time of diversification and are restricted to the Early Devonian. This group may represent an evolutionary divergent trend from group A (Fig. 4A), but may as well rightfully belong with group A based upon other features (e.g. denticle pattern) not considered here.

Low diversity characterizes the early history of group C (Fig. 4C) until the late Early Devonian, but the greatest diversification occurred in the Middle Devonian. This rapid diversification may represent an adaptive radiation into newly available subenvironments. The origin of spur-less *Icriodus* from those with posterior spurs (groups A, B, Fig. 4) was undoubtedly a neotenic event and the spur-less grade was probably polyphyletic (Bultynck 1972). Very small juveniles of *I. woschmidti* from west-central Tennessee lack the posterior spur, which apparently developed during later ontogeny.

Nonetheless, the existence of many species of group C that possess a large prominent cusp (e.g. *I. angustoides* subspp.) suggests a possible close affinity to species of *Pelekysgnathus*, which characteristically has a large cusp. Close examination of interrelationships among individual species of *Icriodus* is

beyond the scope of this report, so that exact relationships, if any, to *Pelekysgnathus* have not been suggested. Evolutionary relationships among species of *Icriodus* have been suggested by several authors including Klapper & Ziegler (1967), Bultynck (1972) and Weddige & Ziegler (1979); the last suggested a demonstrable relationship between environmental factors and evolution of Middle Devonian species derived from *I. corniger* (1979:162–163).

Evolutionary synthesis

Icriodus probably evolved rapidly by neoteny from *Pedavis* during the Late Silurian. Because neotenic development may occur within one or only a few generations, no succession of intermediate morphologies is likely to be found. Species of *Pedavis* existed from the Late Silurian into the Early Devonian and gave rise to *Pelekysgnathus*, *Icriodus* , and *Sannemannia*, but did not diversify rapidly at any time. Rapid diversification of spur-bearing species of *Icriodus* in the Early Devonian and of spur-less species in the Middle Devonian attests to the selective advantage of both morphologies. Differences in denticulation development commonly characterize species and are more likely to be observed as temporally or geographically continuous changes than the two major shifts in morphology discussed here.

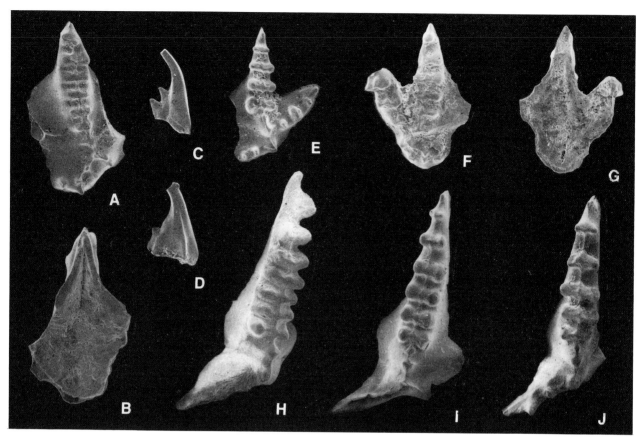

Fig. 3. Late Silurian – Early Devonian conodonts from west-central Tennessee. Localities, see appendix. All figures magnified ×45. □A–B, E–G. *Pedavis* n.sp. E. □A–B. Upper and under sides of Pa element with broken anterior lateral processes (Locality 1-D, SUI 49406). □E. Upper side of Pa element with broken outer lateral process (Locality 1-A, SUI 49407). □F–G. Upper and under sides of Pa element with broken outer lateral process (Locality 1-C, SUI 49408). □C–D, H–J. *Icriodus woschmidti.* □C. Lateral view of S element (Locality 1-D, SUI 499409). □D. Lateral view of S element (Locality 1-C, SUI 49410). □H. Upper view of Pa element (Locality 2, SUI 49411). □I. Upper view of Pa element (Locality 1-B, SUI 49412). □J. Upper view of Pa element (Locality 1-C, SUI 49413).

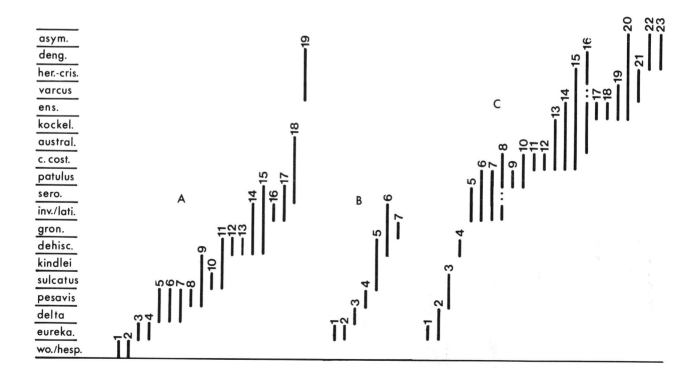

Pedavis n.sp. E

Fig. 3A, B, E–G.

Systematic affinity.–Family Icriodontidae Müller & Müller 1957; Genus *Pedavis* Klapper & Philip 1971.

Material. – Three incomplete Pa elements from west-central Tennessee, USA.

Derivation of name. – The small number of specimens and incomplete preservation of Pa elements necessitates placing this species in open nomenclature.

Diagnosis. – Basal cavity broadly expanded, not extended posteriorly from main part of platform; posterior spur angled from cusp with respect to main platform; denticle ridges smooth, with poorly developed median ridge.

Occurrence. – Rockhouse Limestone (Lower Devonian, *I. woschmidti* Zone), west-central Tennessee, USA.

Icriodus woschmidti Ziegler 1960

Fig. 3C, D, H–J

Systematic affinity. – Family Icriodontidae Müller & Müller 1957; Genus *Icriodus* Branson & Mehl 1938.

Material. – More than 200 complete and fragmental Pa elements plus denticulate and nondenticulate coniform elements (after Serpagli 1982) from west-central Tennessee, USA.

Remarks.– These specimens differ from those illustrated by Ziegler (1960) in possessing more regularly developed dentition and a larger number of denticle rows.

Fig. 4. Stratigraphic distribution of Early to early Late Devonian (*Icriodus woschmidti* Zone to lower *Polygnathus asymmetricus* Zone) species of *Icriodus*. □A. Species with well-developed, denticulate lateral process: □1. *I. woschmidti*, □2. *I. w. hesperius*, □3. *I. postwoschmidti*, □4. *I. rectangularis*, □5. *I. r. lotzei*, □6. *I. fallax*, □7. *I. vinearum*, □8. *I. simulator*, □9. *I. steinachensis*, □10. *I. curvicauda*, □11. *I. claudiae*, □12. *I. sigmoidalis*, □13. *I. latus*, □14. *I. bilatericrescens*, □15. *I. nevadensis*, □16. *I. beckmanni beckmanni*, □17. *I. b. sinuatus*, □18. *I. latericrescens robustus*, □19, *I. l. latericrescens* (termination of morphologic character). □B. Species with poorly-developed adenticulate to poorly denticulate posterior lateral process: □1. *I.* n.sp. G of Klapper (1977), □2. *I. eolatericrescens*, □3. *I. angustoides alcolae*, □4. *I. a. castilianus*, □5. *I. celtibericus*, □6. *I. trojani*, □7. *I.* n.sp. O of Johnson & Klapper (1981). □C. Species lacking a posterior lateral process: □1. *I. hadnagyi*, □2. *I. angustoides bidentatus*, □3. *I. a. angustoides*, □4. *I. taimyricus*, □5. *I. fusiformis*, □6. *I. corniger corniger*, □7. *I. c. rectirostratus*, □8. *I. culicellus*, □9. *I. corniger leptus*, □10. *I. werneri*, □11. *I. norfordi*, □12. *I.* n.sp. 1 of Klapper & Ziegler (1967), □13. *I. angustus*, □14. *I. regularicrescens*, □15. *I. struvei*, □16. *I. expansus*, □17. *I. arkonensis*, □18. *I. lindensis*, □19. *I. obliquimarginatus*, □20. *I. difficilis*, □21. *I. brevis*, □22. *I. symmetricus*, □23. *I. subterminus*, (morphotype continues to end of Devonian).

Occurrence. – Decatur Limestone (Upper Silurian? – Lower Devonian, *I. woschmidti* Zone) upper 2 m, and Rockhouse Limestone (Lower Devonian, *I. woschmidti* Zone), west-central Tennesse.

Acknowledgements. – We thank Gilbert Klapper of the University of Iowa and James E. Barrick of Texas Tech University for comments during several stages of development of this paper, and Charles S. Harris for preparation of Fig. 3. Valuable criticism of the manuscript by Pierre Bultynck of the Koninklijk Belgisch Instituut voor Natuurwetenschappen is gratefully acknowledged. All specimens figured herein are reposited in the collections of the University of Iowa, Iowa City, Iowa, USA, and bear catalog numbers (SUI) of that institution.

References

Branson, E. B. & Mehl, M. G. 1938: The conodont *Icriodus* and its stratigraphic distribution. *Journal of Paleontology 12*, 156–166.

Bultynck. P. 1970: Révision stratigraphique et paléontologique (brachiopodes et conodontes) de la coupe type du Couvinien. *Mém. Inst. Géol. Univ. Louvain 26*, 1–152.

Bultynck, P. 1972: Middle Devonian *Icriodus* assemblages (Conodonta). *Geologica et Palaeontologica 6*, 71–86.

Carls, P. 1969: Die Conodonten des tieferen Unter-Devons der Guadarrama (Mittel-Spanien) und die Stellung des Grenzbereiches Lochkovium/Pragium nach der rheinischen Gliederung. *Senckenbergiana lethaea 50*, 303–355.

Carls, P. 1975: Zusätzliche Conodonten-Funde aus dem tieferen Unter-Devon Keltiberiens (Spanien). *Senckenbergiana lethaea 56*, 399–428.

Chatterton, B. D. E. & Perry, D. G. 1977: Lochkovian trilobites and conodonts from northwestern Canada. *Journal of Paleontology 51*, 772–796.

Gould, S. J. 1977: *Ontogeny and Phylogeny.* 501 pp. Belknap, Cambridge, Massachusetts.

Klapper, G. & Johnson, J. G. 1980: Endemism and dispersal of Devonian conodots. *Journal of Paleontology 54*, 400–455.

Klapper, G. & Murphy, M. A. 1975: Silurian–Lower Devonian conodont sequence in the Roberts Mountains Formation of central Nevada. *University of California Publications in Geological Sciences 111*, 1–62.

Klapper, G. & Philip, G. M. 1971: Devonian conodont apparatuses and their vicarious skeletal elements. *Lethaia 4*, 429–452.

Klapper, G. & Philip, G. M. 1972: Familial classification of reconstructed Devonian conodont apparatuses. *Geologica et Palaeontologica SB 1*, 97–114.

Klapper, G. & Ziegler, W. 1967: Evolutionary development of the *Icriodus latericrescens* group (Conodonta) in the Devonian of Europe and North America. *Palaeontographica Abt. A 127*, 68–83.

Nicoll, R. S. 1982: Multielement composition of the conodont *Icriodus expansus* Branson & Mehl from the Upper Devonian of the Canning Basin, Western Australia. *BMR Journal of Australian Geology & Geophysics 7*, 197–213.

Serpagli, E. 1982: Remarks on the apparatus of *Icriodus woschmidti* Ziegler (Conodonta). *In* Jeppsson, L. & Löfgren, A. (eds.): Third European Conodont Symposium (ECOS III) Abstracts. *Publications from the Institutes of Mineralogy, Paleontology and Quaternary Geology, University of Lund, Sweden, 238*, 21.

Walliser, O. H. 1971: Conodont biostratigraphy of the Silurian of Europe. *In* Sweet, W. C. & Bergström, S. M. (eds.): Symposium on conodont biostratigraphy. *Geological Society of America, Memoir 127*, 195–206.

Weddige, K. & Ziegler, W. 1979: Evolutionary patterns in Middle Devonian conodont genera *Polygnathus* and *Icriodus*. *Geologica et Palaeontologica 13*, 157–164.

Ziegler, W. 1960: Conodonten aus dem Rheinischen Unterdevon (Gedinnium) des Remscheider Sattels (Rheinisches Schiefergebirge). *Paläontologische Zeitschrift 34*, 169–201.

Appendix: Localities

1 – Parsons Plant Quarry of the Vulcan Materials Company, 4.9 km north of Parsons, Tennessee, USA, 35°42′N 88°6′W. Excavation in the quarry has revealed the Decatur Limestone (*O. eosteinhornensis* Zone, *I. woschmidti* Zone 2 m from top of formation) and the Rockhouse Limestone (*I. woschmidti* Zone) and Birdsong Shale (zonation as yet undetermined). Sample 1A is from 0.1 m above the base of the Rockhouse Limestone, 1B is from 1.3 m above the base, 1C is from 3.4 m above the base and 1D is from 3.7 m above the base.

2 – Roadcut on the east side of Tennessee state highway 69, 9.9 km north of Parsons, Tennessee, USA, 35°44′N 88°6′W. The contact between the Decatur Limestone and overlying Rockhouse Limestone is exposed at the top of the cut. Sample 2 is from 0.4 m above the base of the Rockhouse.

The conodont apparatus of *Icriodus woschmidti woschmidti* Ziegler

ENRICO SERPAGLI

FOSSILS AND STRATA

ECOS III

A contribution to the Third European
Conodont Symposium, Lund, 1982

Serpagli, Enrico 1983 12 15: The conodont apparatus of *Icriodus woschmidti woschmidti* Ziegler.
Fossils and Strata, No. 15, pp. 155–161. Oslo. ISSN 0300-9491. ISBN 82-0006737-8.

Study of Sardinian conodonts close to the Silurian–Devonian boundary supports the idea that the
basic number of morphologically distinct constituents in the apparatus of *Icriodus woschmidti
woschmidti* Ziegler was six and not two as hitherto supposed, and included several ramiform
elements. □*Conodonta*, Icriodus, *conodont apparatus, Lower Devonian.*

Enrico Serpagli, Istituto di Paleontologia, Via Università No 4, I-41100 Modena, Italy; 24th July, 1982.

Icriodus woschmidti is certainly one of the most representative
conodonts, both because of its stratigraphic importance as a
useful index for defining the Silurian–Devonian boundary and
the fact that it is always easily recognizable, even when present
in not quite complete specimens. Moreover, it is very widespread geographically, being found in various areas of Europe,
Asia, North Africa, North America, and Australia. This
species can therefore be used over a wide geographic area for
correlation.

Four subspecies of *Icriodus woschmidti* have been proposed:
Icriodus woschmidti woschmidti Ziegler 1960, *I. w. postwoschmidti*
Mashkova 1968, *I. w transiens* Carls & Gandl 1969, and *I. w.
hesperius* Klapper & Murphy 1975. Of these, *I. w. transiens* is
now considered a junior synonym of *I. w. postwoschmidti*
whereas the other two, although having more or less the same
stratigraphic distribution, are both distinct taxa. *I. w. woschmidti* occurs on all continents studied whereas *I. w. hesperius*
is widespread but more limited in its distribution.

Previous work

The apparatus of *Icriodus* has been interpreted as having
mainly an icriodiform element and a simple cone, called
Acodina in element-based taxonomy. The possible occurrence
of elements referred to as *Icriodus* and *Acodina* in the same
conodont apparatus was first proposed by Lange (1968), but
he reported a frequency ratio between these elements types (2
'*Icriodus*' to 30 '*Acodina*' elements) very different from that
proposed here. Klapper & Philip (1971) proposed an apparatus belonging to their 'Type 4 apparatuses' with the same
elements, an 'icriodontan' element (I) and an 'acodinan'
element outnumbering the 'icriodontan' one. This point of
view was put forward at about the same time also by Ziegler
(1972) and has been accepted by the majority of subsequent
authors, with the exception of Bultynck (1972:72) and Boo-

gaard & Kuhry (1979:15) who stated that 'it is extremely
unlikely that these forms belonged to a common apparatus'.
Recently it has been suggested by Klapper & Ziegler (*in*
Ziegler 1975:67), Nicoll (1977:222) and Johnson & Klapper
(1981) that an additional M_2 element similar to that of
Pelekysgnathus may also belong to the *Icriodus* apparatus.

The apparatus of *Icriodus woschmidti woschmidti*

From the study of collections from some Lower Devonian
sections very close to the Silurian–Devonian boundary in
southern Sardinia I conclude that the basic number of
morphologically distinct constituents in the apparatus of *I. w.
woschmidti* was six and not two as hitherto supposed, since
several ramiform elements seem to be an integral part of the
apparatus.

In three different sections (Fig. 1), some of which were
already reported by Serpagli & Mastandrea (1980) and by
Olivieri, Mastandrea & Serpagli (1981), the conodont faunas
of at least five samples are composed of the different elements
of only two taxa, *Ozarkodina remscheidensis remscheidensis* and
Icriodus woschmidti woschmidti, plus some ramiform elements
which I am inclined to associate with the apparatus of *I. w.
woschmidti* for the following reasons:

(1) It is hard to imagine that the same types of ramiform
elements could have been reworked in at least 5 different
samples from three sections situated in areas separated from
each another by as much as 56 km.

(2) A study of literature makes it clear that ramiform elements,
in all respects similar to those which I maintain may belong to
the apparatus of *I. w. woschmidti* or related forms, in several
cases have been found associated with the icriodiform elements. Such occurrences include: (a) From the lowermost part
of the Rhenish Gedinnian Ziegler (1960) reports two ramiform

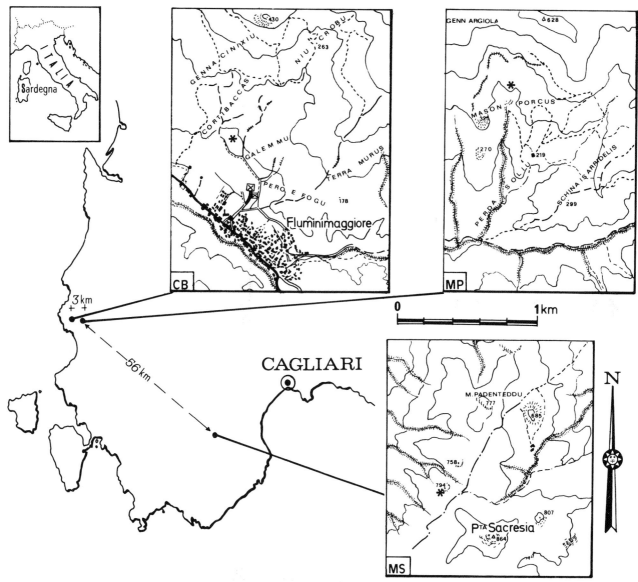

Fig. 1. Location map of conodont localities (asterisks); Corti Baccas (CB), Mason Procus (MP) and Monte Santo – Monte Padenteddu (MS).

elements (*Cordylodus* n. sp. and *Paltodus* n. sp. C), as well as an acodiniform element with denticles on the anticusp (*Drepanodus* sp. a) (Fig. 2C), which are strikingly similar to those from Sardinia; (b) Adrichem Boogaert (1967) reports five ramiform elements, two of which are illustrated as 'n. gen. A n. sp. a' (Fig. 2A), from the Lebanza Formation (Gedinnian) of the Cantabrian Mountains; (c) From the Lower Devonian of Podolia Mashkova (1971) reports two ramiform elements referred to as '*Gen.* et sp. nov. A' and '*Gen.* et sp. nov. B' (Fig. 2B) respectively; (d) Barnett (1971), in a biometric study of some conodonts from the Ravena Member of the Coeymans Formation of eastern North America, reports two ramiform components (*Neoprioniodus brevirameus* and *Oepikodus* sp.) besides the icriodiform (*I. woschmidti*) and acodinan elements (*Acodus* sp.) (Fig. 2 F).

(3) It has also been noted that other ramiform elements, referred to as *Rotundacodina dubia* (e.g. Carls & Gandl 1969), are associated with both the icriodiform element (*I. w. transiens*) and various acodiniform elements (*Acodina plicata, A. retracta,*

A. aragonica) (Fig. 2 D); Drygant (1974) indicated that such elements are associated only with the acodiniform element (*A. plicata*) (Fig. 2 E). However, in his 1974 paper, Drygant described only the simple conodonts although he hinted that in the association there are also more complex forms such as the 'platform' ones.

(4) Bearing in mind that the basic plan of the 'Type 4 apparatuses' of Klapper & Philip (1971) is that of *Icriodella superba* Rhodes, as reconstructed by Sweet & Bergström (1970, 1972), the apparatus of *I. w. woschmidti*, as reconstructed up to now, lacks the corresponding ramiform elements that are probably the ones I describe in the present paper.

Structure of the apparatus.–In accordance with the scheme proposed by Barnes *et al.* (1979) for 'Type IV A Ordovician conodont apparatuses', the elements of the apparatus of *Icriodus woschmidti woschmidti* are grouped into two transition series, both consisting of three morphotypes (Fig. 6). The morphotypes of the first transition series are:

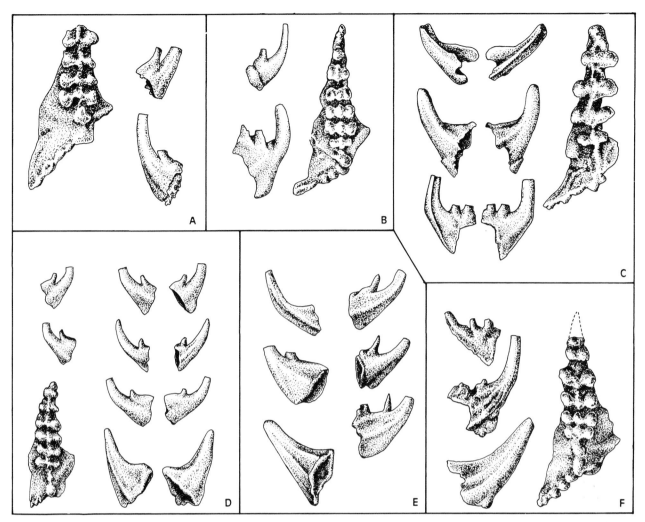

Fig. 2. Previously described ramiform elements occurring with the acodiniform and icriodiform ones. Redrafted from: □A. Adrichem Boogaert (1967): *I. woschmidti* (Pl. 2:1), n. gen. A n. sp. *a* (Pl. 3:28, 29). □B. Mashkova (1971): Gen. et sp. nov. A (Pl. 2:9), Gen. et sp. nov. B (Pl. 2:10), *I. woschmidti* (Pl. 2:2). □C. Ziegler (1960): *Paltodus* sp. c (Pl. 14:11), *Drepanodus* sp. a (Pl. 13:17), *Cordylodus* n. sp. (Pl. 14:19), *I. woschmidti* (Pl. 15:16). □D. Carls & Gandl (1969): *Rotundacodina dubia* (Pl. 20:13–16), *Acodina plicata* (Pl. 19:28), *I. woschmidti transiens* (Pl. 15:1). □E. Drygant (1974): *Rotundacodina dubia* (Pl. 2:26–29), *Acodina retracta* (Pl. 2:30), *Acodina plicata* (Pl. 2:30). □F. Barnett (1971): *Neoprioniodus bevirameus* (Pl. 37:6), *Oepikodus* sp. (Pl. 37:9), *Acodus* sp. (Pl. 37:4), *I. woschmidti* (Pl. 37:17).

a = a slightly asymmetrical element with a denticulated posterior process (cordylodiform; Fig. 3A–F);

b = an asymmetrical element similar to a but bearing one adenticulated lateral costa (gothodiform; Fig. 3G–L);

c = an almost symmetrical element similar to b but with two symmetrically (or almost symmetrically) developed adenticulated lateral costae (trichonodelliform; Fig. 3 M–R);

d = element not recovered.

(Note that rare adenticulated specimens of the a, b, and c morphotypes can also be present).

The morphotypes of the second transition series are:

e = a slightly asymmetrical element with a downwardly directed adenticulate anterior edge and a sharp posterior edge. Its cusp is commonly deflected (triangular; Fig. 4 A–F). Simple cones with a widely flaring basal cavity could belong to this morphotype (Fig. 4G–I);

f = a moderately asymmetrical element with two or three sharp edges (costae), one of which may or may not bear denticles, situated on the posterior, anterior and lateral margins (acodiniform or sagittodontiform; Fig. 4J–R);

g = a platform element (icriodiform; Fig. 5A–E).

The inferred ratio of the several morphotypes is given in Table 1.

Table 1. Abundance and grouping of the different elements of the *Icriodus woschmidti woschmidti* apparatus in Sardinian samples.

Samples	Morphotypes					
	a	b	c	e	f	g
CB	6	6	8	26	15	43
MP$_6$	1	–	–	2	3	22
MP$_9$	4	1	1	4	6	19
MS$_0$	3	3	1	28	32	99
MS$_1$	3	3	1	7	5	42
Total	17	13	11	67	61	225
Inferred ratio	2	1	1	6	6	20

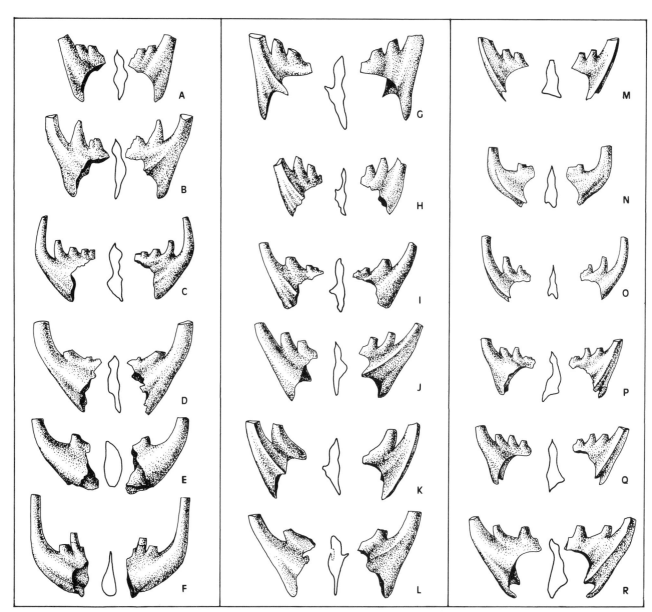

Fig. 3. Apparatus of *Icriodus woschmidti woschmidti*. Ramiform elements of the first transition series. □A–F. *a* morphotype. □G–L. *b* morphotype. □M–R. *c* morphotype. Specimens A, D, H, L are from Monte Santo (MS 1); specimens B, C, I, R are from Monte Santo (MS O); specimens E, G, J, K, M, O, P, Q are from Corti Baccas (CB); specimens F, N are from Mason Porcus (MP). All specimens are about ×52.

By comparing the apparatus of *I. w. woschmidti* as reconstructed here (Fig. 6) with the traditional two-element model, it will be readily apparent that the former model is more nearly complete and closer to the basic plan already established, for instance, for some Ordovician and Silurian apparatuses. It demonstrates also the persistence of only a few structural apparatus types in the Palaeozoic.

It is of interest to note that here is another case where certain ramiform elements that have long been considered little more than part of the general 'background noise', prove to be significant morphological structures of the species they represent.

Objections to the present hypothesis

There are two main objections to my hypothesis regarding the composition of the *I. woschmidti* apparatus: (1) Why have ramiform elements not been identified as elements of *Icriodus* up to now? (2) Why do not stratigraphically younger species, belonging to the same genus or to closely related genera (e.g. *Pedavis*), preserve the six-element structural plan?

The first objection has been partly answered already in point 2 on page 155, where I have shown how other authors noted the elements *a*, *b*, *c*, without, however, including them in the apparatus of *I. woschmidti*. Furthermore, it is quite likely that the *g* (icriodiform) morphotype was preferred in the picking process, both because it is the most characteristic element and because its finding enables immediate solution of stratigraphic problems. Lastly, there is no doubt that the ramiform elements are more fragile than element *g* and, to a lesser extent, than elements *e* and *f*.

To answer to the second objection is by no means so simple, even though a depletion in number of element types can be invoked. This could also agree with the observation by Dzik

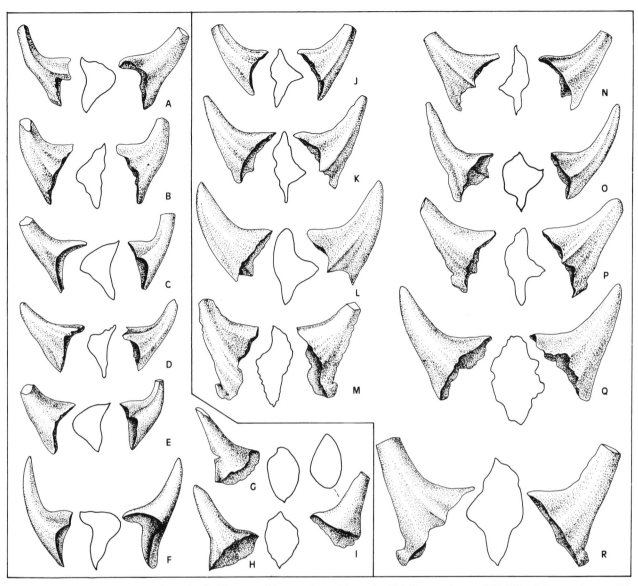

Fig. 4. Apparatus of *Icriodus woschmidti woschmidti.* Elements of the second transition series. □A–F. *e* morphotype. □G–I. Possible *e* morphotype. □J–R. *f* morphotype. Specimens C, F, I are from Corti Baccas (CB); specimens A, B, Q are from Monte Santo (MS 1); specimens D, E, J, K, M–P are from Monte Santo (MS O); specimens H, L, R are from Mason Porcus (MP 6); specimen G is from Mason Porcus (MP 9). All specimens are about ×52.

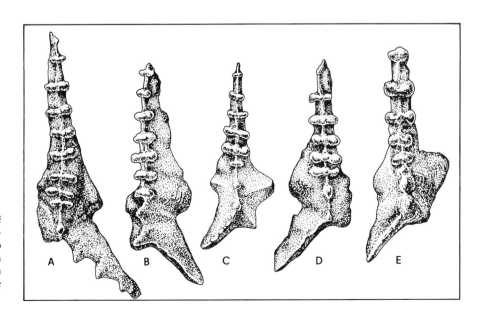

Fig. 5. Apparatus of *Icriodus woschmidti woschmidti.* Platform elements. Specimens A, B, C are from Monte Santo (MS O); specimen D is from Mason Porcus (MP 6); specimen E is from Corti Baccas (CB). All specimens are about ×50.

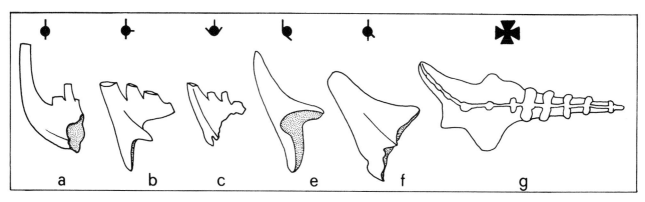

Fig. 6. Structure of *Icriodus woschmidti woschmidti* skeletal apparatus which includes elements arranged in both a first (ramiform) and second transition series. These transition series are represented by morphotypes *a–c* and *e–g* respectively. Morphotype *d* is missing. Symbols after Barnes *et al.* (1979).

(1976:411) who notes that a trend of decrease in the number of elements is typical of the evolution of Icriodontidae.

However, in the apparatus of various species of *Pedavis* (*P. pedavis*, *P. latialata*, *P. marianne*, *P. thorsteinssoni*, P. n. sp. A Uyeno 1980), and of at least one species of *Pelekysgnathus* (*P.* n. sp. B Uyeno 1980), a denticulate element (M$_2$) called a striate or costate cone appears which could be the *a* morphotype of the first transition series. Furthermore, the M$_2$ element is represented by a rather variable set of drepanodiform cones, both smooth and costate; these, properly arranged, can be fitted into the two transition series proposed above, according to the following scheme:

Pedavis		*Pelekysgnathus*	
M$_2$ denticulate	⟶ *a*	M$_2$ denticulate	⟶ *a*
M$_2$ adenticulate	⟶ *e*	M$_2$ adenticulate	⟶ *e*
S$_1$ adenticulate	⟶ *e*	S$_2$ adenticulate $\Big\{$	⟶ *e*
S$_1$ denticulate	⟶ *f*		⟶ *f*
I platform	⟶ *g*	I platform	⟶ *g*

Conclusions

My conclusion about the composition of 'type 4 apparatuses' differs somewhat from that of Klapper & Philip (1971), in that I recognize six (or four), rather than two (or three), components arranged in two symmetry transition series. My reconstruction is in agreement with the statement by several authors (e.g. Sweet & Schönlaub 1975:43) who suggest that six is the basic number of morphologically distinct constituents in most multielement apparatuses from the Ordovician to Triassic.

Acknowledgements.–Special thanks are due to Prof. S. M. Bergström (Ohio State University, Columbus, Ohio) and to Dr. T. W. Broadhead (The University of Tennessee, Knoxville, Tennessee) for linguistic improvements and for the critical reading of the manuscript even if this is not to imply that they necessarily agree with all my conclusions. Drs. M. Gnoli, R. Olivieri and A. Mastandrea shared the field work and kindly allowed me to use co-sampled material. The artist G. Leonardi has carefully re-drafted all the figures on the basis of my camera lucida drawings. Field studies and printing of the paper were supported financially by Consiglio Nazionale delle Ricerche and Ministero della Pubblica Instruzione (Rome).

References

Adrichem Boogaert, H. A. van 1967: Devonian and Lower Carboniferous conodonts of the Cantabrian Mountains (Spain) and their stratigraphic application. *Leidse Geol. Med. 39*, 129–192.

Barnes, C. R., Kennedy, D. J., Mc Cracken, A. D., Nowlan, G. S. & Tarrant, G. A. 1979: The structure and evolution of Ordovician conodont apparatuses. *Lethaia 12*, 125–151.

Barnett, S. G. 1971: Biometric determination of the evolution of *Spathognathodus remscheidensis:* a method for precise intrabasinal time correlations in the northern Appalachians. *J. Paleontol. 45*, 274–300.

Boogard, M. van den & Kuhry, B. 1979: Statistical reconstruction of the *Palmatolepis* apparatus (Late Devonian Conodontophorids) at the generic, subgeneric, and specific level. *Scripta Geol. 49*, 1–57.

Bultynck, P. 1972: Middle Devonian *Icriodus* Assemblanges (Conodonta). *Geol. Palaeontol. 6*, 71–86.

Carls, P. & Gandl, J. 1969: Stratigraphie und Conodonten des unter-Devons des Östlichen Iberischen Ketten (NE-Spanien). *N. Jb. Geol. Paläont. Abh. 132*, 155–218.

Drygant, D.M. 1974: Simple conodonts from the Silurian and Lowermost Devonian of the Volyno-Podolian. *Paleontol. Sbornik 10*, 64–70.

Dzik, J. 1976: Remarks on the evolution of Ordovician conodonts. *Acta Palaeontol. Polonica 21*, 395–455.

Johnson, D. B. & Klapper G. 1981: New Early Devonian conodont species of Central Nevada. *J. Paleontol. 55* 1237–1250.

Klapper, G. & Murphy, M. A. 1975: Silurian – Lower Devonian conodont sequence in the Roberts Mountains Formation of Central Nevada. *Univ. California Publ. Geol. Sci. 111*, 1–62.

Klapper, G. & Philip, G. M. 1971: Devonian conodont apparatuses and their vicarious skeletal elements. *Lethaia 4*, 429–452.

Lange, F. G. 1968: Conodonten-Gruppenfunde aus Kalken des tieferen Oberdevon. *Geol. Palaeontol. 2*, 37–57.

Mashkova, T. V. 1968: Konodonty roda *Icriodus* Branson & Mehl, 1938, iz Borshchovskogo i Chortkovskogo gorizontov Podolii. *Dokl. Akad. Nauk SSSR. 182*, 941–944. (In Russian.)

Mashkova, T. V. 1971: Zonal conodont assemblages from boundary beds of the Silurian and Devonian of Podolia. *In:* Granitsa silura i devona i biostratigrafija silura, III. *Int. symp. Leningrad 1968, 1*, 147–164, Akad. Nauk SSSR, Leningrad (In Russian.)

Nicoll, R. S. 1977: Conodont apparatuses in an Upper Devonian palaeoniscoid fish from the Canning Basin, Western Australia. *Bur. Mineral. Resour. Jour. of Austral. Geol. & Geophysic 2*, 217–228.

Olivieri, R., Mastandrea, A. & Serpagli, E. 1981: Riconoscimento di alcune zone a conodonti del Devoniano inferiore nei calcari di Monte Padenteddu nella Sardegna Meridionale. *Atti Soc. Nat. Mat. di Modena. 111*, 15–26.

Serpagli, E. & Mastandrea, A. 1980: Conodont assemblages from the Silurian–Devonian boundary beds of southwestern Sardinia (Italy). *N. Jb. Geol. Paläont. Mh. 1980:1*, 37–42.

Sweet, W. C. & Bergström S. M. 1970: The generic concept in conodont taxonomy. *Proceedings North American Paleont. Convention Chicago, 1969, Part C*, 157–173.

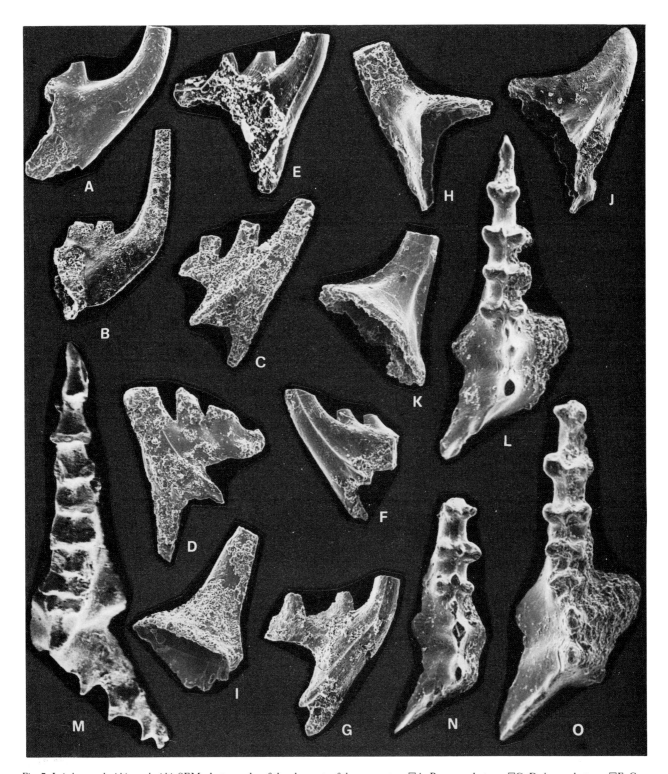

Fig. 7. Icriodus woschmidti woschmidti. SEM photographs of the elements of the apparatus. □A, B. *a* morphotype. □C, D. *b* morphotype. □E–G. *c* morphotype. □H, I. *e* morphotype. □J, K. *f* morphotype. □L–O. *g* morphotype. Specimens C–F, H, I, K, L, N, O are from Corti Baccas (CB); specimens A, B are from Mason Porcus (MP 9); specimens G, J, M are from Monte Santo (MS O). All specimens are about ×88.

Sweet, W. C. & Bergström, S. M. 1972: Multielement taxonomy and Ordovician conodonts. *Geol. Palaeontol. SB1*, 29–42.

Sweet, W.C. & Schönlaub H. P. 1975: Conodonts of the genus *Oulodus* Branson & Mehl, 1933. *Geol. Palaeontol. 9*, 41–59.

Uyeno, T. T. 1980: Stratigraphy and conodonts of Upper Silurian and Lower Devonian rocks in the environs of the Boothia Uplift, Canadian Arctic Archipelago. Part. II. Systematic study of conodonts. *Geol. Survey of Canada, Bull. 292*, 39–75.

Ziegler, W. 1960: Conodonten aus dem Rheinischen Unterdevon (Gedinnium) des Remscheider Sattels (Rheinisches Schiefergebirge). *Paläontol. Z. 34*, 169–201.

Ziegler, W. 1972: Über devonische Conodonten-Apparate. *Geol. Palaeontol. SB1*, 91–96.

Ziegler, W. (ed.) 1975: *Catalogue of Conodonts, II* E. Scweizerbart'sche Verlangsbuchhandlung, 1–404, Stuttgart.

Origin and development of the conodont genus *Ancyrodella* in the late Givetian – early Frasnian

PIERRE BULTYNCK

FOSSILS AND STRATA

ECOS III

A contribution to the Third European Conodont Symposium, Lund, 1982

Bultynck, Pierre 1983 12 15: Origin and development of the conodont genus *Ancyrodella* in the late Givetian – early Frasnian. *Fossils and Strata*, No. 15, pp. 163–168. Oslo. ISSN 0300-9491. ISBN 82-0006737-8.

Early Frasnian *Ancyrodella* species are descended from a late Givetian – early Frasnian *Ozarkodina* stock, as demonstrated by the ontogenetic development of the platform elements in the early *Ancyrodella* species *A. binodosa* and *A. rotundiloba*. Furthermore, one morphotype of *A. binodosa* (δ) is linked to *Ozarkodina* sp. A by a number of morphological intermediates; another morphotype (α) may be derived from *Ozarkodina insita*. This evolution is observed in a near-shore environment over a large geographical area (N. America and W. Europe). The γ morphotype of *A. binodosa* linked to the α morphotype gives rise to *A. rotundiloba rotundiloba*; this transition corresponds with a dispersion of the genus in the off-shore environment. □*Conodonta, biostratigraphy, palaeoecology, evolution, Devonian.*

Pierre Bultynck, Department of Paleontology, Koninklijk Belgisch Instituut voor Natuurwetenschappen, Vautierstraat 29, B-1040 Brussel, Belgium; 1 st September, 1982.

According to Klapper & Philip (1972:99) the apparatus of *Ancyrodella* Ulrich & Brassler 1926, consists of ancyrodellan platform or P elements and various compound elements. The apparatus of *Ozarkodina* Branson & Mehl 1933, to which the ancestral stock of *Ancyrodella* belongs, is of the same type, the platform element being spathognathodontan. I consider *Pandorinellina* Müller & Müller 1957 to be a junior synonym to *Ozarkodina*. The platforms are the most diagnostic element in these apparatuses. The present discussion is, therefore, only based on the spathognathodontan and ancyrodellan elements.

Three hypotheses have been proposed for the origin of the two earliest *Ancyrodella* species, *A. binodosa* Uyeno 1967 and *A. rotundiloba rotundiloba* (Bryant 1921). Ziegler (1962:146) regarded *Polygnathus asymmetricus* Bischoff & Ziegler 1957 as the direct ancestor of *A. rotundiloba*. He later advocated (1973:26) a more vague origin in stating that 'the species has evolved from the wide-plated *Polygnathus* stock'.

Uyeno (1967:11, 1974:18, 44) admitted a probable phylogenetic sequence from '*Spathognathodus*' *insitus* (Stauffer 1940) to '*Spathognathodus*'? sp. and finally to *A. rotundiloba binodosa*.

Ziegler (1973:35) doubted the generic and specific affinity of *A. rotundiloba binodosa* and concluded (1973:26) that if *A. rotundiloba binodosa* is recognized, the origin of the genus may lie in some '*Spathognathodus*' stock, preferably '*S.*' *sannemanni* Bischoff & Ziegler 1957. Druce (1976:52, 53) considered *Ancyrodella* as polyphyletic, having its origin in two subspecies of '*S.*' *sannemanni* both established by Pollock (1968), '*S.*' *sannemanni* subsp. A and '*S.*' *sannemanni variabilis*. Bultynck (1979, Pl. 27:18–20) and Bultynck & Coen (1982:40–42) proposed a slightly different origin in establishing a phylogenetic sequence *Ozarkodina sannemanni adventa* (Pollock 1968) – *O. sannemanni proxima* (Pollock 1968) – *A. binodosa*.

Morpho-phylogenetic aspects

Transitional forms, in stratigraphic sequence, demonstrating the descent of *A. rotundiloba* from *P. asymmetricus* have never been figured, whereas this has been done for the *Ozarkodina insita–A. binodosa* lineage by Uyeno (1974, Pl. 8:8, 10) and for the *O. sannemanni adventa–A. binodosa* lineage by Bultynck (1979, Pl. 27:18–20). Several authors demonstrated that early growth stages of *A. binodosa*, *A. rotundiloba rotundiloba* and the still younger *A. curvata* (Branson & Mehl 1934) are spathognathodontiform in outline (Ethington & Furnish 1962; Müller & Clark 1967; Uyeno 1967; Bultynck & Jacobs 1981). These two arguments, the ontogenetic development, and transitional forms in stratigraphic sequence, strongly support an evolution from an *Ozarkodina* stock rather than from a *Polygnathus* Hinde 1879. The question arises whether *A. binodosa* derives from one *Ozarkodina* species or whether the species and the genus have a polyphyletic origin.

Relevant to this question is the presence of two main morphotypes, α and δ, recognized by Bultynck & Jacobs (1981:16) within *A. binodosa* material from S. Morocco. They are also represented in the type material from the Waterways Formation in northeastern and central Alberta, in the basal part of the Frasnes group in the Ardennes (Bultynck & Jacobs 1982) and the Beaulieu Formation from the Boulonnais (N. France) (Brice, Bultynck, Deunff, Loboziak & Streel 1979).

The α morphotype (Fig. 1:21–27) is characterized by a triangular asymmetric platform, with raised, denticulated rim. The platform is widest at or near its anterior end. In the δ morphotype (Fig. 1:12–20) the platform is rounded or oval-shaped, widest at about its midlength, with thick, rounded, smooth platform borders. In the earliest δ morphotypes the platform ornamentation is restricted to two large nodes (Fig. 1:12–15); in later forms the number of nodes increases (Fig. 1:16–20).

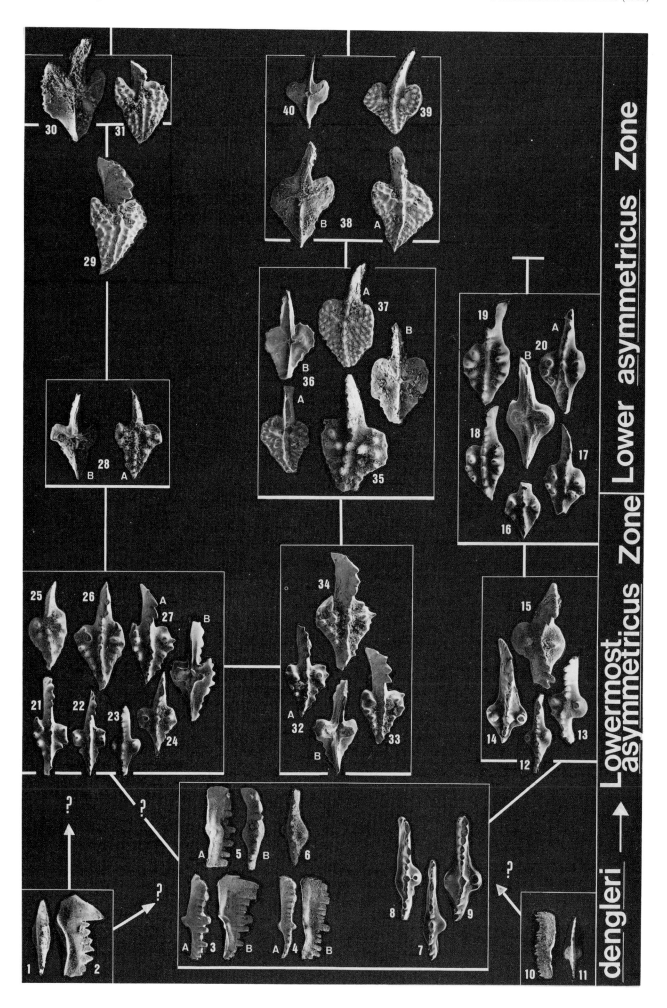

Material from the Boulonnais (N. France) and the Ardennes (Fig. 1:3–9; 12–15) demonstrates that the δ morphotype derives from *Ozarkodina* sp. A (Fig. 1:4), developing later a flat, smooth platform like extension on each side of the blade (Fig. 1:4, 5, 7). Forms herein called *Ozarkodina* aff. *adventa* and *Ozarkodina* aff. *proxima* (Fig. 1:6, 8, 9) represent a further evolution to *Ancyrodella binodosa* by development of the lateral extensions and the presence of a node on one or both of these. All the forms assigned to this lineage share common characteristics of the blade-carina, identical outline of the upper and lower margins in lateral view, the twisted posterior end, and the presence of a cusp. *Ozarkodina* aff. *adventa* and *O.* aff. *proxima* were previously identified as *O. sannemanni adventa* and *O. sannemanni proxima* by Bultynck (1979) and Bultynck & Coen (1982). The different outline of the lower margin, the twisted posterior end and the presence of a cusp permit distinction from, respectively, *O. sannemanni adventa* and *O. sannemanni proxima*. *Ozarkodina* sp. A is somewhat similar to *O. sannemanni sannemanni* (Fig. 1:10–11) in the denticulation of the blade and the outline of the upper margin. It differs however in the arched outline of the lower margin posterior of the cavity, the cavity without marked lateral expansions, and the twisted posterior end. Transitional specimens have not been found and the phylogenetic relationship of the two taxa is questioned. Specimens from the lineage *Ozarkodina* sp. A–*A. binodosa* δ morphotype resemble *O. insita* (Fig. 1:1–2) in the high denticle(s) at the anterior end of the blade and in the profile, especially of the lower margin (Fig. 1:2 and 3B). In some sections *O. insita* occurs below *O.* sp. A–*A. binodosa* δ (Fig. 2 section 1 Nismes). Intermediate forms have not been collected, and the relationship between *O. insita* and the lineage *O.* sp. A–*A. binodosa* δ morphotype is uncertain. On Fig. 1 the δ morphotype is considered to be a branch which became extinct.

Two possibilities are discussed here concerning the α morphotype of *Ancyrodella binodosa*. *O.* sp. A could function as the ancestral form. An objection to this view is that *O.* aff. *adventa* and *O.* aff. *proxima* can not be considered as intermedi-

---

Fig. 1. Possible origins and phylogeny of late Givetian to early Frasnian *Ancyrodella* taxa. All specimens ×22.5. Each rectangle contains a species, subspecies or morphotype. The lower horizontal bold line of each rectangle indicates the approximately earliest occurrence of the taxon with reference to the conodont zonation. The upper horizontal line has no stratigraphic significance and does not correspond to the latest occurrence of the taxon. The distance between rectangles is not to scale. The position of the figured specimens does not necessarily correspond to their position in the given sections. All figured specimens are deposited at the *Koninklijk Belgisch Instituut voor Natuurwetenschappen – Institut royal des Sciences naturelles de Belgique*, Department of Paleontology, Micropaleontology–Paleobotany Section.

☐1–2. *Ozarkodina insita* Sample 14 from the Upper Member of the Fromelennes Formation, Sourd d'Ave, Ardennes (Belgium), section described by Bultynck & Jacobs (1982, Fig. 4). ☐1. Upper view. I. R. Sc. N. B., specimen No. b1431. ☐2. Lateral view. I. R. Sc. N. B., specimen No. b1430. ☐3–6. Lineage *Ozarkodina* sp. A. – *Ozarkodina* aff. *adventa* – *Ozarkodina* aff. *proxima*. All specimens are from sample 42 from the lower part of the Frasnes Group, Nismes, Ardennes (Belgium), section described by Bultynck & Jacobs (1982, Fig. 2). ☐3 A. Oblique upper view of *O.* aff. *adventa*. I. R. Sc. N. B., specimen No. b1437. ☐3 B. Lateral view of the same specimen as in 3 A. ☐4 A. Oblique upper view of *O.* sp. A. I. R. Sc. N. B., specimen No. b1436. ☐4 B. Lateral view of same specimen as in 4 A. ☐5 A. Lateral view of *O.*aff. *adventa*. I. R. Sc. N. B., specimen No. b1439. ☐5 B. Upper view of same specimen as in 5 A. ☐6. Upper view of *O.* aff. *proxima*. I. R. Sc. N. B., specimen No. b1440. ☐7–9. Lineage *Ozarkodina* aff. *adventa* – *O.* aff. *proxima* – *Ancyrodella binodosa*. Specimen in 7 is from sample 3 and specimens in 8 and 9 from sample 4 from the Cambresèque Member of the Beaulieu Formation, Banc Noir, Boulonnais (France), section described by Brice, Bultynck, Deunff, Loboziak & Streel (1977, Table 3). ☐7. Upper view of *O.* aff. *adventa*. I. R. Sc. N. B., specimen No. b1124. ☐8. Upper view of *O.* aff. *sannemanni*, I. R. Sc. N. B., specimen No. b1125. ☐9. Upper view of *O.* aff. *sannemanni*, specimen transitional to *A. binodosa*. I. R. Sc. N. B., specimen No. b1522. ☐10–11. *Ozarkodina sannemanni*. Sample BT43 from the Bouia Formation, Bou Tchrafine, Tafilalt (S. Morocco), section described by Bultynck & Jacobs (1981, Fig. 3). ☐10. Lateral view. I. R. Sc. N. B., specimen No. b1523. ☐11. Upper view. I. R. Sc. N. B., specimen No. b1524. ☐12–15. Earliest δ morphotypes of *Ancyrodella binodosa*. ☐12. Upper view. Sample 45, same section as for specimens in 3–6. I. R. Sc. N. B., specimen No. b1453. ☐13. Upper view. Sample 45, same section as for specimens in 3–6. I. R. Sc. N. B., specimen No. b1455. ☐14. Upper view. Sample 4, same section as for specimens in 7–9. I. R. Sc. N. B., specimen No. b1126. ☐15. Upper view. Sample 43, same section as for specimens in 3–6. I. R. Sc. N. B., specimen No. 1454. ☐ 16–20. Ontogenetic series of δ morphotypes of *Ancyrodella binodosa*. All specimens are from sample AB5 from the Upper Member of the Bou Dib Formation, Ma'der (S. Morocco), section described by Bultynck & Jacobs (1981, Fig. 5). ☐16. Upper view. I. R. Sc. N. B., specimen No. b1409. ☐17. Upper view. I. R. Sc. N. B., specimen No. b1408. ☐18. Upper view. I. R. Sc. N. B., specimen No. b1407. ☐19. Upper view. I. R. Sc. N. B., specimen No. b1411. ☐20 A. Upper view. I. R. Sc. N. B., specimen No. b1410. ☐20 B. Lower view of the same specimen as in 20 A. ☐21–23, 25–27. *Ancyrodella binodosa* α morphotype. Specimens in 23, 22, 21, 27 and 26 representing an ontogenetic series are from the same sample as those in 16–20. Specimen in 25 from sample 46, same section as for specimens in 1:3–6. ☐21. Upper view. I. R. Sc. N. B., specimen No. b1525. ☐22. Upper view. I. R. Sc. N. B., specimen No. b1393. ☐23. Upper view. I. R. Sc. N. B., No. b1391. ☐25. Upper view. I. R. Sc. N. B., specimen No. b1448. ☐26. Upper view. I. R. Sc. N. B., specimen No. b1526. ☐27 A. Upper view. I. R. Sc. N. B., specimen No. b1527. ☐27 B. Lower view of same specimen as in 27 A. ☐24. *Ancyrodella binodosa*, β morphotype transitional between the asymmetric α morphotype and the nearly symmetric γ morphotype. Upper view. Sample 41 same section as for specimens in 3–6. I. R. Sc. N. B., specimen No. b1450. ☐28. *Ancyrodella rotundiloba rotundiloba*, asymmetric form. Sample 52 C, same section as for specimens in 3–6. ☐28 A. Upper view. I. R. Sc. N. B., specimen No. b1459. ☐28 B. Lower view of same specimen as 28 A. ☐29. *Ancyrodella rotundiloba rotundiloba* – *Ancyrodella rugosa*. Oblique upper view. Arrangement of nodes on the upper surface as in *Ancyrodella rugosa*; lower surface without secondary keels. Sample 65, same section as for specimens in 3–6. I. R. Sc. N. B., specimen No. b1469. ☐30–31. *Ancyrodella rugosa*. Sample 30 from the lower part of the Frasnes Group, same section as for specimens in 1–2. ☐30. Lower view. I. R. Sc. N. B. specimen No. b1472. ☐31. Upper view. I. R. Sc. N. B., specimen No. b1472. ☐32–34. *Ancyrodella binodosa*, γ morphotype. All specimens from the same sample as those in 16–20. ☐32 A. Upper view. I. R. Sc. N. B., specimen No. b1404. ☐32 B. Lower view of same specimen as in 32 A. ☐33. Upper view. I. R. Sc. N. B., specimen No. b1528. ☐34. Upper view. I. R. Sc. N. B., specimen No. b1406. ☐35. *Ancyrodella rotundiloba rotundiloba*. Early form. Upper view. Sample 41, same section as for specimens in 3–6. I. R. Sc. N. B., specimen No b1457. ☐36. *Ancyrodella rotundiloba rotundiloba*. Typical form. Sample AB 8, same section as for specimens in 16–20. ☐36 A. Upper view. I. R. Sc. N. B., specimen No. b1424. ☐36 B. Lower view of same specimen as in 36 A. ☐37. *Ancyrodella rotundiloba rotundiloba*. Typical form. Upper view. Sample 66, same section as for specimens in 3–6. I. R. Sc. N. B., specimen No. b1461. ☐38. *Ancyrodella rotundiloba alata*. Early form. Sample 65, same section as for specimens in 3–6. ☐38 A. Upper view. I. R. Sc. N. B., specimen No b1467. ☐38 B. Lower view of same specimen as in 38 A. ☐39–40. *Ancyrodella rotundiloba alata*. typical form. Samples 49 and 32, same section as for specimens in 1–2. ☐39. Upper view. I. R. Sc. N. B., specimen No. b1463. ☐40. Lower view. I. R. Sc. N. B., specimen No. b1466.

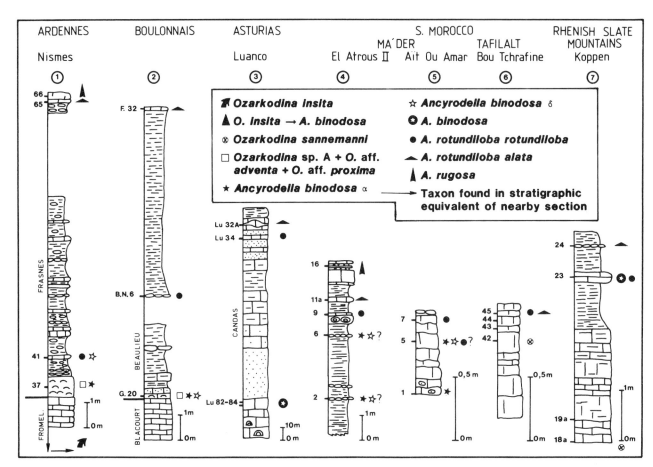

Fig. 2. Sequence of late Givetian to early Frasnian *Ancyrodella* taxa and their possible ancestors in selected western European, North African and North American sections.

Original references to conodont faunas, conodont biofacies, dominant megafauna, environment (near-shore and off-shore), and sample numbers for each of the numbered sections: □1. Ardennes, Nismes (Belgium) – Bultynck, including contributions by Jacobs (1982, Figs. 2, 3, Table 1). Benthic megafauna, brachiopods and corals. *Icriodus–Polygnathus* biofacies. Near-shore. □2. Boulonnais (northeastern France) – Brice, Bultynck, Deunff, Loboziak & Streel (1979:332, Table 3) Benthic megafauna, brachiopods, bryozoans and corals. *Icriodus–Polygnathus* biofacies. Near-shore. □3. Asturias, Luanco (north Spain) – Garcia Lopez (1982, Fig. 53). Benthic megafauna, brachiopods, corals, trilobites. *Icriodus–Polygnathus* biofacies with rare *P. asymmetricus*. Near-shore. □4. Ma'der, El Atrous II (Southern Morocco) – Hollard (1974:41, Fig. 7). Samples 2 to 9 benthic megafauna, brachiopods; sample 11a pelagic fauna, goniatites and styliolinids. Samples 2 to 9 *Icriodus–Polygnathus* biofacies; sample 11 a, *P. asymmetricus*. □5. Ma'der, Ait ou Amar (Southern Morocco) – Bultynck & Jacobs (1981, Fig. 5). Sample 1, benthic megafauna,

brachiopods; samples 3–7, pelagic fauna, styliolinids and goniatites. Samples 1–5, *Polygnathus–Ancyrodella* biofacies; sample 7, *P. asymmetricus* and *P. dengleri* Bischoff & Ziegler 1957. □6. Tafilalt, Bou Tchrafine (Southern Morocco) – Bultynck & Jacobs (1981, Fig. 3). Pelagic fauna, goniatites, orthoceratites, tentaculites. *Polygnathus–Ancyrodella–Palmatolepis* biofacies. Off-shore. □7. E. Rhenish Slate Mountains Koppen (Western Germany) – Ziegler, Klapper & Johnson (1976, Fig. 2, Table 13). Pelagic fauna, tentaculites and styliolinids. *Polygnathus–Ancyrodella–Palmatolepis* biofacies. Off-shore. □8 and 9. Northeastern and central Alberta (Canada) – Uyeno (1974, Tables 1 and 4b) and Norris & Uyeno (1981, Figs. 2, 4). Benthic megafauna, brachiopods. *Icriodus–Polygnathus* biofacies; insitus biofacies. Near-shore. □10 and 11. Northern Antelope Range, Nevada (U. S. A.) – Johnson, Klapper & Trojan (1980, Fig. 4, Tables 18, 22). Benthic megafauna, brachiopods, corals, crinoids. Diverse conodont association. Off-shore. □12. New York, Fall Brook (U.S.A.) – Huddle, assisted by Repetski (1981, Table 1, Sheets 1–5). Mainly pelagic, goniatites, styliolinids, small brachiopods. Diverse conodont association. Distal basin.

ate forms. Both have flat platform-like extensions with smooth rounded rims as observed in the δ morphotype of *A. binodosa*, whereas the α form has raised, denticulated rims. 'Spathognathodus'? sp., described by Uyeno (1967, 1974) from the Firebag Member of the Waterways Formation of northeastern and central Alberta, has an incipient platform with erect lateral denticles and fits well as a morphological intermediate between *O. insita* and *A. binodosa* α morphotype. Two branches are recognized in the further development of the last mentioned form. An asymmetrical form of *A. rotundiloba rotundiloba* with one rounded anterior lobe and another angular anterior lobe (Fig. 1:28) derives from the α morphotype of *A. binodosa* by reduction of the cavity to a small rhombic pit and by increase of

number of nodes on the platform surface. Similar asymmetric forms from the Squaw Bay Formation of Michigan have been figured by Müller & Clark (1967, Figs. 5e and 6n, x, i). The progressive development of two secondary keels extending from the pit in an anterior direction, and the arrangement of the upper surface nodes in two or four longitudinal rows gave rise to *Ancyrodella rugosa* Branson & Mehl 1934 (Fig. 1:29–31).

A nearly symmetrical γ morphotype of *A. binodosa* (Fig. 1:32–34) is linked to the α morphotype by intermediate forms (Bultynck & Jacobs 1981:16). It is characterized by a combination of a triangular platform with slightly raised, denticulated rims and two rounded anterior lobes. *A. binodosa* γ is ancestral to typical *A. rotundiloba rotundiloba* specimens with

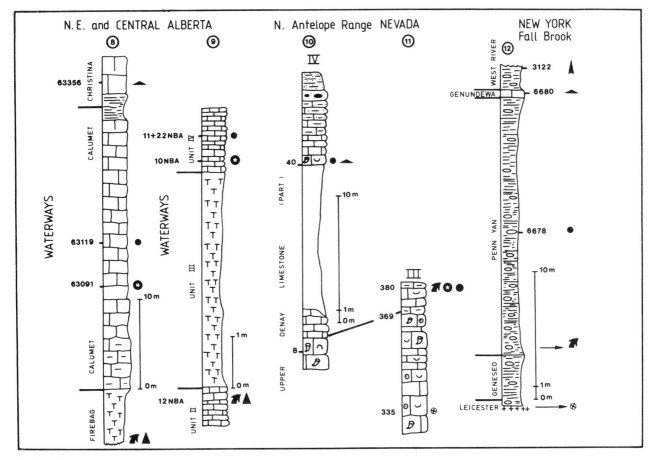

Fig. 2 (cont.).

broad, rounded anterior lobes. There is also an increase in the number of nodes on the platform surface and reduction of the cavity to a small rhombic pit (Fig. 1:35–37). The transverse development of the anterior lobes and the originating of two secondary keels, one extending anteriorly from the pit, the other laterally or slightly posteriorly, leads to *A. rotundiloba alata* Glenister & Klapper 1966 (Fig. 1:38–40 herein).

Stratigraphic and paleogeographic aspects

The postulated development of early *Ancyrodella* species (*A. binodosa*, *A. rotundiloba* and *A. rugosa*) from an *Ozarkodina* stock as outlined here took place in the upper *P. dengleri* Subzone – lowermost *P. asymmetricus* Zone and the lower *P. asymmetricus* Zone, that is during the late Givetian to early Frasnian according to the conodont succession in the type area of these stages.

Fig. 2 represents twelve sections in Europe, N. Africa and N. America showing the earliest stratigraphic occurrence of the *Ancyrodella* taxa discussed here and of the possible ancestral *Ozarkodina* taxa. From the conodont distribution it is concluded that *Ozarkodina insita*, *O. insita*→*Ancyrodella binodosa*, *O. sannemanni*, *O.* sp. A., *O.* aff. *adventa* and *O.* aff. *proxima* appear earlier than or coexist with *A. binodosa*. All the successions where *A. binodosa* is linked to the possible ancestral *Ozarkodina* taxa by intermediate forms belong to the near-shore facies: (1) Ardennes, (2) Boulonnais, (8) and (9) northeastern and central Alberta. This is also true for sections where *A. binodosa*

appears alone and earlier than *A. rotundiloba rotundiloba*. Conodont faunas of all these sections belong either to the *O. insitus* biofacies or to the icriodid–polygnathid biofacies and contain polygnathids possessing narrow platforms with simple ornamention (*P. dubius* Hinde 1879–*P. webbi* Stauffer 1938 type). Polygnathids with a broad and heavy tuberculated platform surface (*P. cristatus* Hinde 1879–*P. asymmetricus*) are absent. From these observations it is evident that the development of *A. binodosa* from an *Ozarkodina* stock took place in a near-shore environment which extended over a large geographic area including at least western Europe and North America. *O. sannemanni sannemanni* occurs mainly in the offshore facies and it is unlikely therefore that it is the ancestral form of *A. binodosa*. This confirms the conclusion from the part on the morpho-phylogenetic aspects (p. 165).

The evolution of *Ancyrodella rotundiloba rotundiloba* from *A. binodosa* also corresponds to a dispersal of the genus in the offshore facies. This is illustrated by the occurrence of *A. rotundiloba rotundiloba* in all the successions in Fig. 2.

From a biostratigraphic point of view it seems that *A. rotundiloba alata* appeared later than *A. rotundiloba rotundiloba*, and *A. rugosa* later than *A. rotundiloba alata*. This succession provides a valuable tool for subdivision of the Lower *P. asymmetricus* Zone into three: lower with *A. rotundiloba rotundiloba* and eventually *A. binodosa;* middle with *A. rotundiloba rotundiloba* and *A. rotundiloba alata;* upper with *A. rotundiloba rotundiloba*, *A. rotundiloba alata* and *A. rugosa*. Huddle & Repetski (1981:B4–B5) observed the same sequence in the Genesee Formation in western New York.

References

Bischoff, G. & Ziegler, W. 1957: Die Conodontenchronologie des Mitteldevons und des tiefsten Oberdevons. *Abh. Hess. L.-Amt Bodenforsch. 22*, 1–136.

Branson, E. B. & Mehl, M. G. 1933: Conodont studies number 1. *Missouri Univ. Studies 8*, 1–72.

Branson, E. B. & Mehl, M. G. 1934: Conodont studies number 3. *Missouri Univ. Studies 8*, 172–260.

Brice, D., Bultynck, P., Deunff, J., Loboziak, S. & Streel, M. 1979: Données biostratigraphiques nouvelles sur le Givétien et le Frasnien de Ferques (Boulonnais, France). *Ann. Soc. Géol. Nord 98*, 325–344.

Bryant, W. L. 1921: The Genesee conodonts. *Bull. Buffalo Soc. Nat. Sc. 13, 2*, 1–59.

Bultynck, P. 1979: Les conodontes. *In* Brice, D. *et al.:* Données biostratigraphiques nouvelles sur le Givétien et le Frasnien de Ferques (Boulonnais, France). *Ann. Soc. Géol. Nord 98*, 331–335.

Bultynck, P. & Coen, M. 1982: Распространение конодонтов в свите фромелен и нижней части „слоев фрасн" (граница среднего и верхнего девона в Арденнах). [Conodont distribution in the Fromelennes Formation and the lower part of the 'Assise de Frasnes' (Middle/Upper Devonian boundary of the Ardennes).] *In* Соколов, Б.С. & Ржонсницкая, М.А. (ред.): Биостратиграфия пограничных отложений нижнего и среднего девона: Труды полевой сессии Международной подкомиссии по стратиграфии девона, Самарканд, 1978. „Наука", Ленинград.

Bultynck, P. & Jacobs, L. 1981: Conodontes et sédimentologie des couches de passage du Givétien au Frasnien dans le nord du Tafilalt et dans le Ma'der (Maroc présaharien). *Bull. Inst. R. Sci. Nat. Belg. Ser. Sci. Terre 53*, 1–24.

Bultynck, P. & Jacobs, L. 1982: Conodont succession and general faunal distribution across the Givetian–Frasnian boundary beds in the type area. *In:* Papers on the Frasnian–Givetian boundary. *Geological Survey of Belgium, Spec. Vol.*, 34–59.

Druce, E. C. 1976: Conodont biostratigraphy of the Upper Devonian reef complexes of the Canning Basin, Western Australia. *Bureau Mineral Resourc., Geol. and Geoph. Bull. 158*, 1–301.

Ethington, R. L. & Furnish, W. M. 1962: Silurian and Devonian conodonts from Spanish Sahara. *J. Paleont. 36*, 1253–1290.

[Garcia Lopez, S. 1982: Los Conodontos y su aplicacion al estudio de las divisiones cronostratigraficas mayores del devonico Asturleones. – España. Unpublished Dr. Thesis, Facultad de Geologia Universidad de Oviedo, 1–292. Oviedo.]

Glenister, B. F. & Klapper, G. 1966: Upper Devonian conodonts from the Canning Basin, Western Australia. *J. Paleont. 40*, 777–842.

Hinde, G. J. 1879: On conodonts from the Chazy and Cincinati Group of the Cambro-silurian, and from the Hamilton and Genesee-Shale divisions of the Devonian, in Canada and the United States. *Quart. J. Geol. Soc. London 35*, 351–369.

Hollard, H. 1974: Recherches sur la stratigraphie des formations du Devonien Moyen, de l'Emsien supérieur au Frasnien, dans le Sud du Tafilalt et dans le Ma'der (Anti-Atlas Oriental). *Notes Serv. Géol. Maroc 36*, 7–68.

Huddle, J. W., assisted by Repetski, J. E. 1981: Conodonts from the Genesee Formation in Western New York. *U.S. Geol. Survey Prof. Paper 1032-B*, B1-B66.

Johnson, J. G., Klapper, G. & Trojan, W. R. 1980: Brachiopod and conodont successions in the Devonian of the northern Antelope Range, central Nevada. *Geologica et Palaeontologica 14*, 77–116.

Klapper, G. & Philip, G. M. 1971: Devonian conodont apparatuses and their vicarious skeletal elements. *Lethaia 4*, 429–452.

Klapper, G. & Philip, G. M. 1972: Familial classification of reconstructed Devonian conodont apparatuses. *Geologica et Palaeontologica, SB1*, 97–114.

Müller, K. J. & Clark, D. L. 1967: Early late Devonian conodonts from the Squaw Bay Limestone in Michigan. *J. Paleont. 41*, 902–919.

Müller, K. J. & Müller, E. M. 1957: Early Upper Devonian (Independence) conodonts from Iowa, part 1. *J. Paleont. 31*, 1069–1108.

Norris, A. W. & Uyeno, T. T. 1981: Stratigraphy and paleontology of the lowermost Upper Devonian Slave Point Formation on lake Claire and the lower Upper Devonian Waterways Formation on Birch River, northeastern Alberta. *Bull. Geol. Surv. Canada, 334*, 1–53.

Pollock, C. A. 1968: Lower Upper Devonian conodonts from Alberta, Canada. *J. Paleont. 42*, 415–443.

Stauffer, C. R. 1938: Conodonts of the Olentangy Shale. J. Paleont. 14, 417–435.

Stauffer, C. R. 1940: Conodonts from the Devonian and associated clays of Minnesota. *J. Paleont. 14*, 417–435.

Ulrich, E. O. & Bassler, R. S. 1926: A classification of the toothlike fossils, conodonts, with descriptions of American Devonian and Mississippian species. *U. S. Nat. Mus. Proc. 68:12*, 1–63.

Uyeno, T. T. 1967: Conodont zonation, Waterways Formation (Upper Devonian) of northeastern and central Alberta. *Geol. Surv. Canada Paper 67–30*, 1–20.

Uyeno, T. T. 1974: Conodonts of the Waterways formation (Upper Devonian) of northeastern and central Alberta. *Geol. Surv. Canada Bull. 232*, 1–93.

Ziegler, W. 1962: Phylogenetische Entwicklung stratigraphisch wichtiger conodonten-Gattungen in der *Manticoceras*-Stufe (Oberdevon, Deutschland). *N. Jb. Geol. Paläont. Abh. 114*, 142–168.

Ziegler, W. (ed.) 1973: *Catalogue of conodonts 1*, 1–504. Stuttgart.

Ziegler, W., Klapper, G. & Johnson, J. G. 1976: Redefinition and Subdivision of the *varcus*-Zone (Conodonts, Middle–?Upper Devonian) in Euorpe and North America. *Geologica et Palaeontologica 10*, 109–140.

A Late Mississippian conodont faunule from area of proposed Pennsylvanian System stratotype, eastern Appalachians

JOHN E. REPETSKI AND THOMAS W. HENRY

FOSSILS AND STRATA

Repetski, John E. & Henry, Thomas W. 1983 12 15: A Late Mississippian conodont faunule from area of proposed Pennsylvanian System stratotype, eastern Appalachians. *Fossils and Strata*, No. 15, pp. 169–170. Oslo. ISSN 0300-9491. ISBN 82-0006737-8.

A late Chesterian (Late Mississippian) conodont faunule assigned to the *Adetognathus unicornis* or lower *Rhachistognathus muricatus* Conodont Assemblage Zone was recovered from the Bramwell Member of the Bluestone Formation in southern West Virginia, thereby extending the geographic range of these assemblage zones into the eastern Appalachians. The Bramwell Member thus correlates with, or may even be younger than, the Grove Church Shale, the highest unit traditionally placed in the type Chesterian sequence, and it correlates with the uppermost Namurian A strata in western and central Europe. □*Conodonta, biostratigraphy, Carboniferous, Appalachians.*

ECOS III

A contribution to the Third European Conodont Symposium, Lund, 1982

John E. Repetski and Thomas W. Henry, U.S. Geological Survey, E-501, U.S. National Museum, Washington, D.C. 20560, USA; 16th July, 1983.

Upper Mississippian and Lower Pennsylvanian rocks form a thick sequence in the eastern part of the Appalachian basin. The succession near Bluefield, Virginia – West Virginia, is one of the few in eastern North America in which deposition was virtually continuous across the Mississippian–Pennsylvanian boundary. The U.S. Geological Survey is proposing that stratotypes for the Mississippian–Pennsylvanian boundary and the Lower Pennsylvanian Series be established in this area.

The Bramwell Member of the Bluestone Formation is the youngest unit included in the Mississippian (Fig. 1). About 30 m thick at its type locality, it consists primarily of a sequence of fine-grained terrigenous clastic rocks, which coarsens upward. It was deposited as a set of nearshore marine beds during a transgression over coastal-plain and near-coastal nonmarine deposits of the red member of the Bluestone Formation. The Bramwell Member generally is overlain conformably either by the upper member (near-coastal nonmarine beds) of the Bluestone or by the lower sandstone member of the Pocahontas Formation. The lower sandstone member was formed by a series of distributaries prograding over the upper member and Bramwell Member of the Bluestone. As shown by Englund (1979) and by Englund *et al.* (1981), the lower sandstone member of the Pocahontas interfingers and intertongues with the upper member of the Bluestone in interdistributary areas; however, near distributary axes, the lower sandstone member of the Pocahontas truncates subjacent beds and rests directly on the upper part of the red member of the Bluestone.

Studies of the macrofloras by Gillespie & Pfefferkorn (1979) and Pfefferkorn & Gillespie (1981) have indicated that the Mississippian–Pennsylvanian boundary, defined in this sequence at the top of the Bramwell Member of the Bluestone

Formation, corresponds exactly with the Namurian A – Namurian B boundary of western Europe.

The Bramwell Member contains a locally diverse marine invertebrate fauna. Preliminary studies by Gordon & Henry (1981) have suggested that this molluscan-dominated fauna consists of at least 45 macroinvertebrate taxa and that it is closely similar to the faunas of the upper Chesterian of the Ozark Plateaus region of north-central Arkansas.

A 5 kg sample of fossiliferous silty limestone from the upper part of the Bramwell Member at Freeman, Mercer County, West Virginia (U.S. Geological Survey collection 26789-PC), yielded a small collection of conodonts, including Pa elements of *Adetognathus unicornis* (Rexroad & Burton 1961), *Cavusgnathus convexus* Rexroad 1957, *C. naviculus* (Hinde 1900), and *Gnathodus bilineatus* (Roundy 1926) morphotype δ (Fig. 2). This co-occurrence establishes a late Chesterian age, limited to the *A. unicornis* through at least the lower *Rhachistognathus muricatus* Zones *sensu* Lane & Straka (1974).

Although further sampling may or may not allow ultimate assignment of these strata more precisely to either the *A. unicornis* or the *R. muricatus* Zone, some important points can be made from this discovery:

(1) If the upper part of the Bramwell Member represents the *A. unicornis* Zone, then the Mississippian–Pennsylvanian boundary as used in this area at present is somewhat different from that used elsewhere in North America by most conodont workers, i.e. at the *R. muricatus* – *R. primus* Zone boundary (Lane & Straka 1974; Lane 1977; Lane & Baesemann 1982).

(2) If these strata can be shown to represent the *R. muricatus* Zone, then the Bramwell Member would be slightly younger

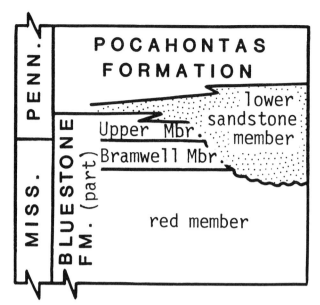

Fig. 1. Sketch of stratigraphic relationships of Mississippian–Pennsylvanian boundary sequence in area of proposed Pennsylvanian System stratotype near Bluefield, Virginia – West Virginia.

than the Grove Church Shale at the top of the type Chesterian sequence in the Illinois basin.

(3) This faunule from the Bramwell Member is the first record of Chesterian-age conodonts as young as the *A. unicornis* Zone from the Appalachian basin. It therefore is the youngest Mississippian conodont fauna yet known from eastern North America and represents a significant extension of geographic range for this Midcontinent fauna.

Fig. 2. SEM photomicrographs of conodont Pa elements from upper part of Bramwell Member of the Bluestone Formation at Freeman, West Virginia (USGS collection 26789-PC). Specimens, photographed coated lightly with carbon (later removed), reposited in collections of Department of Paleobiology, U.S. National Museum of Natural History, Washington, D.C., U.S.A. □A. *Cavusgnathus convexus* Rexroad. Inner lateral view, ×50, USNM 347251. □B. *Cavusgnathus naviculus* (Hinde). Inner lateral view, ×50, USNM 347252. □C, D, E. *Gnathodus bilineatus* (Roundy) morphotype δ. Lower, ×50; outer lateral, ×50; and upper (stereopair, ×45) views, respectively, USNM 347253. □F, G. *Adetognathus unicornis* (Rexroad & Burton). Lateral, ×100, and upper (stereo-pair, ×105) views, respectively, of specimen having most of free blade missing, USNM 347254.

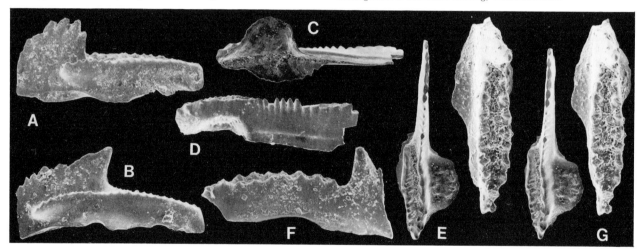

References

Englund, K. J. 1979: Mississippian System and Lower Series of the Pennsylvanian System in the proposed Pennsylvanian System stratotype area. *In* Englund, K. J., Arndt, H. H. & Henry, T. W. (eds.): Proposed Pennsylvanian System stratotype, Virginia and West Virginia. *American Geological Institute, Selected Guidebook Series No. 1*, 69–72.

Englund, K. J., Henry, T. W. & Cecil, C. B. 1981: Upper Misissippian and Lower Pennsylvanian depositional environments, southwestern Virginia and southern West Virginia. *In* Roberts, T. G. (ed.): *GSA Cincinnati 1981, Field Trip Guidebooks. Vol. 1, Stratigraphy, Sedimentology,* 171–175. American Geological Institute, Washington, D. C.

Gillespie, W. H. & Pfefferkorn, H. W. 1979: Distribution of commonly occurring megafossils in the proposed Pennsylvanian System stratotype, Virginia and West Virginia. *American Geological Institute, Selected Guidebook Series No. 1*, 86–96.

Gordon, Mackenzie, Jr. & Henry, T. W. 1981: Late Mississippian and Early Pennsylvanian invertebrate faunas, east-central Appalachians – A preliminary report. *In* Roberts, T. G. (ed.): *GSA Cincinnati 1981, Field Trip Guidebooks. Vol. 1, Stratigraphy, Sedimentology,* 165–171. American Geological Institute, Washington, D. C.

Hinde, G. J. 1900: *In* Smith, J.: Conodonts from the Carboniferous limestone strata of the west of Scotland. *Glasgow Natural History Society Transactions 5, new series,* 336–346.

Lane, H. R. 1977: Morrowan (Early Pennsylvanian) conodonts of northwestern Arkansas and northeastern Oklahoma. *In* Suther-

land, P. K. & Manger, W. L. (eds.): Mississippian–Pennsylvanian boundary in northeastern Oklahoma and northwestern Arkansas. *Oklahoma Geological Survey, Guidebook 18,* 177–180.

Lane, H. R. & Baesemann, J. F. 1982: A Mid-Carboniferous boundary based on conodonts and revised intercontinental correlations. *In* Ramsbottom, W. H. C., Saunders, W. B. & Owen, B. (eds.): *Biostratigraphic Data for a Mid-Carboniferous boundary; International Union of Geological Sciences, Subcommission on Carboniferous Stratigraphy, Proceedings of Leeds Symposium,* 6–12.

Lane, H. R. & Straka, J. J., III 1974: Late Mississippian and Early Pennsylvanian conodonts, Arkansas and Oklahoma. *Geological Society of America, Special Paper 152.* 144 pp.

Pfefferkorn, H. W. & Gillespie, W. H. 1981: Biostratigraphic significance of plant megafossils near the Mississippian–Pennsylvanian boundary in southern West Virginia and southwestern Virginia. *In* Roberts, T. G. (ed.): *GSA Cincinnati 1981, Field Trip Guidebooks. Vol. 1, Stratigraphy, Sedimentology,* 159–164. American Geological Institute, Washington, D. C.

Rexroad, C. B. 1957: Conodonts from the Chester Series in the type area of southwestern Illinois. *Illinois Geological Survey, Report of Investigations 199.* 43 pp.

Rexroad, C. B. & Burton, R. C. 1961: Conodonts from the Kinkaid Formation (Chester) in Illinois. *Journal of Paleontology 35,* 1143–1158.

Roundy, P. V. 1926: *In* Roundy, P. V., Girty, G. H. & Goldman, M. I. (eds.): Mississippian formations of San Saba County, Texas; Part II, The micro-fauna: *U. S. Geological Survey Professional Paper 146,* 5–17.

Paleoenvironmental factors and the distribution of conodonts in the Lower Triassic of Svalbard and Nepal

DAVID L. CLARK AND ERIC W. HATLEBERG

FOSSILS AND STRATA

ECOS III

A contribution to the Third European
Conodont Symposium, Lund, 1982

Clark, David L. & Hatleberg, Eric W. 1983 12 15: Paleoenvironmental factors and the distribution of conodonts in the Lower Triassic of Svalbard and Nepal. *Fossils and Strata*, No. 15, pp. 171–175. Oslo. ISSN 0300-9491. ISBN 82-0006737-8.

Lower Triassic sections in Svalbard and Nepal contain markedly different conodont faunas. Each of the sections has yielded approximately 20 species but there are only two species in common. Associated megafauna confirms that a more or less complete Lower Triassic sequence exists in both sections, but if comparison of the sections were based exclusively on conodonts, little reliable correlation would be possible. The problem is solved by correlation of the conodonts of the Svalbard and Nepal sections with a third section, the comprehensive Lower Triassic of Utah and Nevada. Both Arctic and Tethyan sections have 7 species in common with that of Western U.S. and demonstration of equivalency of stages and zones is possible through use of the intermediate U.S. section. Possible explanations for lack of common species in two time-equivalent sections (Svalbard and Nepal) include such things as sampling and/or preservational problems at either or both sites, paleoclimatic differences for the two sections including latitudinal differences and a range of ecologic factors produced by different climates, provincialism because of paleogeographic barriers, and lithofacies and biofacies factors. Field and laboratory data are compelling evidence that the differences between conodont faunas in Svalbard and Nepal are due to lithofacies factors. Comparison of the conodonts with a previously established Lower Triassic biofacies model tends to confirm that faunal differences are related to distinct biofacies and suggests that there may be real value for the predictive power of conodonts for paleoenvironmental definition. □ *Conodonta, biogeography, ecology, Nepal, Svalbard, Triassic.*

David L. Clark, Department of Geology and Geophysics, University of Wisconsin, Madison, Wisconsin, 53706, U.S.A.; Eric W. Hatleberg, Arco Exploration Co., Lafayette, Louisiana, 70505, U.S.A.; 20th May, 1982.

Lower Triassic rocks of Spitsbergen average approximately 450 m in thickness and consist of clastics with minor carbonates that have been organized into the Vardebukta, Sticky Keep and part of the Botneheia Formations.

In contrast, Lower Triassic rocks of Nepal average approximately 25 m in thickness, consist of shales and nodular carbonates and have been referred to the Panjang and Thinigaon Formations.

Both Arctic (Svalbard) and Tethyan (Nepal) sections include megafossils that indicate a more or less complete Lower Triassic sequence and demonstrate equivalency of the two geographically separated sections (Buchan *et al.* 1965 for Svalbard; Waterhouse 1977, Bassoullet & Colchen 1977, Bordet *et al.* 1971, for Nepal). Of some 100 samples taken from the two areas, 42 % yielded conodonts. Some 20 species are represented in the Nepal collections, whereas 18 species were identified in the collections from Spitsbergen (Table 1). The two Lower Triassic sections have only two clearly identifiable species in common, however. In spite of the general usefulness of conodonts in stratigraphic studies for most part of the Paleozoic and Triassic, it seems apparent that if comparison of the sections was based exclusively on conodonts, little reliable correlation would be possible. The absence of common species that could be used in precise correlation raises several classical stratigraphic problems.

For this study we have considered two questions: (1) How do we establish precise stratigraphic correlations between the two sections? and, (2) What are the best possible explanations for lack of common species in two time-equivalent sections?

Comparisons, correlations and coefficients

Lower Triassic sections containing conodonts are widespread. In addition to the sections discussed in this report, sections in the Salt and Trans-Indus Ranges of West Pakistan and at the Guryul Ravine of Kashmir have been discussed by Sweet (1970a, b). Considered on a world-wide basis, these sections and that from Nepal are so closely spaced geographically (and so obviously part of the same Tethyan belt) that similarities are expected. In fact, the Tethyan sections are lithologically and faunally the most similar of the worldwide sections available for this study. Most of the species represented in Nepal were described originally from Pakistan. For these reasons, the central Tethyan sections are not contrasted in the discussion that follows.

However, because so few species were found to be common to the Lower Triassic of Nepal and Spitsbergen, we decided to compare these faunas with Lower Triassic conodonts from other parts of the world. Faunas from a few important Lower

Table 1. Lower Triassic conodonts of Svalbard and Nepal. From Hatleberg (1982); Svalbard section from Spitsbergen, including southern shoreline of Van Keulenfjorden at Reinodden, Pitnerodden and Ahlstravdodden; Nepal section from Thakkhola Valley in central Himalaya above Jamsom (Hatleberg & Clark, in press).

Species	Svalbard	Nepal
Hindeodus typicalis		x
Neogondolella carinata		x
N. milleri		x
N. jubata	x	x
N. aff. *mombergensis*	x	x
N. aff. *timorensis*		x
N. sp.		x
Neospathodus waageni		x
N. aff. *waageni*		x
N. conservativus		x
N. discretus		x
N. aff. *spathi*		x
N. homeri	x	x
N. triangularis		x
N. sp. a		x
N. sp. b		x
N. sp. c		x
N. sp. d		x
Ellisonia sp.		x
Ellisonia triassica	x	
Neogondolella elongata	x	
N. nevadensis	x	
N. regale?	x	
N.? mombergensis	x	
N. sp. a	x	
N. sp. b	x	
Neospathodus dieneri	x	
N. cristagalli	x	
N. svalbardensis	x	
N. pakistanensis	x	
N. peculiaris	x	
N. collinsoni	x	
N. aff. *triangularis*	x	
Xaniognathus? sp.	x	

Table 2. Lower Triassic conodonts from Australia. From McTavish (1973).

Neogondolella carinata
N. elongata
N. jubata
N. planata
Neospathodus bicuspidatus
N. conservativus
N. dieneri
N. homeri
N. novaehollandiae
N. pakistanensis
N. timorensis
N. waageni
N. aff. *discretus*
Ellisonia sp.
Xaniognathus sp.

Table 3. Lower Triassic conodonts from Western United States. From Carr (1981) and Paull (1980).

Isarcicella sp.
Hindeodus typicalis
Ellisonia gradata
Ellisonia triassica
Neogondolella carinata
N. milleri
N. jubata
N. timorensis
N. n. sp. a
Neospathodus peculiaris
N. dieneri
N. bicuspidatus
N. waageni
N. triangularis
N. collinsoni
N. homeri
Furnishius triserratus
Pachycladina sp.
Platyvillosus asperatus
Parachirognathus ethingtoni

Triassic sections are shown on Tables 1–4, and include those described in reports from western North America (Paull 1980; Carr 1981; Clark *et al.* 1979), South Primorye of the U.S.S.R. (Burij 1979), and Australia (McTavish 1973). Comparison of number of species from these areas plus Japan (Koike 1979) is shown in Figure 1. These sections provide a good data base for world-wide Lower Triassic conodont distribution.

Because we have not examined all of the faunas reported and have relied on published illustrations and descriptions for species identification, some subjectivity is inherent in our comparisons (Tables 1–4). We have not included the Japanese report for these data. In order to establish some kind of uniform baseline for the comparisons, coefficients of similarity were calculated using the equation $2w/a+b$, in which w = number of conodont species common to two areas and $a+b$ = total species for the two areas. The results for the 5 areas (Tables 1–4) are shown in Fig. 2.

Both the Nepal and Spitsbergen sections have their greatest similarity with the section in the western U.S. Both the Arctic and the Tethyan sections have 7 species in common with the U.S. section.

Even this low similarity coefficient (0.35–0.36) is adequate for demonstration of equivalency of several stages and zones by use of the intermediate U.S. section.

Table 4. Lower Triassic conodonts from the U.S.S.R. From Burij (1979).

Ellisonia triassica
Ellisonia teicherti
Ellisonia gradata
Ellisonia robusta
Ellisonia delicatula
Ellisonia torta
Ellisonia clarki
Hindeodus typicalis
Neogondolella carinata
N. elongata
N. jubata
Xaniognathus curvatus
Xaniognathus defectens
Xaniognathus elongatus
Isarcicella sp.
Neospathodus kummeli
N. dieneri
N. cristagalli
N. peculiaris
N. pakistanensis
N. waageni
N. spathi
N. triangularis
N. homeri
N. timorensis
'Prioniodella' prioniodellides

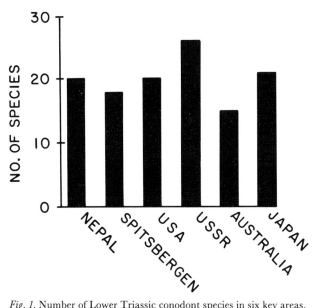

Fig. 1. Number of Lower Triassic conodont species in six key areas.

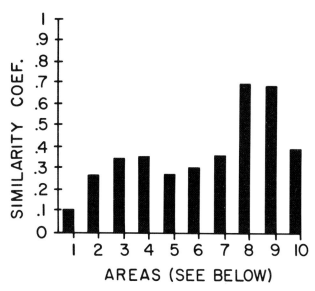

Fig. 2. Similarity coefficients for Lower Triassic conodonts in 5 areas. Data from Japan were not used. 1: Nepal–Svalbard. 2: Nepal–USSR. 3: Nepal–Australia. 4: Nepal–western US. 5: Svalbard–USSR. 6: Svalbard–Australia. 7: Svalbard–western US. 8: Western US–USSR. 9: Western US–Australia. 10: Australia–USSR.

Greatest similarity for Lower Triassic conodonts is that indicated for the U.S.S.R.–western U.S. sections. A similarly strong coefficient exists for the Australia–U.S. sections. The lowest similarity of all Lower Triassic faunas is that between the Nepal and Spitsbergen sections. We address this problem in the following pages.

Interpretations

The extremes in similarity coefficients among the various Lower Triassic sections pose fundamental stratigraphic questions: Are there sampling and/or preservation problems? Are there latitudinal differences that could account for the range of ecologic factors produced by different climates? What is the role of provincialism, and what are the biofacies factors?

Sampling and preservation

The section in Spitsbergen and that in Nepal are difficult to study although exposures are good to excellent. In both sections every exposed carbonate bed was sampled. The interbedded clastics were also sampled but not uniformly. The clastic beds did not yield conodonts, therefore, we conclude that all exposed rocks with potential conodonts were sampled. Preservation of conodont elements in the two sections is similar. In both Nepal and Spitsbergen, CAI values of 1.5 to 5 were obtained. Element abundances are higher in Nepal and CAI values range in both sections. The question of sampling or preservations problems for these sections may be moot.

Latitude difference

Knowledge of the general paleogeography of Earth 220 to 200 million years ago (Early Triassic) is still sketchy but the best evidence at hand (e.g., Ziegler 1981) indicates that Nepal was part of the Tethyan sea and probably within 10° to 15° of the Triassic equator. Spitsbergen paleogeography is less well known. Although Birkenmajer (1977) suggests that Svalbard may have been between 28°–30° North, probably most of Svalbard was at least 60° North, and perhaps even higher,

although there is no widespread agreement on this (e.g., Frakes 1979). Therefore there may have been a minimum of 50° difference between the two sections during the Early Triassic. Although this latitude spread is significant by modern climate standards, there are suggestions that during the Triassic this may not have been as important. Invertebrate and plant assemblages from many Lower Triassic sections suggest that there were broad, warm climate zones for this period. Subtropical to warm, temperate conditions probably extended from 15 to 55° North Latitude. Most of the Earth probably had average temperatures 20° warmer than at present (Frakes 1979). The other Lower Triassic faunas (Tables 2–4) had more uniform latitude addresses than those of Nepal and Spitsbergen. Clearly the generally uniform world wide climate patterns interpreted for the Early Triassic do not indicate that latitude differences were the most important factors to explain faunal differences and the low similarity coefficient for Nepal and Svalbard.

Provincialism and paleogeographic barriers

There are presently known at least 40 conodont species for the Lower Triassic. Most of the species are known at several localities. In addition, similarity coefficient calculations do not support provincialism for the Lower Triassic. In fact there is no distinctive fauna with elements unique to any one area. The general withdrawal of seas from the shallow shelf areas during the earliest Triassic may have produced paleogeographic barriers that were not present before eustatic lowering. Nevertheless, during the Early Triassic there apparently was more or less free interchange of species in the oceans and it is safe to assume that geographic barriers were not a significant factor producing faunal differences.

Biofacies

The Lower Triassic section of Spitsbergen consists of clastics, many coarse, and a few thin interbedded carbonates. In

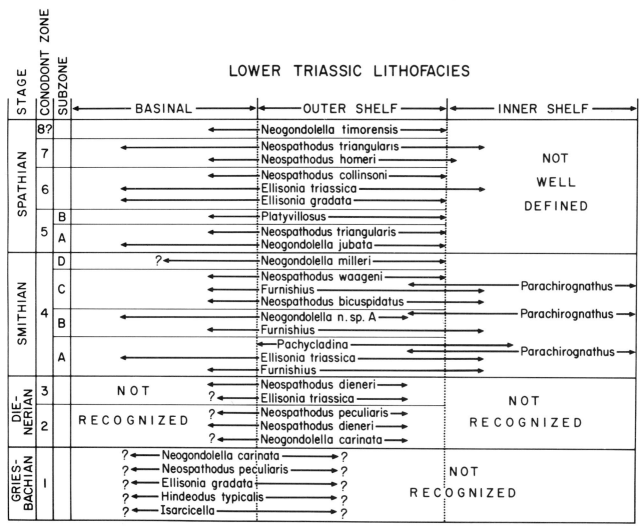

Fig. 3. Basinal, outer-shelf and inner-shelf conodont biofacies for western U.S. Lower Triassic. Environmental range of most species is approximate. Key species for biostratigraphy appear for intervals representing stratigraphic range. This figure is the basis for the ecologic calculations referred to in text.

general, the section includes a greater proportion of clastics in the younger parts. Middle and Upper Triassic sections are predominantly clastic. Sixty-six samples were taken, of which 28 yielded conodont elements. The carbonates that yielded conodonts averaged 50 % insoluble material. Overall, 75 % insoluble residue was the average for all samples. The section has been studied by sedimentologists who generally have argued for a deltaic–tidal zone interpretation for environment of deposition (Buchan *et al.* 1965; Birkenmajer 1977; Gazdzicki & Trammer 1977). Our field and laboratory observations on percentage of coarse clastic material and rareness of open-marine invertebrates support these interpretations. In addition, we have attempted to classify the conodont species present in Spitsbergen by comparison with a 'North American Biofacies Standard' (Fig. 3), based on extensive collections whose paleogeography and carbonate petrology have been determined (Paull 1980; Carr 1981; Carr & Paull 1981). Six of the Spitsbergen species are the same as those interpreted to range (in the western North America section) from the outer shelf to the basin; a total of 46 % of the ecologically classifiable Spitsbergen species are the same that elsewhere are interpreted to range from outer shelf to basin. A single species from

Spitsbergen is the same as a western North America species that has been interpreted to range onto the inner shelf.

In contrast, the Lower Triassic section in Nepal consists of shales at the base and has thin nodular carbonates throughout. Most of the carbonates are fine grained, and ammonites and other pelagic and benthic fossils are present. The nodular carbonates have been interpreted to represent an environment that may include water depths of 375–500 m (Hatleberg 1982). Some 56 % of the samples yielded conodonts and less than 10 % insoluble material remained after acidization. Comparison of the species present shows that 78 %, are the same as those in western North America that have been interpreted to range from the outer shelf into the basin. No inner-shelf species are present.

The lithofacies clearly suggest different environmental parameters for the two sections. The Spitsbergen section probably represents shallow-marine (10–30 m) to deltaic deposits with little carbonate but abundant clastics. The Nepal section represents deeper basinal deposition (375–500 m) and consists of thin nodular carbonates with abundant pelagic and benthic fossils.

Comparison of the percentage of species classifiable with

North American species is more subjective but indicates that 46 % of the Spitsbergen species and 78 % of the Nepal species represent outer-shelf to basinal taxa. These different percentages are interpreted to indicate important biofacies difference. Taken together, the petrologic and faunal data argue for differences in environment of deposition for the sections; similarly coefficient calculations confirm that different conodont biofacies characterize the different lithofacies.

Conclusions

The faunal comparisons for several major Lower Triassic sections provide data that indicate the sections in Nepal and Svalbard are most dissimilar. Biofacies differences may be sufficient to account for this dissimilarity in spite of obvious latitudinal and possible climatic differences in the two sections. We suggest that the Svalbard section represents the inner part of the marine outer shelf environment during the Early Triassic, while the Nepal secton represents more outer-shelf to basinal conditions.

The coefficient of similarity probably is a good expression of biofacies similarity. Sections that represent a basinal environment and an inner-shelf environment would have a similarity coefficient of 0 (i.e. no species in common) for the Lower Triassic. Sections from basinal and outer shelf edges would show higher coefficients. The USSR–US–Australia coefficients are very strong (Fig. 2) and probably indicate that very similar biofacies existed in the sampled areas during the Early Triassic. The very low coefficient for Nepal–Svalbard (0.10, Fig. 2), as is suggested in this report, indicates more dissimilar biofacies, perhaps toward the extremes of basinal and shelf.

Although documentation of biofacies differences in different sections is of considerable importance in understanding such things as low coefficient of similarity of the contained fauna, a second and perhaps more important conclusion results from these observations. This is the possibility that conodonts may have predictive power in biofacies definition. If a matrix could be arranged that included the biofacies range of various species based on reliable paleogeographic, faunal and petrologic studies, these species might then be used in biofacies interpretation of sections for which little petrologic or paleogeographic data are available. For a group such as conodonts whose members were predominantly parts of the pelagic realm, understanding of biofacies may account for most or all of the differences that are suggested by differences in similarity coefficients.

Clearly better data are needed for the basic biofacies matrix for each series or stage of the various periods during which conodonts were part of the marine fauna. Quantitative data for the various intervals could allow the use of species assemblages in biofacies definition of relatively poorly studied or understood sections. This is a well-defined challenge for the serious conodont student.

References

Bassoulet, J. P. & Colchen, M. 1977: La limite Permien–Trias dans le domaine tebetian de l'Himalaya du Nepal. *Coll. Intern. C.N.R.S. Ecologie et Geologie de l'Himalaya, Paris, 268,* 41–50.

Birkenmajer, K. 1977: Triassic sedimentary formations of the Hornsund area, Spitsbergen. *Stud. Geol. Polonica 51,* 7–74.

Bordet, P., Colchen, M., Krummenocher, D., LeFort, P., Mouterde, R. & Remy, M. 1971: *Recherches geologiques dans l'Himalaya du Nepal, region de la Thakkhola.* 279 pp. Centre Nat. Recher., Sci., Paris.

Buchan, S. H., Challinor, A., Harland, W. B. & Parker, J. R. 1965: The Triassic stratigraphy of Svalbard. *Norsk Polar. Skrifter, N.R. 135,* 1–93.

Burij, Galina (Бурий, Галина) 1979: Нижнетриасовые конодонты южного Приморья [*Lower Triassic Conodonts of South Primorye*]. 144 pp. Akademia Nauk SSSR, Sibirskoe Otdelenie. Moskva.

[Carr, T. R. 1981: Paleogeography, depositional history and conodont paleoecology of the Lower Triassic Thaynes Formation in the Cordilleran Miogeosyncline. Unpub. Ph. D. dissertation, Univ. Wiconsin–Madison.]

Carr, T. R. & Paull, R. K. 1981: Early Triassic stratigraphy and depositional history of the Cordilleran Miogeosyncline. *Geol. Soc. Am., Abstracts with Programs 12:7,* 399.

Clark, D. L., Paull, R. K., Solien, M. A. & Morgan, W. A. 1979: Triassic conodont biostratigraphy in the Great Basin. *In* Sandberg, C. A. & Clark, D. L. (eds.): Conodont biostratigraphy of the Great Basin and Rocky Mountains, *Brigham Young Univ. Geol. Studies 26:3,* 179–185.

Frakes, L. A. 1979: *Climates Throughout Geologic Time.* 310 pp. Elsevier Sci. Pub. Co. New York.

Gazdzicki, A. & Trammer, J. 1977: The Sverdrupi Zone in the Lower Triassic of Svalbard. *Acta Geol. Polonica 27,* 349–356.

[Hatleberg, E. W. 1982: Conodont biostratigraphy of the Lower Triassic at Van Keulenfjorden, Spitsbergen and the Thakkhola Valley, Nepal. Unpub. M. S. thesis, Univ. Wisconsin–Madison.]

Hatleberg, E. W. & Clark, D. L. (In press): Lower Triassic conodonts from Svalbard and Nepal. *Geol. et Palaeont. 17.*

Koike, Toshio 1979: Conodont biostratigraphy in the Taho Limestone (Triassic), Shirokawa-cho, Higashiuwa-gun, Ehime Prefecture. *In: Professor Mosaburo Kanuma Memorial Volume,* 115–137.

McTavish, R. A. 1973: Triassic conodont faunas from Western Australia. *Neues Jahrb. Geol. Paläont., Abh. 143,* 275–303.

[Paull, R. K. 1980: Conodont biostratigraphy of the Lower Triassic Dinwoody Formation in northwestern Utah, northeast Nevada, and southeastern Idaho. Unpub. Ph. D. dissertation, Univ. Wisconsin–Madison.]

Sweet, W. C. 1970a: Uppermost Permian and Lower Triassic conodonts of the Salt Range and Trans-Indus Ranges, West Pakistan. *In* B. Kummel & C. Teichert (eds.): Stratigraphic boundary problems: Permian and Triassic of west Pakistan. *Univ. Kansas Dept. Geol. Sp. Pub. 4,* 207–275.

Sweet, W. C. 1970b: Permian and Triassic conodonts from a section at Guryul Ravine, Vihi District, Kashmir. *Univ. Kansas Paleont. Institute Paper 4a.* 10 pp.

Waterhouse, J. B. 1977. The Permian rocks and faunas of Dolpo, northwest Nepal. *Coll. Intern. C.N.R.S. Ecologie Geol. de l'Himalaya, Paris, 268,* 479–496.

Ziegler, A. M. 1981: Paleozoic paleogeography. *In* M. W. McElhinny & D. A. Valencio (eds.): Paleoreconstruction of the continents, 31–37. *Geodynamics Series, v. 2, Amer. Geoph. Union.*

Epigondolella populations and their phylogeny and zonation in the Upper Triassic

MICHAEL JAMES ORCHARD

FOSSILS AND STRATA

Orchard, Michael J. 1983 12 15: *Epigondolella* populations and their phylogeny and zonation in the Upper Triassic. *Fossils and Strata*, No. 15, pp. 177–192. Oslo. ISSN 0300-9491. ISBN 82-0006737-8.

Eight distinctive populations of the conodont *Epigondolella* occur in the Norian (Upper Triassic) of British Columbia. Each is associated with rich ammonoid faunas. Intercalibration of the faunal succession through a remarkably complete Norian sequence is documented. The range of morphological variety within each conodont population is presented, a central morphotype is selected, and its growth series described. The problems arising from an inadequate appreciation of growth stages and of homeomorphy are stressed. New criteria for species determination are outlined, including the nature of microreticulation. The following succession is recognized: *E. primitia* – restricted to the *S. kerri* Zone; *E. abneptis* subsp. A – upper *S. kerri* and lower *M. dawsoni* Zones; *E. abneptis* subsp. B – upper *M. dawsoni* and *J. magnus* Zones; *E. multidentata – D. rutherfordi* Zone; *E.* n. sp. C – lower lower *M. columbianus* Zone; *E. postera* – upper lower *M. columbianus* Zone; *E.* n. sp. D – lower upper *M. columbianus* Zone; and *E. bidentata – G. cordilleranus* and *C. amoenum* Zones. □*Conodonta*, Epigondolella, *biostratigraphy, evolution, Triassic, British Columbia.*

Michael J. Orchard, Geological Survey of Canada, 100 West Pender Street, Vancouver, B.C., V6B 1R8, Canada; 3rd September, 1982.

ECOS III

A contribution to the Third European Conodont Symposium, Lund, 1982

The Pardonet Formation (McLearn 1960) of northeastern British Columbia in western Canada is a thin (maximum of about 135 m) but remarkably complete succession of Norian (Upper Triassic) strata deposited near the edge of the Cordilleran miogeosyncline. Now exposed in the Rocky Mountain foothills, particularly in the area of Peace River, the formation consists of dark calcareous siltstones and shales with subordinate coquinoid limestones (see Gibson 1971:22–3 for details) of relatively deep-water origin that contrast markedly with the Carnian dolomites of the underlying Baldonnel Formation.

The Pardonet Formation contains rich ammonoid faunas that have been central to the development of a biochronological standard for late Triassic time (Tozer 1967). New exposures of the formation, resulting from damming of the Peace River, have been studied and sampled in collaboration with E.T. Tozer, who has recognized sequences of ammonoid faunas embracing the zones of *Stikinoceras kerri*, *Malayites dawsoni*, *Juvavites magnus* (Lower Norian), *Drepaniceras rutherfordi*, *Mesohimavatites columbianus* (Middle Norian) and *Gnomohalorites cordilleranus* (Upper Norian). Abundant conodont faunules have been recovered from all of these zones. Elsewhere in western Canada, rare conodonts have also been found in strata assigned to the *Cochloceras amoenum* Zone (Upper Norian).

This paper outlines the succession of *Epigondolella* species, establishes criteria for recognizing and distinguishing eight principle species-complexes, and presents their relationship with each other and with the ammonoid standard.

Illustrated specimens are deposited with the Geological Survey of Canada in Ottawa.

Epigondolella populations

Conodont faunules from the Pardonet Formation include abundant representatives of the discrete platform element of *Epigondolella*. Eight successive *Epigondolella* populations covering the Norian ammonoid zones of *S. kerri* through *C. amoenum* (Fig. 1) are recognized. Within each population examples of the diverse array of morphotypes, including representatives of different growth stages, have been chosen for illustration (Figs. 2, 4, 6, 8, 10, 11, 13, 14). Of these, the most common element type has been selected as a *central morphotype* (shown shaded in the figs.).

Each population comprises specimens in addition to the central morphotype that have characteristic features in common with it, yet in some respects appear markedly different, particularly if considered in isolation. To a large extent individual conodont elements form points in a broad morphological continuum that I have attempted to illustrate and describe below (see also Fig. 14). Many individual morphotypes of a population can be hypothetically derived from different growth stages (frequently the earlier ones) of the central morphotype by the retention of certain characteristics or the exaggeration of certain trends. Most members of a population may be linked as shown in Fig. 14, which shows these relationships within *Epigondolella bidentata*, the youngest and structurally least complex of the epigondolellids. In this way, stratigraphically restricted, but less common morphotypes are generated in each population in addition to the central morphotype. These specimens are often homeomorphic with structurally comparable morphotypes of a different age.

	Ammonoid Zones (Tozer 1967, 1979, 1981)		Conodont Zones (Orchard, this paper)
UPPER	CRICKMAYI		?
	AMOENUM		– – – – – – – –
	CORDILLERANUS		E. bidentata
MIDDLE	COLUMBIANUS	4	– – – – – ? – – – – –
		3	E. n. sp. D.
		2	E. postera
		1	E. n. sp. C.
	RUTHERFORDI		E. multidentata
LOWER	MAGNUS		E. abneptis subsp. B
	DAWSONI		E. abneptis subsp. A.
	KERRI		E. primitia

Fig. 1. Zonation of the Norian based on ammonoids and conodonts.

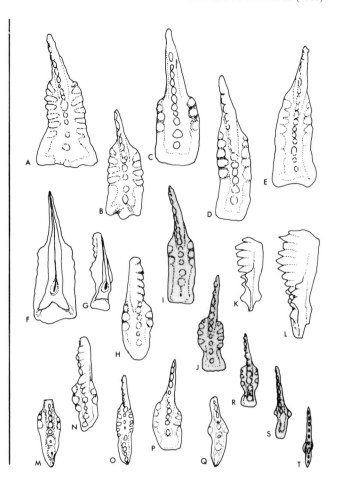

Fig. 2. Central morphotype growth series (shaded specimens) and morphological variation in *Epigondolella primitia* populations. The scale bar is 1 mm long.

In some species of *Epigondolella*, early growth stages are quite different from later ones, but may resemble later growth stages of other species. This phenomenon of neoteny, recognized in *Epigondolella* by Mosher (1970:740), demands that an accurate determination of growth stages be made, which requires a full appreciation of scale. This work attempts to define the growth series of each *Epigondolella* species as exemplified by its central morphotype. Put into this perspective, it becomes clear that many *Epigondolella* species, and their stratigraphic range, have been misinterpreted.

There follows an outline of the eight *Epigondolella* species-complexes recognized herein, and their age. New systematics, numerical and stratigraphic detail will be presented in a future paper.

The *Epigondolella primitia* population

Figs. 2, 3 A, B, F, 7 A, L, X, 15 A, B, C

Epigondolella primitia was originally described by Mosher (1970). The type material was recovered from the matrix of *S. kerri* Zone ammonoids from Brown Hill on Peace River. Abundant collections from the type locality and elsewhere demonstrate the extent of variation within populations of the species. This is shown in Fig. 2, and described below.

Platform shapes. – Generally elongate, length to breadth ratio of between 2:1 and 3:1, commonly 5:2. Small specimens have a characteristic mid-platform constriction (Fig. 2 R, J). Late

growth stages may have either subparallel margins (Fig. 2 B, C, D, I) or wedge-shaped posterior platforms (Fig. 2 A). Some specimens are tapered to a narrowly rounded posterior termination (Fig. 2 H), which is pointed in rare specimens (Fig. 2 M, N, O). Generally, the posterior margin is squared-off; rarely it is irregular due to undulations of the platform margins (Fig. 2 B). Specimens sometimes have a linguiform posterior when one postero-lateral margin is more strongly developed and the axis of the conodont is flexed: this condition occurs in younger epigondolellids too.

Platform ornament. – Almost entirely restricted to the anterior platform margin. Three to five, round to transversely elongate nodes occur on each side, except in the earliest growth stages, and in rare larger specimens, in which there is only one (Fig. 2 Q). Nodes are generally subequal in size, well differentiated and discrete (Fig. 2 A, B) but may be subdued and coalescing (Fig. 2 E); rarely, they are only feebly developed. Posterior platforms are characteristically free of relief; occasionally they may be marginally undulose, but they are never serrate.

Carina. – Nearly always subterminal. Small specimens display a prominent subterminal cusp (Fig. 2 K), which becomes subdued with additional platform growth. Uncommon morphotypes (Fig. 2 M, O) have a posteriorly continuous carina as well as a narrow, pointed platform. Large specimens may develop secondary carinae.

Free blade. – Between one-third and one-half total unit length, decreasing with growth, arcuate in profile and becoming progressively lower onto the platform.

Lower-side morphology. – A pit with a distinct lip is situated posterior of the platform midlength. The basal attachment area may be bifurcate in those large specimens with marked postero-lateral growth (Fig. 2 F).

Microreticulae. – Present over the whole of the platform except for the carina and an area surrounding it on all sides. This is manifest as a distinct marginal band that contrasts sharply with a smooth adcarinal area. The reticulae consist of closely-shaped, equidimensional pits that have a diameter of about 10 µm (Fig. 3 A, B, F).

Central morphotype growth series. – Growth proceeds through an early bidentate condition (Fig. 2 S) by increase in the size of the anterior platform and in the number of marginal nodes. The posterior platform expands laterally less quickly, but the medial part shows least growth initially so that a mid-platform constriction arises. Thereafter, subparallel platform margins are developed by medial filling-in (Fig. 2 I): this is the condition of the holotype. Later growth may be uniform (Fig. 2 D) or the posterior platform may outstrip anterior growth to produce wedge-shaped outlines (Fig. 2 A).

Comparisons. – Uncommon specimens that are relatively short (Fig. 2 C, P) mimic *E. abneptis* but may be distinguished by the relatively low anterior nodes, which are also more numerous, and the rounded postero-lateral platform margins. Further, microreticulae are more uniformly developed marginally, including over the nodes. Specimens that have a posterior carina (Fig. 2 M, N, O) vaguely resemble *E. multidentata* but both the carina and anterior nodes of the latter species are much more prominent and the microreticulae are totally different, as is the case also with bidentate (one denticle on each side of the carina) specimens (Fig. 2 Q) that resemble *E. bidentata* (see below).

Age. – The *E. primitia* population described above appears at the base of the Norian and occurs throughout its range with *S. kerri* Zone ammonoids. Late Carnian *E. primitia*? occurs within faunules dominated by gondolellids of the *Paragondolella polygnathiformis – Metapolygnathus nodosa* group but its morphological variation within those faunules has yet to be assessed.

The *Epigondolella abneptis* subsp. A population

Figs. 3 D, E, G, 4, 7 B, M, N, Y, 9 A, 15 D, E, F

Epigondolella abneptis was originally described by Huckriede (1958) from the Hallstatt Limestone at Sommeraukogel, Austria. The type material was recovered from strata containing an ammonoid association referred to the *Cyrtopleurites bicrenatus* Zone. Krystyn (1980:90) elucidated the faunal

succession at Sommeraukogel and demonstrated that the *C. bicrenatus*-Fauna proper is more restricted than originally thought. The precise stratigraphic origin of the type of *E. abneptis* is therefore in doubt.

The species is the most generalised of the epigondolellids. All previous authors have regarded it as ranging throughout most of the Norian as well as the late Carnian. Previous attempts to subdivide it have not substantially improved its stratigraphic utility. I recognize three distinct *abneptis*-like homeomorphs, of Late Carnian, Early Norian and Middle Norian age. Each arose independently from non-'abneptid' predecessors. Additional species also occur that might fall within the existing, rather broad concept of the species and this has undoubtedly led to the long range attributed to *E. abneptis*.

Pending a study of the type material and a rationalized taxonomy, I retain the name *E. abneptis* for the early Norian species (e.g. Mosher 1973). Two broad groups may be distinguished, based on the morphology of the posterior platform. In addition, several distinct morphotypes of stratigraphic value are recognized. Variability within *E. abneptis* subsp. A populations is illustrated in Fig. 4.

Platform shapes. – Relatively squat elements with an average platform length to breadth ratio of about 3:2. This contrasts markedly with *E. primitia* populations as shown graphically in Fig. 5. Posterior margins are generally quadrate, although the 'linguiform condition' distorts this symmetry (Fig. 4 O, Q). Some forms become increasingly wedge-shaped posteriorly, and this reaches an extreme in specimens that have sharp, strongly extended, postero-lateral corners (Fig. 4 F). Other specimens, generally more elongate, have rounded posterior terminations (Fig. 4 K, L).

Platform ornament. – Nodes or denticles are generally restricted to the anterior but are fewer in number (usually two or three) and have greater relief than those of *E. primitia*. The posterior platform is generally smooth or has incipient nodes developed marginally, particularly at the postero-lateral corners or at the posterior border in line with the carina (Fig. 4 E, F), or as secondary carinae (Fig. 4 B). Specimens with distinct posterior ornament are rare and never have nodes as strongly developed as those anteriorly.

Carina. – Always subterminal, even in early growth stages. Rare specimens with crenulated posterior margins may have a marginal node aligned with the carina and in some late growth stages, one or both secondary carinae may extend to near the postero-lateral corner.

Free blade. – Generally one-third unit length, with convex profile.

Lower-side morphology. – A pit with a distinct lip is situated at or slightly posterior of platform midlength. A bifurcate keel is common.

Microreticulae. – Present over much of the platform but receding from the more prominent anterior nodes. For the most part comparable to that in *E. primitia* but showing elongation, enlargement and decreasing relief toward the carina (Fig. 3 D,

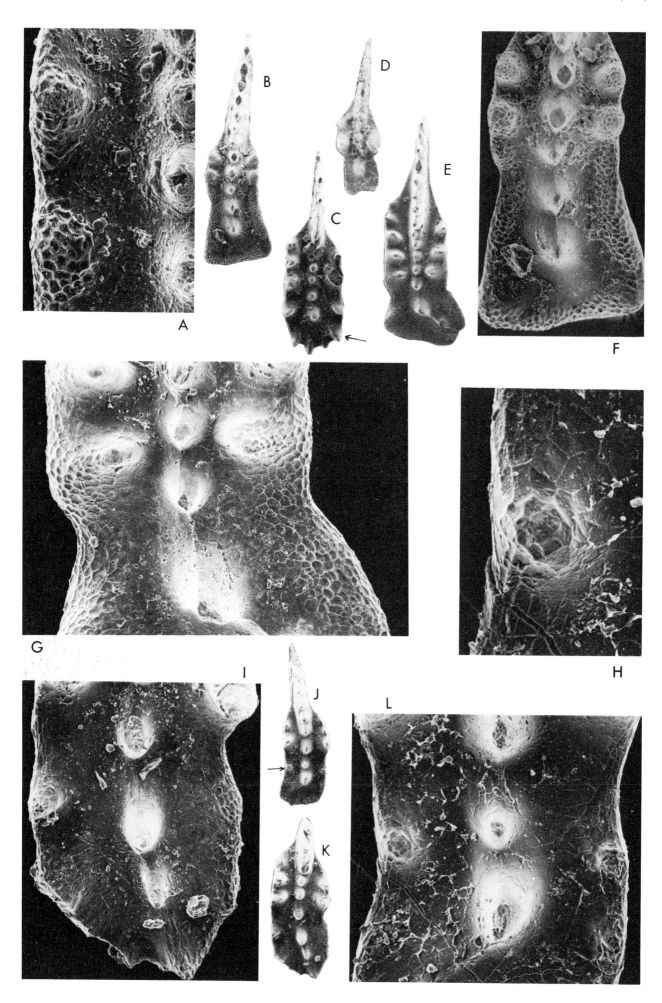

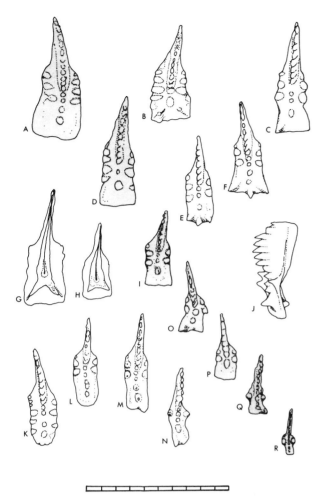

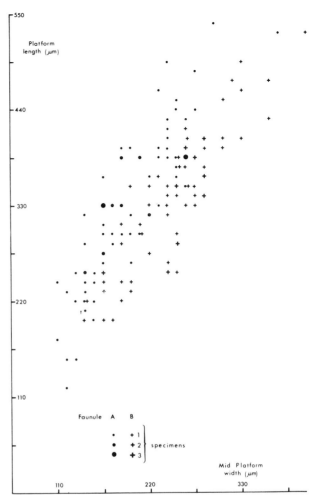

Fig. 4. Central morphotype growth series (shaded specimens) and morphological variation in *Epigondolella abneptis* subsp. A populations. The scale bar is 1 mm long.

Fig. 5. Graphical illustration of the change in platform proportions between faunule A = *E. primitia* from GSC Loc. No. C-87908, and faunule B = *E. abneptis* subsp. A from GSC Loc. C-87909. The collections, from Brown Hill, were made 1.5 m apart.

E, G). This is an initial stage in the development of the more subdued, open reticulae that characterize Middle Norian epigondolellids.

Central morphotype growth series. – Growth proceeds uniformly from an early stage that strongly resembles later stages except that they have fewer anterior nodes (Fig. 4 R, Q, P, O, I, D, A).

Comparisons. – See *E. primitia* and *E. abneptis* subsp. B.

Age. – *E. abneptis* subsp. A appears abruptly within the late *S. kerri* Zone. At Pardonet Hill, its appearance corresponds to a distinctive *Aulacoceras* belemnite–ammonoid bed. *Epigondolella* faunules from low in the *M. dawsoni* Zone are not abundant but are probably referable to *E. abneptis* subsp. A. Thereafter, representatives of this complex, if present, are strongly subordinate to *E. abneptis* subsp. B. Rare specimens resembling the central morphotype persist throughout the remainder of the Lower Norian, but not beyond it.

Fig. 3. Microreticulation on the platforms of Lower Norian epigondolellids. Arrows indicate positions of illustrated detail. □A, B, F. *E. primitia.* Note uniform, compact reticulation. □A. GSC 68846 (= Fig. 15 B) from Pardonet Hill, GSC Loc. No. O-98514. ×500. □B, F. GSC 68869 from Brown Hill, GSC Loc. No. C-87908. ×80, ×210. □C. *E. abneptis* subsp. B, late form. GSC 68870 (= Fig. 9 I) from McLay Spur, GSC Loc. No. O-98537. ×80. □D, E, G. *E. abneptis* subsp. A. □D. GSC 68871 (= Fig. 9 A) from McLay Spur, GSC Loc. No. O-98538. ×80. □E, G. GSC 68872 from Pardonet Hill, GSC Loc. No. O-98509. ×80, ×350. Note more open, finer reticulae toward carina. □H, I, J, K, L. *E. abneptis* subsp. B, early forms. Note that compact reticulae are confined to nodes whereas open form characterizes platform. □H, J, L. GSC 68873 from Brown Hill, GSC Loc. No. C-87912. ×1200, ×80, ×500. □I, K. GSC 68874 from Brown Hill, GSC Loc. No. C-87913. ×300, ×80.

The *E. abneptis* subsp. B population

Figs. 3 C, H, I, J, K, L, 6, 7 C, O, P, Q, R, Z, 9 I, 15 G, H, I

The upper half of the Lower Norian is dominated by strongly ornate epigondolellids. Although similar in relative dimensions to *E. abneptis* subsp. A, the species-complex is characterized by a distinctive growth series and by morphotypes that allow further subdivision of the interval. Morphological variation is illustrated in Fig. 6.

Platform shapes. – Relatively squat elements with a length to breadth ratio comparable to *E. a.* subsp. A, but with fewer

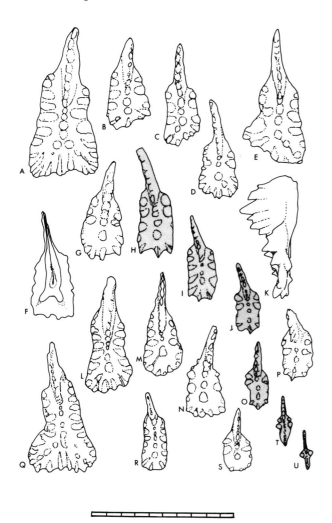

Fig. 6. Central morphotype growth series (shaded specimens) and morphological variation in *Epigondolella abneptis* subsp. B populations. The scale bar is 1 mm long.

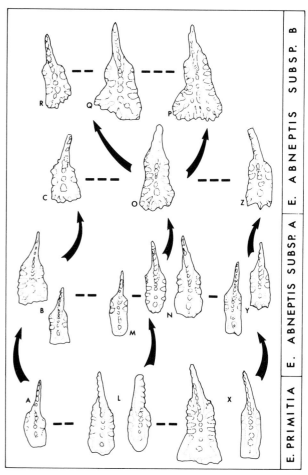

Fig. 7. Some intraspecific morphological variability (horizontal links) and phylogenetic trends (vertical arrows) in some Lower Norian epigondolellids. □A, B, C. Relatively squat morphotypes. □L, M, N. Morphotypes with rounded posterior platforms. □O, P. Morphotypes with increasing latero-posterior growth. □Q, R. Morphotypes with asymmetric posterior growth. □X, Y, Z. Relatively elongate morphotypes.

elongate morphotypes, and including elements with virtually equidimensional platforms. Most are subquadrate apart from the early growth stages, which are narrow and pointed. Irregular outlines are produced by strong marginal ornament. Elements with round posterior margins, present throughout the Lower Norian, become progressively more expanded laterally producing flask-shaped outlines (Fig. 7 L–P). This trend appears to have stratigraphic utility, as does a further trend toward increasing asymmetry of the posterior platform whereby one postero-lateral lobe is developed more strongly posteriorward (Fig. 6 B, D, E). In the upper *J. magnus* Zone, rare elongate elements with one posterior lobe suppressed (Fig. 6 C) appear; these are similar to some Middle Norian epigondolellids.

Platform ornament. – All but the smallest growth stages bear strong marginal nodes around their entire perimeter. At first, the anterior denticles are stronger but subsequent growth introduces and enlarges the posterior nodes until they too are very prominent. As observed by Mosher (1973:159), these tend to radiate from the terminal node of the carina, a feature exaggerated by additional nodes introduced as secondary carinae.

Carina. – In small specimens, the carina persists to the posterior tip (Fig. 6 T, U) but with continued growth the terminal platform node and the carina become separated (Fig. 6 O) although rare specimens retain the continuity into later growth stages (Fig. 6 P). Secondary carinae are common and in the stratigraphically youngest representatives of the group (see above) the central (terminal) node of the primary carina may be connected to the tip of the enlarged posterior lobe (Fig. 6 C).

Free blade. – Between one-quarter and one-third unit length, with a strongly convex profile.

Lower-surface morphology. – A pit with a strong lip is situated at or slightly anterior of platform midlength. Bifurcate keels are common.

Microreticulae. – Less common on the strongly ornate specimens. When present they are marginally developed and generally restricted to the posterior platform. For the most part they are of the open, irregular type (see *E. multidentata*), but the compact style is often preserved on denticle ridges and platform edges (Fig. 3 C, H, I, J, K, L).

Central morphotype growth series. – The smallest elements are bidentate (Fig. 6 U). Initially, nodes are added anteriorly with only a minor increase in posterior platform width (Fig. 6 T). Subsequently, nodes are added in postero-lateral positions giving the elements their characteristic subquadrate outline, and thereafter marginally as the elements grow more uniformly (Fig. 6 O, J, I, H).

Comparisons. – Their strongly ornate character separates *E. abneptis* subsp. B from other Lower Norian conodonts. Similar elements occur in the Middle Norian but they are characterized by one or more of the following features: a continuous carina, a more elongate form, more asymmetrically disposed nodes, a longer free blade, a more anteriorly situated slit-like pit without a distinct lip and/or a style of microreticulae that does not include the close-packed form.

Age. – *E. abneptis* subsp. B ranges from within the *M. dawsoni* Zone through the *J. magnus* Zone, up to the Lower/Middle Norian boundary.

The *Epigondolella multidentata* population

Figs. 8, 9 E, H, 12 A, B, L, X, Y, 15 J, K, L

The type of *E. multidentata* Mosher 1970 was recovered from the *D. rutherfordi* Zone at the Crying Girl locality of the Pardonet Formation. Rich collections from that zone at the type locality and along the Peace River demonstrate a range of variation illustrated in Fig. 8.

Platform shapes. – Six, partly intergradational morphological categories are recognized. The central morphotype (e.g. Fig. 8 I) is the most common and compares closely to the holotype. It is characteristically elongate having a length to breadth ratio of about 3:1. The platform generally has subparallel anterior margins and thereafter tapers, often sinuously (the linguiform condition), to a posterior point, although there is frequently a node adjacent to the posterior tip, which produces a narrow, squared-off termination: this is the case in the holotype. An extension of the latter trend produces specimens that bear relatively broad posterior outlines (Fig. 8 D). A further, less common morphotype is relatively short, and is characterized by rounded posterior platform margin (Fig. 8 M to O). A rare morphotype is plano-convex in upper view (Fig. 8 C). Two additional morphotypes (Fig. 8 E, H) are distinguished on the basis of platform ornament but have platform proportions similar to the central morphotype.

Platform ornament. – The majority of specimens that occur in faunules in the lower half of the *D. rutherfordi* Zone are characterized by strong anterior denticles numbering between two and four. Rare morphotypes have only a single denticle on one margin (Fig. 8 H), a morphology that becomes very common in the upper Middle Norian. The posterior platform is typically smooth although one marginal node may occur, particularly on the posterior border. In the upper part of the *D. rutherfordi* Zone, specimens with strong marginal

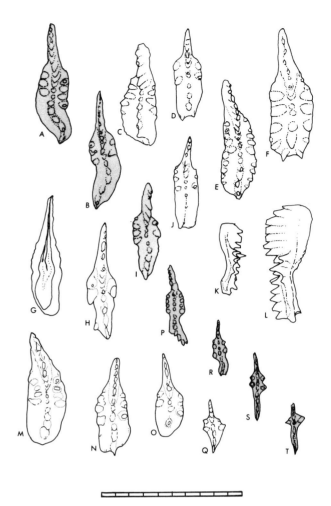

Fig. 8. Central morphotype growth series (shaded specimens) and morphological variation in *Epigondolella multidentata* populations. The scale bar is 1 mm long.

ornament throughout their length (Fig. 8 E), which are uncommon earlier, become dominant whilst the predominantly smooth morphotypes give way to forms with an increasing number of marginal nodes, usually on one side. This imparts a strong asymmetry to the elements, and represents a trend that culminates in *Epigondolella* n. sp. C (Fig. 12 A to C).

Carina. – From the beginning of Middle Norian time, epigondolellids are characterized by a prominent carina that extends, and generally rises, to the posterior tip of the elements (Fig. 8 L). In a few elements the continuity is not complete (Fig. 8 D, F) but the carina is nevertheless prominent. This morphology appears suddenly and is most evident in the early *D. rutherfordi* Zone. Subsequently, the carina becomes less prominent in *E.* n. sp. C but most other Middle and Upper Norian epigondolellids are characterized by a posteriorly extended carina, and thus differ from most of those in the Lower Norian.

Free blade. – This is between one-quarter and one-third unit length and has an arcuate profile. Some younger representatives develop a shorter blade than is typical of the central morphotype.

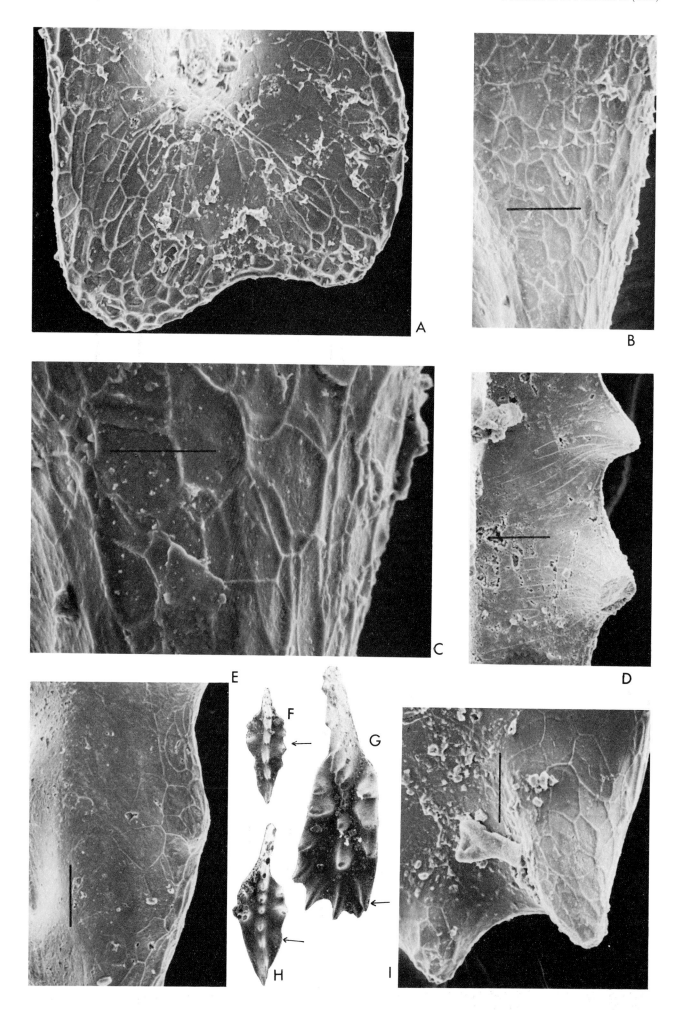

Lower-side morphology. – The pit is slit-like and is not surrounded by a prominent lip. The basal attachment scar is elongate and extends far posterior of the pit, which is situated anterior of the platform midlength.

Microreticulae. – The development of micro-ornament in post-Lower Norian epigondolellids continues the trend that began with the development of *E. abneptis. E. multidentata* is characterized by a total absence of compact reticulae. Instead, platform margins may be covered (they are not always present) by a relatively faint, very irregular anastomosing network in which subdued ridges are commonly up to 20 μm apart (Fig. 9 E, H). This style of reticulae has completely replaced the compact form.

Central morphotype growth series. – As exemplified in Fig. 8 T, S, R, P, I, B, A, growth begins with a very elongate form. Anterior nodes are added prior to the enlargement of the posterior platform.

Comparisons. – The prominent carina and elongate form separates these specimens from all others. Rare squat forms with rounded margins (Fig. 8 O) resemble elements of *E. postera* but the anterior denticles are more numerous and the whole element is more robust. See also *E.* n. sp. D.

Age. – *E. multidentata* ranges through the *D. rutherfordi* Zone, but the ornate morphotypes replace the central morphotype in the upper part of that zone (and allow subdivision of it), and typical specimens are not found later. Elements resembling early growth stages of the central morphotype do occur in late Middle Norian faunas but they are referred to *E.* n. sp. D.

The *Epigondolella* n. sp. C population

Figs. 9 B, C, G, 10, 12 C, D, Z, 15 M, N, O

At the beginning of *M. columbianus* Zone time, conodont faunas become dominated by *Epigondolella* n. sp. C, a homeomorph of *E. abneptis* subsp. B. At the moment, there is uncertainty as to which species should bear the name *E. abneptis*, but they are clearly different and I do not regard the younger as having developed directly from the older. Rather, *E.* n. sp. C developed from the central morphotype of *E. multidentata* by way of a general reduction in length and in posterior carina prominence and an increase in posterior ornamentation

Fig. 9. Microreticulation of some Lower and Middle Norian epigondolellids. Arrows indicate positions of illustrated detail. Bar = 20 μm, except C = 10 μm. □A. *E. abneptis* subsp. A. GSC 68871 (= Fig. 3 D) from McLay Spur, GSC Loc. No. O-98538. ×910. Note marginal compact reticulae passing into subdued irregular reticulae toward carinae. □B, C, G. *E.* n. sp. C. GSC 68876 from Crying Girl, GSC Loc. No. O-83835. ×1000, ×2700, ×80. Note superimposition of two generations of subdued, irregular reticulae. □D, F. *E.* n. sp. D. GSC 68875 from Black Bear Ridge, GSC Loc. No. O-98548. ×800, ×80. Note irregular, discontinuous striae replacing reticulae. □E, H. *E. multidentata.* GSC 68877 from Brown Hill, GSC Loc. No. C-87921. ×810, ×80. Note subdued, irregular reticulae covering platform margin. □I. *E. abneptis* subsp. B. GSC 68870 (= Fig. 3 C) from McLay Spur, GSC Loc. No. O-98537. ×900.

coupled with an asymmetric development of the postero-lateral margins. Fig. 10 illustrates morphological variation within the population.

Platform shapes. – Compared with the platform element of *E. multidentata*, that of *E.* n. sp. C is relatively squat and some elements have rectangular platforms (Fig. 10 S, T). A more common shape is distinctly asymmetric with one margin laterally expanded at midlength and then incurved to meet the posteriormost point of the element on the opposite margin (Fig. 10 J). Many of the elements have very irregular outlines due to the outgrowth of strong marginal nodes. Some specimens have symmetrical, subparallel or weakly convex margins and squared-off posterior borders (Fig. 10 F, G, M), others have rounded posterior outlines (Fig. 10 N, R).

Platform ornament. – As in all Middle Norian epigondolellids, anterior denticles are strongly developed. These number between one and three, relatively fewer on average than in *E. multidentata;* two nodes on one side and one on the other, a condition seen only rarely in populations of the latter, is more common. Posteriorly, early growth stages and rare large specimens are smooth (Fig. 10 N, R) whereas other subsymmetrical elements have either completely nodose posterior margins (Fig. 10 F), or one lateral margin that is ornate (Fig. 10 G), or bears nodes only on the posterior border (Fig. 10 M). To an extent, all of these specimens intergrade with the central morphotype in which nodes are developed initially, and thereafter more strongly, on one postero-lateral margin, whereas the opposite, less expanded, margin carries fewer or sometimes no nodes. The postero-lateral corner of the expanded margin bears the strongest node, which is ultimately responsible for the subquadrate outline of some platforms (e.g. Fig. 10 O). In such specimens it is common for the anteriormost platform nodes to be unpaired.

Carina. – Carina development is variable in this group of conodonts. In many elements it extends to the posterior end, but it is not especially prominent. In the central morphotype, the carina tends to become progressively retarded through growth in conjunction with the increased development of one postero-lateral corner. Ultimately, the terminal node of the carina becomes indistinguishable from adjacent marginal nodes, while anteriorly the carina is represented by discrete, often large nodes occupying a central position on the platform (Fig. 10 S, T).

Free blade. – Rather variable. In some specimens relatively long, like that of *E. multidentata*, in others shorter, like that of *E. postera* (see below).

Lower-surface morphology. – A slit-like pit is situated anterior, sometimes far anterior of platform midlength. There is no distinct protuberance surrounding it. The basal scar may be bifurcate but secondary branches are developed laterally from the principal scar rather than arising from its division, as may be envisaged in *E. abneptis.*

Microreticulae. – Often difficult to detect on the strongly ornate elements but present marginally and on the sides of platform

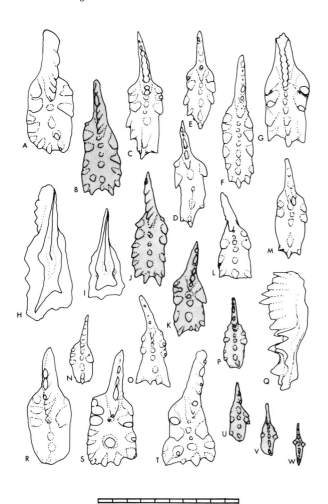

Fig. 10. Central morphotype growth series (shaded specimens) and morphological variation in *Epigondolella* n. sp. C populations. The scale bar is 1 mm long.

denticles. It is of the subdued, irregular-network type, like that of all post-Lower Norian epigondolellids. The phenomenon of superimposed reticulae was first observed in a specimen of *E.* n. sp. C, although I have subsequently recognized it in other species that bear the irregular style of reticulae. In these specimens, the latest reticulae development appears to be superimposed completely randomly over an earlier set that, although faint, can still be seen at high magnification. It is possible that the greater relief of the early Norian compact reticulae arises from the superimposition of later sets exactly over earlier ones, whereas the pattern that characterized younger species remains subdued because it is never duplicated.

Central morphotype growth series. – Asymmetry of the posterior platform arises at an early stage in the growth of *E.* n. sp. C (Fig. 10 V). This morphology is essentially that retained by the next younger species *E. postera.* Growth proceeds with the addition of postero-lateral nodes on the expanded margin as well as generally even growth elsewhere (Fig. 10 U, P). Later growth tends to emphasize the earlier asymmetry as one postero-lateral margin expands to rival the posterior carina in position and development (Fig. 10 K, J, B). This is often accompanied by a posterior shifting of the anterior denticles on the expanded side producing anterior asymmetry as posterior symmetry increases (Fig. 10 S, T).

Comparisons. – Specimens of *E.* n. sp. C that develop posterior symmetry strongly resemble *E. abneptis* subsp. B but specimens of the latter are generally more symmetrically developed anteriorly. In young *E. abneptis* subsp. B populations, in which posterior asymmetry develops, one postero-lateral lobe is developed in a relatively medial position. In general, *E. abneptis* bears nodes that have a stronger vertical component of growth than *E.* n. sp. C. Hence, in upper view, elements of *E. abneptis* subsp. B are not as marginally serrate. Other criteria that may be used in distinguishing the two species are the relative blade lengths (compare Figs. 6 K, 10 Q), the frequent suggestion of posterior carina continuity in *E.* n. sp. C, the position and shape of the pit and its surrounding area, and the presence of 'residual' compact reticulae in *E. abneptis* subsp. B. Morphological variability within populations of the two species are quite different too.

Age. – *E.* n. sp. C characterizes the lowermost division of the *M. columbianus* Zone of the Middle Norian. It is replaced toward the top of that interval (*M. columbianus* 1) and especially in the zone of *M. columbianus* 2, by *E. postera.*

The *Epigondolella postera* population

Figs. 11, 12 M, N, O, P, Q, R, S, T, 15 P, Q, R

The type of *E. postera* (Kozur & Mostler 1971) came from the Middle Norian of Sommeraukogel, Austria. Available illustrations of this specimen correspond closely to elements that dominate collections from the *M. columbianus* 2 Zone in NE British Columbia. Fig. 11 illustrates representative morphologies of the *E. postera* population.

Platform shapes. – The platform elements of *E. postera* are smaller than those of older species of *Epigondolella* and mark the beginning of a general diminution in size of the genus that continued until its extinction. This trend is evident even within the range of *E. postera* and provides the means to subdivide the interval. The oldest faunules are characterized by relatively elongate specimens (Fig. 11 D, I, cf. Fig. 10 C, D) that often gradually taper to a point, although the margins may be very irregular due to node development. Rather squat elements, often with lobate outlines like the holotype, characterize the younger faunules. They may taper to a point (Fig. 11 U) but are frequently posteriorly rounded (Fig. 11 V) or, with node development, squared-off. The abruptly terminated morphotypes (Fig. 11 G, M) in association with bidentate elements (see below) characterize the youngest faunules.

Platform ornament. – The common anterior morphology of *E. postera* comprises three prominent denticles: two on one platform margin and one on the other. Less commonly, there are as many as three on one margin, but unequal development of the two margins is normal. In the upper part of the range, bidentate morphotypes appear (Fig. 11 P, T, Y, Z, AA). The posterior platform is commonly unornamented but single nodes may be developed on one lateral margin (Fig. 11 D, F, Y), on one postero-lateral margin (Fig. 11 K, L, M, T), or on both sides of the posterior carina (Fig. 11 H, G). Less common

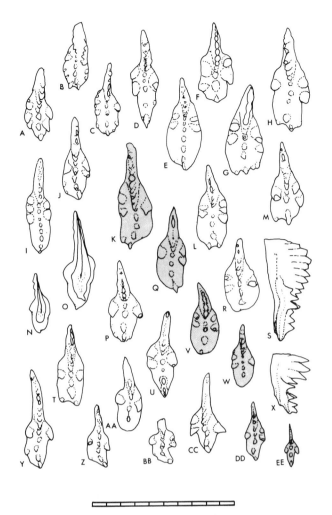

Fig. 11. Central morphotype growth series (shaded specimens) and morphological variation in *Epigondolella postera* populations. The scale bar is 1 mm long.

specimens bear multiple nodes on one (Fig. 11 J) or both (Fig. 11 A, B, C) margins.

Carina. – Generally continuous to the posterior tip but sometimes terminating in front of this point (Fig. 11 R, W, AA). It is often prominent posteriorly (Fig. 11 X), especially in small specimens, but with platform growth it may become relatively submerged (Fig. 11 S).

Free blade. – Many specimens of *E. postera* have a short high blade with relatively few stout denticles (Fig. 11 X). This type of blade is characteristic of this and all younger representatives of the genus whereas a longer blade is more common in *D. rutherfordi* and basal *M. columbianus* Zone faunas. Some specimens (Fig. 11 S, Y) still retain a long blade, but they are rare.

Lower-side morphology. – A slit-like pit without a lip is situated anterior of platform midlength.

Microreticulae. – Not frequently observed but when present are marginal, subdued and irregular.

Central morphotype growth series. – As with most if not all epigondolellids, small specimens of *E. postera* are bidentate. General enlargement of the platform is accompanied by the

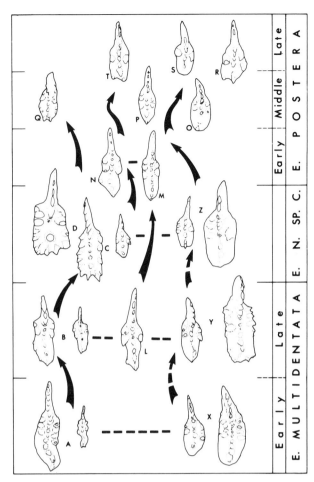

Fig. 12. Some intraspecific morphological variability (horizontal links) and phylogenetic trends (vertical arrows) in some Middle Norian epigondolellids. □A, B, C. Development of *E.* n. sp. C from *E. multidentata* via an intermediate morphology. □X, Y, Z. Uncommon morphotypes in successive faunules characterized by an *E. postera*-like morphology. □L. Elongate, 'tridentate' element with *E. postera*-like morphology. □M, N, O, P, Q. *E. postera* morphotypes showing possible derivation by retention of 'juvenile' characteristics (N) of *E.* n. sp. C, and/or development from *E. postera*-like variants L, X, Y, Z. Specimen Q shows a 'dwarfed' development of 'mature' *E.* n. sp. C morphology. □R, S, T. Late stage *E. postera* morphotypes with abruptly terminated posterior platforms, and development of 'bidentate' morphology.

formation of an additional node on one anterior margin. This general morphology is retained thereafter as the specimen simply enlarges in size, commonly developing one postero-lateral margin more strongly than the other.

Comparisons. – Small specimens of other species may resemble *E. postera* but the rather lobate posterior outline is distinctive. Such an outline characterizes early growth stages of *E.* n. sp. C but that species quickly developed postero-lateral nodes and attained a larger size. Rare specimens in older Middle Norian faunules mimic *E. postera* in possessing smooth, rounded posterior platforms but their anterior nodes are characteristically more numerous and they do not exhibit the same lobate outline (Fig. 12 X, Y, Z): isolated specimens may be difficult to determine, however. Large bidentate specimens (Fig. 11 P, T, Y, Z, AA) retain the anterior morphology developed at an early growth stage (Fig. 11 EE) but have broad platforms that are identical to that of the central morphotype of *E. postera*, and

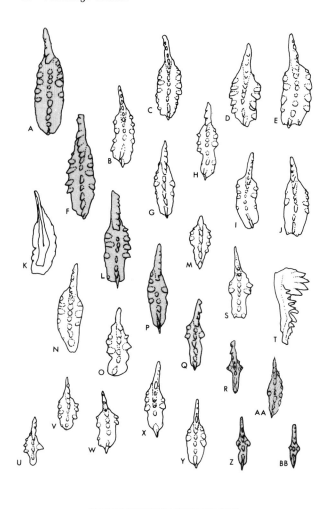

their length, although some specimens have a medial constriction (Fig. 13 H, O), and others taper markedly in their posterior half (Fig. 13 M, N). Anterior platform terminations are often less abrupt than in older species, the margins meeting the blade in an even curve (Fig. 13 F). Posterior terminations are pointed to narrowly square depending on posteriormost ornamentation; a few specimens have round, lobate posteriors (Fig. 13 O, U, V).

Platform ornament. – Rather variable but commonly smooth posteriorly. Anterior nodes are relatively numerous, small and spike-like. They increase to a maximum of five in the largest specimens available, although the development may be noticeably unequal on the two margins (Fig. 13 P, AA). As in other Middle Norian species, posterior ornament may be present on one margin (Fig. 13 H), both margins (Fig. 13 C, E) or only terminally (Fig. 13 J, S).

Carina. – Usually continuous to the posterior tip of the elements and often projecting beyond it as strong spike-like extension (Fig. 13 L, P). Less commonly terminating short, as is the case in posteriorly lobate specimens (Fig. 13 O, U, V).

Free blade. – Distinctive cockscomb-like, much shorter than lower Middle Norian epigondolellids. Commonly one-quarter or one-fifth unit length.

Lower-side morphology. – Slit-like pit on featureless attachment scar is medial or anterior of the medial point.

Microreticulae. – When present, marginal, subdued and irregular, sometimes extending onto denticles as striations.

Central morphotype growth series. – Initially bidentate, becoming tridentate and enlarging platform progressively in all directions, particularly antero-posteriorly, with the irregular introduction of additional nodes.

Comparisons. – The small size yet highly ornate character of *E.* n. sp. D populations is distinctive. The short blade and the spike-like nature of the ornament differ from that in *E. multidentata*. The posteriorly lobate forms and those with medial constrictions are unknown in older Middle Norian faunules but bear some resemblance to *E. primitia* which generally has relatively subdued nodes, a short carina, longer blade and totally different microreticulae.

Age. – Restricted to the *M. columbianus* 3 Zone. Only sparse conodont faunules have been recovered from the *M. columbianus* 4 Zone, so it is uncertain whether *E.* n. sp. D extends nearer to the top of the Middle Norian.

Fig. 13. Central morphotype growth series (shaded specimens) and morphological variation in *Epigondolella* n. sp. D populations. The scale bar is 1 mm long.

thus differ from *E. bidentata,* which is a narrow, relatively elongate species.

Age. – The central morphotype of *E. postera* is characteristic of the *M. columbianus* 2 Zone. Rare morphotypes (e.g. Fig. 12 L) that appear within the *D. rutherfordi* Zone and continue into the *M. columbianus* Zone have much in common with *E. postera* but are for the present regarded as extremes within the *E. multidentata* and *E.* n. sp. C populations. Elongate morphotypes dominate faunules in the upper *M. columbianus* 1 Zone, whereas bidentate and abruptly terminated specimens characterize the upper part of the *M. columbianus* 2 Zone.

The *Epigondolella* n. sp. D population

Figs. 9 D, F, 13, 15 S, T, U

Epigondolellids from *M. columbianus* 3 Zone are markedly different from the preceding *E. postera* populations. They resemble those of *E. multidentata* but are noticeably smaller and more delicate. Fig. 13 illustrates the variety of form.

Platform shapes. – Generally elongate, average length to breadth ratio of 5: 2. Margins are often subparallel for much of

The *Epigondolella bidentata* population

Figs. 14, 15 V, W, X

E. bidentata Mosher 1968 is based on Upper Norian material from the Hallstatt Limestone at Steinbergkogel by Hallstatt Salzberg, Austria. It is recognized that all platform elements of

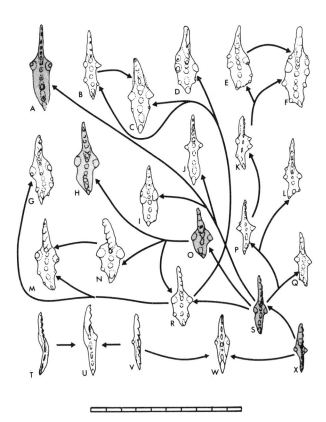

Fig. 14. Central morphotype growth series (shaded specimens) and morphological variation in *Epigondolella bidentata* populations. The arrowed lines link morphotypes that could be theoretically derived through growth. This illustrates the complex relationships typical of an *Epigondolella* population. The scale bar is 1 mm long.

Epigondolella pass through a 'bidentate' morphology early in their growth but most do not retain it. Large bidentate specimens are known, however, within the Lower and Middle Norian too, so it is important to examine bidentate forms closely. Small isolated growth stages may be difficult to determine. Fig. 14 illustrates the variety of platform elements within the Lower Upper Norian *E. bidentata* faunules.

Platform shapes. – Always small, generally slender, occasionally almost totally reduced to a blade-like condition (Fig. 14 T, U, V). Platforms may be subparallel, biconvex, plano-convex, sagittate or spindle-shaped in outline. Posterior terminations generally pointed, occasionally squared-off by node development.

Platform ornament. – A single strong node on either side of the anterior platform is typical, but these may be supplemented by an additional, generally smaller denticle anterior of one of these (Fig. 14 D, E, I, K, L, P); specimens with additional anterior nodes (Fig. 14 F) are rare. The posterior platform is smooth in the holotype, but incipient nodes (as occur in the type of *E. mosheri* Kozur & Mostler 1971) or distinct nodes may be developed symmetrically (Fig. 14 B, C, D, R) or asymmetrically (Fig. 14 E, F, G, K, L, M, N). In the younger *E. bidentata* populations, specimens with one or no platform denticles occur (Fig. 14 T, U, V), and their size demonstrates that they are not simply early growth stages of the bidentate forms.

Carina. – Extends to posterior tip of elements and is often very prominent at the posterior end (Fig. 14 W).

Free blade. – Identical to that of *E.* n. sp. D.

Lower-surface morphology. – Slit-like pit situated within narrow attachment area beneath or slightly posterior of principal anterior nodes, that is anterior of platform midlength.

Microreticulae. – Often absent.

Central morphotype growth series. – The bidentate condition is introduced at an early stage and persists with uniform or slightly asymmetrical platform enlargement.

Comparisons. – The holotype of *E. bidentata* is about 570 µm in length, which is about the size of the largest specimen in my collection (Fig. 14 A). Comparable morphology is known up to about 250 µm in *E. abneptis* subsp. B, up to about 300 µm in *E.* n. sp. C, and up to about 350 µm in *E.* n. sp. D. This suggests a progressively longer retention of the bidentate condition through time. On the other hand, bidentate morphotypes in both the *E. primitia* (Fig. 2 Q) and *E. postera* (Fig. 11 Y) populations attain a size greater than found in *E. bidentata*. The Lower Norian specimens can be readily distinguished on the basis of microreticulation, whilst *E. postera* is generally broader. Other isolated post-Lower Norian specimens should be regarded as specifically indeterminate unless they attain a length of greater than 350 µm. Alternatively, '*E. bidentata*' would have to be regarded as ranging through much of the Middle and Upper Norian.

Age. – The *E. bidentata* population characterizes the *G. cordilleranus* Zone and at least part of the *C. amoenum* Zone. It is possible that the species appears first in the zone of *M. columbianus* 4. Higher *G. cordilleranus* Zone and younger collections may be characterized by a greater number of morphotypes with strongly reduced platforms.

A summary of *Epigondolella* phylogeny

Eight *Epigondolella* populations are recognized in the Norian. They follow one upon the other with only one sparsely productive interval of uncertain content, that is the uppermost Middle Norian (*M. columbianus* 4). Each of the species complexes is bounded by a faunal break of varying abruptness. The most profound of these occurs at the Lower–Middle Norian boundary between the *J. magnus* and *D. rutherfordi* Zones, and within the late Middle Norian between the Zones of *M. columbianus* 2 and 3. The change that occurs near the top of the Lower Norian *S. kerri* Zone is also abrupt, but the derivation of *E. abneptis* subsp. A from *E. primitia* appears straightforward. The evolutionary event is more significant, however, because the microreticulation of the platform began to change at that point. This criterion may be useful in separating the two species at the generic level since '*E.*' *primitia* is characterized by what might be regarded as a *Metapolygnathus* microreticulation (Orchard, in preparation).

Morphological changes between the *E. abneptis* subspp. populations, between the *E. multidentata*, *E.* n. sp. C and *E. postera* faunas, and between those characterized by *E.* n. sp. D and *E. bidentata* are also marked, but rare morphotypes within

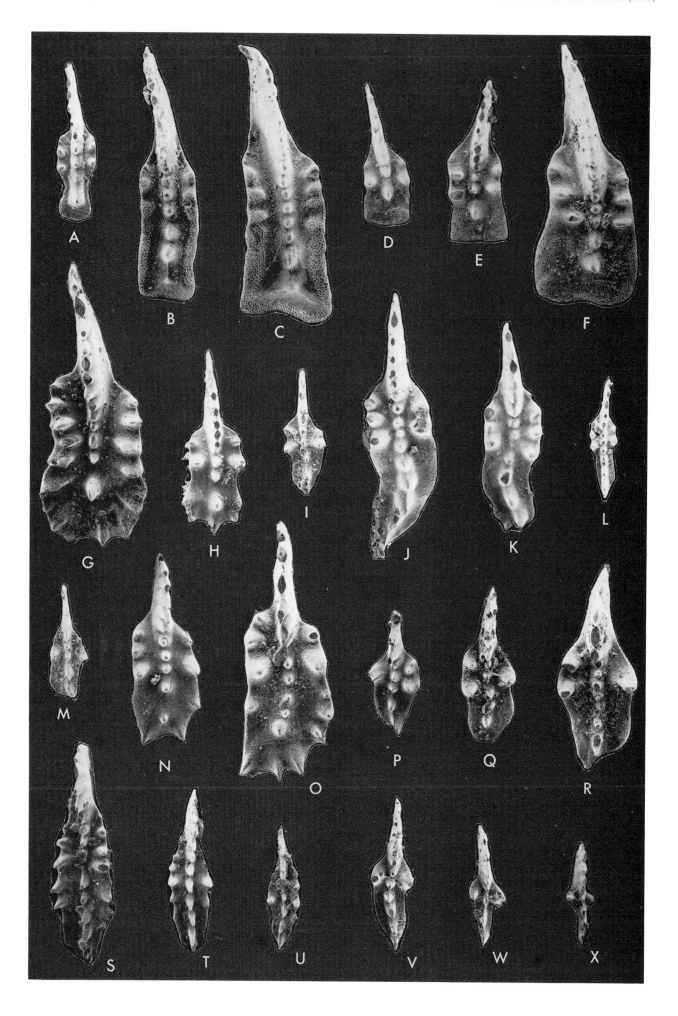

the ancestor populations predict the changes that find common expression in the descendant faunas. In the case of both *E. postera* and *E. bidentata,* the central morphotypes show a retention of 'juvenile' characteristics (neoteny).

Upper Lower Norian faunules of *E. abneptis* subsp. B include strongly asymmetric elements that become less squat and more elongate as one postero-lateral lobe (and secondary carina) is extended and the other is suppressed. This development is the continuation of a trend that begins with the bifurcation of the basal attachment scar in some large early Norian epigondolellids. The morphological change at the Lower–Middle Norian boundary appears to have consisted of a shift of emphasis from dual lobe development to posterior carina growth. Hence the derivation of the uni-lobed, strongly carinate *E. multidentata* from the bi-lobed, weakly carinate *E. abneptis.* The functional advantages of postero-lateral enlargement were presumably great, however, because *E.* n. sp. C evolved by a secondary outgrowth of one platform margin (Fig. 12 A, B, C). This development of homeomorphic characteristics in the mid Middle Norian was superseded by a general diminution both in overall size and in the degree of platform development, particularly of the anterior denticles. *E. postera* thus became posteriorly attenuated and anteriorly bidentate (one denticle on each side of the carina) prior to disappearing in the upper Middle Norian (Fig. 12 R, S, T). The appearance of *E.* n. sp. D marked a second major shift, this time away from broad platforms and a few large denticles to relatively elongate elements with more numerous but more delicate denticles. The general diminution continued into the Upper Norian as *E. bidentata* emerged as the last remnant of platform conodont development.

References

Gibson, D. W. 1971: Triassic stratigraphy of the Sikanni Chief River–Pine Pass Region, Rocky Mountain Foothills, northeastern British Columbia. *Geological Survey of Canada, Paper 1970–31,* 105 pp.

Huckriede, R. 1958: Die Conodonten der Mediterranen Trias und ihr stratigraphischer Wert. *Paläontologische Zeitschrift 32,* 141–175.

Krystyn, L. 1980: Stratigraphy of the Hallstatt region. *Abhandlungen der geologischen Bundesanstalt 35,* 69–98.

Fig. 15. Norian epigondolellids, exemplified by representative growth stages of the central morphotype. All figures ×80. □A, B, G. *E. primitia.* Respectively, GSC 68845, 68846, 68847 from Pardonet Hill, GSC Loc. No. O-98514. □D, E, F. *E. abneptis* subsp. A. Respectively, GSC 68848 from Brown Hill, GSC Loc. No. C-87909, GSC 68849 and 68850 from Pardonet Hill, GSC. Loc. No. O-98509. □G, H, I. *E. abneptis* subsp. B. Respectively, GSC 68851 from McLay Spur, GSC Loc. No. O-98538, GSC 68852 and 68853 from Brown Hill, GSC Loc. No. C-87915. □J, K, L. *E. multidentata.* Respectively, GSC 68854, 68855, 68856 all from Brown Hill, GSC Loc. Nos. C-87921, O-97538, O-97533. □M, N, O. *E.* n. sp. C. Respectively, GSC 68857 from Crying Girl, GSC Loc. No. O-83835, GSC 68858 and 68859 from McLay Spur, GSC Loc. No. O-98878. □P, Q, R. *E. postera.* Respectively, GSC 68860 from McLay Spur, GSC Loc. No. O-98541, GSC 68861 and GSC 68862 from Black Bear Ridge, GSC Loc. Nos. O-98552 and O-98549. □S, T, U. *E.* n. sp. D. Respectively, GSC 68863, 68864, 68865 from Black Bear Ridge, GSC Loc. No. O-98548. □V, W, X. *E. bidentata.*Respectively, GSC 68866, 68867, 68868 from Ne-Parle-Pas Rapids, GSC Loc. No. O-98504.

Kozur, H. & Mostler, H. 1971: Probleme der Conodontenforschung in der Trias. *Geologische Paläontologische Mitteleilungen Innsbruck 1,* 1–19.

McLearn, F. H. 1960: Ammonoid faunas of the Upper Triassic Pardonet formation, Peace River Foothills, British Columbia. *Geological Survey of Canada, Memoir 311,* 118 pp.

Mosher, L. C. 1968: Triassic conodonts from Western North America and Europe and their correlation. *Journal of Paleontology 42,* 895–946.

Mosher, L. C. 1970: New conodont species as Triassic guide fossils. *Journal of Paleontology 44,* 737–742.

Mosher, L. C. 1973: Triassic conodonts from British Columbia and the Northern Arctic Islands. *Geological Survey of Canada, Bulletin 222,* 141–193.

Tozer, E. T. 1967: A standard for Triassic time. *Geological Survey of Canada, Bulletin 156,* 103 pp.

Tozer, E. T. 1979: Latest Triassic ammonoid faunas and biochronology, Western Canada, *Geological Survey of Canada, Paper 79–1B,* 127–135.

Tozer, E. T. 1981: Triassic Ammonoidea: Geographic and stratigraphic distribution. *In* House, M. R. & Senior, J. R. (eds.): The Ammonoidea. *Systematics Association Special Volume 18,* 397–431.

Appendix

The following GSC Locality Numbers refer to conodont collections from the Pardonet Formation (P.F.). All lie within the Halfway River Map Area (Sheet 94 B, 1:250,000 National Topographic Series), British Columbia. Numbers prefixed by O- are from beds containing ammonoid faunas.

Brown Hill

53°06′05″N, 122°53′00″W.

C-87908 6 m above exposed base of P.F. Bracketed by *S. kerri* Zone ammonoids.

C-87909 8 m above base P.F. *S. kerri* Zone ammonoids 2 m below, *M. dawsoni* Zone ammonoids 2.5 m above.

C-87912 19 m above base P.F. Bracketed by *M. dawsoni* Zone ammonoids.

C-87913 22 m above base P.F. Bracketed by *M. dawsoni* Zone ammonoids.

C-87915 24 m above base P.F. Bracketed by *M. dawsoni* Zone ammonoids.

O-97538 33.0 m above base of P.F. Occurring with *D. rutherfordi* Zone ammonoids.

C-87921 34.5 m above base P.F. Bracketed by *D. rutherfordi* Zone ammonoids.

O-97533 38.5 m above base of P.F. Occurring with *D. rutherfordi* Zone ammonoids.

Black Bear Ridge

56°05′10″N, 123°02′25″W. This locality lies immediately to the east of Black Bear Ridge shown on topographic maps.

O-98552 48.5 m below top of P.F. Occurring with *M. columbianus* 2 Zone ammonoids.

O-98549 41 m below top of P.F. Occurring with *M. columbianus* 2 Zone ammonoids.

O-98548 36.5 m below top of P.F. Occurring with *M. columbianus* 3 Zone ammonoids.

Crying Girl Prairie Creek

56°28′N, 122°54′W (White Creek; upper reaches of Graham River).

O-83835 Occurring with *M. columbianus* 1 Zone ammonoids.

McLay Spur

56°06′16″N, 122°43′00″W (east of Childerhose Coulee).

O-98538   About 18.5 m above base of P.F. Occurring with *M. dawsoni* Zone ammonoids.

O-98537   About 48 m above base of P.F. Occurring with *J. magnus* Zone ammonoids.

O-98878   About 54 m above base of P.F. Occurring with *M. columbianus* 1 Zone ammonoids.

O-98541   About 58 m above base of P.F. Occurring with *M. columbianus* 2 Zone ammonoids.

Ne-Parle-Pas Rapids

56°00′53″N, 123°05′05″W (new exposure on south side of Williston Lake).

O-98504   About 25 m below top of Monotis Beds. Occurring with *G. cordilleranus* Zone ammonoids.

Pardonet Hill

56°04′N, 123°02′W.

O-98514   0.5 m above exposed base of P.F., within *Juvavites* Cove. Occurring with *S. kerri* Zone ammonoids.

O-98509   About 30 m above exposed base of P.F. Occurring with belemnites and *S. kerri* Zone ammonoids.